Modernizing Your Windows Applications with the Windows App SDK and WinUI

Expand your desktop apps to support new features and deliver an integrated Windows 11 experience

Matteo Pagani

Marc Plogas

BIRMINGHAM—MUMBAI

Modernizing Your Windows Applications with the Windows App SDK and WinUI

Associate Publishing Product Manager: Sathyanarayanan Ellapulli
Senior Editor: Rohit Singh
Content Development Editor: Tiksha Lad
Technical Editor: Maran Fernandes
Copy Editor: Safis Editing
Project Coordinator: Manisha Singh
Proofreader: Safis Editing
Indexer: Subalakshmi Govindhan
Production Designer: Shyam Sundar Korumilli
Marketing Coordinator: Sonakshi Bubbar

First published: March 2022

Production reference: 1280322

Published by Packt Publishing Ltd.
Livery Place
35 Livery Street
Birmingham
B3 2PB, UK.

ISBN 978-1-80323-566-0

www.packt.com

To my wife, Angela, for being a constant source of love, inspiration, and support. And to my daughter, Giulia, for teaching me to see the world from a different perspective.

– Matteo Pagani

Dedicated to anyone who still holds on to reason and scientific knowledge in today's turbulent times.

– Marc Plogas

Contributors

About the authors

Matteo Pagani is the lead of the Windows App Consult team in Microsoft. In his role, he supports developers and companies all around the world, enabling them to learn and adopt the latest development tools and technologies in the Microsoft 365 and .NET ecosystem. He has a strong passion for client development, which he loves to share with other developers by writing articles, blog posts, and books, and by speaking at conferences all around the world. Before joining Microsoft, he was a Microsoft MVP in the Windows Development category and a Nokia Developer Champion for almost 5 years.

I would like to thank my wife, Angela, and my daughter, Giulia, for their ongoing support and patience during the writing process. Next, I wish to thank my friend Marc for being a partner in crime. Then I would like to thank Sebastien Bovo and Xianghui Li for the excellent feedback they provided during the review process. Lastly, I would like to thank Microsoft and the App Consult team for supporting me on this journey.

Marc Plogas is an Azure AppConsult engineer at Microsoft and was on the Windows AppConsult team in his previous role. He works with start-ups and large companies on IoT, mixed reality, and software architecture. Tinkering with computers has always been his passion – since the tender age of 6. Therefore, he is interested in many development topics, including client development and machine learning, and enjoys all the challenges that can be solved with software engineering. Before joining Microsoft, Marc was a freelance software developer for many years, creating mobile and LoB applications.

First, I would like to thank my wife and daughter for their support, encouragement, and patience. I would also like to thank my friend Matteo, who covered my back more than once and encouraged me with memes and GIFs. I would also like to thank my team lead, David, who helped me free up time to work on the book. I would also like to thank my work-family at Microsoft, namely, both AppConsult teams, for the opportunity they have given me, especially my colleagues, Sebastien Bovo and Xianghui Li, for their feedback.

About the reviewers

Sebastien Bovo has been a member of the worldwide Windows AppConsult team dedicated to the Windows Enterprise Developer product group at Microsoft. His everyday life involves helping customers, partners, and developers in creating Windows and holographic applications. He has worked on Proof of Concept, debugged code, and provided knowledge transfers and workshops for developers.

He loves sharing his experience and knowledge with others to make them better! In particular, he has been a speaker at technical conferences, including BUILD in Seattle, IGNITE in Orlando, and around Europe for the Windows Insider Dev Tour!

Sebastien has just assumed the role of cloud solution architect to help customers modernize and innovate with Microsoft Cloud.

Xianghui Li has solid solution analysis skills and extensive development experience in .NET, web solutions, and distributed systems. He has mastered Azure IaaS and PaaS, as well as IoT deployment, development, and troubleshooting. He is good at unmanaged/managed debugging and has successfully designed, developed, and released several debugging diagnostic tools in Global Azure and Microsoft Internal.

He now works as a Windows Application Consult in Microsoft, focusing on Microsoft 365, Windows development, and cutting-edge software techniques.

Table of Contents

Section 2: Modernization Journey

2

The Windows App SDK for a Windows Forms Developer

3

The Windows App SDK for a WPF Developer

4

The Windows App SDK for a UWP Developer

5

Designing Your Application

6

Building a Future-Proof Architecture

Section 3: Integrating Your App with the Windows Ecosystem

7

Migrating Your Windows Applications to the Windows App SDK and WinUI

8

Integrating Your Application with the Windows Ecosystem

9

Implementing Notifications

10

Infusing Your Apps with Machine Learning Using WinML

Section 4: Distributing Your Application

11

Publishing Your Application

12

Enabling CI/CD for Your Windows Applications

Assessments

Index

Other Books You May Enjoy

Preface

Windows desktop applications continue to play a critical role in many scenarios among consumers and enterprises. Many of them have been the outcome of very long development processes that involved technologies that are now outdated, hard to maintain, and hard to integrate with the latest Windows innovations.

In this book, we'll show how you can modernize the UI and the features of your application using the Windows App SDK and WinUI, the latest iteration of the Windows developer platform.

Who this book is for

This book is for developers who are building Windows applications with Windows Forms, WPF, and UWP with a special focus on **Line-of-Business (LoB)** apps for enterprises. Basic knowledge of Windows app development, .NET/C#, and Visual Studio will assist with understanding the concepts covered in this book.

What this book covers

Chapter 1, Getting Started with the Windows App SDK and WinUI, covers the current status of the developer platform ecosystem and explains how you can start a new Windows App SDK project and how you can integrate its features into an existing Windows application.

Chapter 2, The Windows App SDK for a Windows Forms Developer, covers Windows Forms, which is one of the most popular development platforms for Windows. In this chapter, you'll learn how many concepts from Windows Forms are translated to the Windows App SDK and WinUI.

Chapter 3, The Windows App SDK for a WPF Developer, covers WPF, which has many similarities with the Windows App SDK and WinUI, starting from the shared UI layer based on XAML. In this chapter, you'll learn the key differences between WPF and WinUI.

Chapter 4, The Windows App SDK for a UWP Developer, explains that the Windows App SDK and WinUI are the direct successors to the UWP ecosystem. However, they are based on a different development model. In this chapter, we'll learn the differences we have to take into account.

Chapter 5, Designing Your Application, explains that building the UX of your application in the right way is critical to achieving success. This chapter will focus on important design elements, such as navigation, animations, and responsive layouts.

Chapter 6, Building a Future-Proof Architecture, explains that Windows desktop applications are often built to support critical tasks for many years to come. As such, building a future-proof architecture is critical to ensure you don't have to rebuild your application from scratch every time there's a requirement change. In this chapter, we'll learn how the MVVM pattern can help you build applications that are easy to maintain, test, and evolve over time.

Chapter 7, Migrating Your Windows Applications to the Windows App SDK and WinUI, puts into practice all the knowledge we acquired in the previous chapters to migrate existing Windows applications built with Windows Forms, WPF, and UWP to the Windows App SDK and WinUI.

Chapter 8, Integrating Your Application with the Windows Ecosystem, explains that Windows includes many innovative features that can be integrated into your application, such as Windows Hello and Geolocation. In this chapter, we'll learn how to use them in your Windows applications.

Chapter 9, Implementing Notifications, covers notifications, which play a key role in modern applications, delivering critical information to users. As such, Windows includes a powerful notification ecosystem, which you can integrate into your applications.

Chapter 10, Infusing Your Apps with Machine Learning Using WinML, explains that machine learning and artificial intelligence are two of the most important topics in the technology ecosystem. Windows enables developers to turn their machines into part of the edge ecosystem by supporting the evaluation of machine learning models without an internet connection and using the full power of the device. In this chapter, we'll learn how to integrate this platform, called WinML, into your Windows applications.

Chapter 11, Publishing Your Application, explains that developing a great application isn't enough to be successful. You have also to make it available to your users so that they can acquire it in the easiest possible way. In this chapter, we're going to explore the deployment opportunities offered by MSIX and platforms such as the Microsoft Store and App Installer.

Chapter 12, Enabling CI/CD for Your Windows Applications, covers DevOps, which is one of the key approaches nowadays to be successful in software development. In this chapter, we'll learn how you can use two of the key DevOps pillars, continuous integration and continuous deployment, to deliver your Windows applications in a more reliable, fast, and effective way.

To get the most out of this book

To build Windows desktop applications with the Windows App SDK and WinUI, you will need the following:

Software/hardware covered in the book	Operating system requirements
WPF	Windows
Windows Forms	Windows
Windows App SDK 1.0	Windows
WinUI 3.0	Windows
.NET 6	Windows
Visual Studio 2022	Windows

To follow the last chapter of the book, you will need also a GitHub account.

If you are using the digital version of this book, we advise you to type the code yourself or access the code from the book's GitHub repository (a link is available in the next section). Doing so will help you avoid any potential errors related to the copying and pasting of code.

Download the example code files

You can download the example code files for this book from GitHub at `https://github.com/PacktPublishing/Modernizing-Your-Windows-Applications-with-the-Windows-Apps-SDK-and-WinUI`. If there's an update to the code, it will be updated in the GitHub repository.

We also have other code bundles from our rich catalog of books and videos available at `https://github.com/PacktPublishing/`. Check them out!

Download the color images

We also provide a PDF file that has color images of the screenshots and diagrams used in this book. You can download it here: `https://static.packt-cdn.com/downloads/9781803235660_ColorImages.pdf`.

Conventions used

There are a number of text conventions used throughout this book.

`Code in text`: Indicates code words in text, database table names, folder names, filenames, file extensions, pathnames, dummy URLs, user input, and Twitter handles. Here is an example: "You must use this initialization code before setting the `Source` property in XAML or calling the `EnsureCoreWebView2Async()` method."

A block of code is set as follows:

```
<VariableSizedWrapGrid Orientation="Horizontal"
 MaximumRowsOrColumns="3" ItemHeight="200" ItemWidth="200">
    <Rectangle Fill="Red" />
    <Rectangle Fill="Blue" />
    <Rectangle Fill="Green" />
    <Rectangle Fill="Yellow" />
</VariableSizedWrapGrid>
```

Any command-line input or output is written as follows:

```
winget install Microsoft.Edge
```

Bold: Indicates a new term, an important word, or words that you see onscreen. For instance, words in menus or dialog boxes appear in **bold**. Here is an example: "Click on **Upload files** and wait for the process to finish."

> **Tips or important notes**
> Appear like this.

Get in touch

Feedback from our readers is always welcome.

General feedback: If you have questions about any aspect of this book, email us at customercare@packtpub.com and mention the book title in the subject of your message.

Errata: Although we have taken every care to ensure the accuracy of our content, mistakes do happen. If you have found a mistake in this book, we would be grateful if you would report this to us. Please visit www.packtpub.com/support/errata and fill in the form.

Piracy: If you come across any illegal copies of our works in any form on the internet, we would be grateful if you would provide us with the location address or website name. Please contact us at copyright@packt.com with a link to the material.

If you are interested in becoming an author: If there is a topic that you have expertise in and you are interested in either writing or contributing to a book, please visit authors.packtpub.com.

Share Your Thoughts

Once you've read *Modernizing Your Windows Applications with the Windows App SDK and WinUI 3*, we'd love to hear your thoughts! Scan the QR code below to go straight to the Amazon review page for this book and share your feedback.

https://packt.link/r/1803235667

Your review is important to us and the tech community and will help us make sure we're delivering excellent quality content.

Section 1: Basic Concepts

Section 1 will set the stage for working with the Windows App SDK and WinUI. The chapters will guide developers in understanding the current state of Windows app development, how to start a new project, and the key concepts to know, regardless of the starting technology.

This section contains the following chapter:

- *Chapter 1, Getting Started with the Windows App SDK and WinUI*

1
Getting Started with the Windows App SDK and WinUI

Windows desktop applications have always played a critical role in the enterprise and productivity space. No matter what you do in your day-to-day job, if you are using a computer in a professional capacity, you are using one or more desktop applications to get your job done – Visual Studio, Office, and Photoshop are just a few of the most famous examples. And let's not forget many of the line-of-business applications that we might use in our everyday jobs to perform tasks such as submitting expense reports or creating financial reports.

Web and mobile devices have certainly changed the ecosystem, and many of these tasks can be performed everywhere nowadays. But this doesn't mean that desktop applications aren't relevant anymore. They still play a critical role in our productivity, thanks to better performance, which makes them the best choice for heavy tasks such as video rendering or graphic design. They are optimized for mouse and keyboard, which is still the primary input method for scenarios such as coding, data entry, and data analysis. They can be deeply integrated with every type of external hardware, such as barcode readers, scanners, and blood sample testers.

For all these reasons, Microsoft continues to heavily invest in the Windows desktop space to provide developers the best platform and tools to create powerful experiences for their customers. And with the release of Windows 11, there's a renewed interest among developers to delight their users with applications that take advantage of the latest innovations in the platform.

In this chapter, we're going to explore the following topics:

- What the Windows App SDK is and how it compares to the other existing development platforms for Windows
- The role of the new .NET runtime
- Choosing the right deployment model for your application
- Creating your first Windows App SDK project
- Managing the relationship between the Windows App SDK and Windows
- Building libraries and components

These topics will set the stage for you to get started with the Windows App SDK and **WinUI** (the short name for **Windows UI Library**), which will be useful for the next chapters.

Technical requirements

To build applications with the Windows App SDK, you will need the following:

- A computer with the latest version of Windows 10 or Windows 11.
- Visual Studio 2022 with the following workloads:
- **Universal Windows Platform** (**UWP**) development
    - .NET desktop development
    - Desktop development with C++
- The Windows SDK version 2004 (build 19041) or later. This SDK will be installed with Visual Studio when you enable the UWP development workload.
- The .NET 6 SDK. This SDK will be installed together with Visual Studio when you enable the .NET Framework desktop development workload.
- The Visual Studio extension for the Windows App SDK (if you are using Visual Studio 2022 Update 1 or later, it will already be included).

The code for the chapter can be found at the following URL:

```
https://github.com/PacktPublishing/Modernizing-Your-Windows-
Applications-with-the-Windows-Apps-SDK-and-WinUI/tree/main/
Chapter01
```

A brief history of Windows UI platforms

Over the years, UI guidelines and paradigms have constantly shifted as hardware and platforms evolved. We moved from screens with 640 x 480 resolution to 4K or even 8K screens, from mouse and keyboard only to touch and digital pens. Consequently, Microsoft has created multiple UI platforms over time, with the goal of offering developers the opportunity to build modern applications; each of them represented the state of the art for the time when they were released.

The first platform was called **Microsoft Foundation Class** Library (**MFC**), which was a C++ object-oriented UI library released by Microsoft in 1992. It was a wrapper around most of the Win32 and **Component Object Model** (**COM**) APIs. Thanks to MFC, developers were able to build UIs with the most common Windows controls and build complex interfaces made up of multiple windows, panels, and so on. MFC was a considerable success, and it's still heavily used today by many developers. The following screenshot shows the look and feel of a typical MFC application:

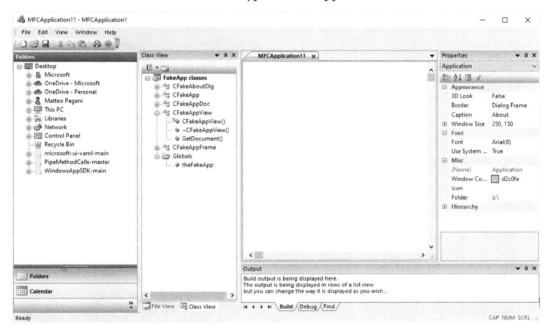

Figure 1.1 – A Windows application that uses MFC as a UI framework

However, as years passed by, it started to show limitations in supporting modern devices and features such as high-resolution screens and touch inputs. Additionally, it can be used only by C++ developers, while many developers over time have migrated to managed languages such as C#, which are easier to learn and support.

In 2002, Microsoft released the first version of .NET Framework with the goal of improving developer productivity. By running applications inside a virtual environment called **Common Language Runtime (CLR)**, developers could get out-of-the-box features such as security, memory, and exception handling that, in the past, needed to be manually managed. Additionally, by introducing languages such as C# and VB.NET, Microsoft reduced the learning curve required to master a programming language and start building software. As part of .NET Framework, Microsoft included a platform to build Windows desktop applications called **Windows Forms**. It's an event-driven platform, which makes it easier to build complex applications by wrapping the existing Windows UI common controls and Windows APIs in managed code. The development experience is mostly UI-based – developers create UIs with a visual designer by dragging and dropping the available controls inside a window. Then, they can write code that reacts to the events exposed by the various controls, such as the click of a button or the selection of an item from a list. The following screenshot shows the development experience provided by Visual Studio to build Windows Forms applications:

Figure 1.2 – The Windows Forms designer in Visual Studio

The platform kept evolving across the various releases of .NET Framework, until it reached full maturity with version 2.0.

With the release of .NET Core 3.0, Windows Forms has been integrated into the modern .NET development stack for the first time. This choice has enabled developers to access all the latest enhancements in the platform, such as newer versions of the C# language, performance improvements, or the latest Windows APIs. However, when it comes to building the UI, it still lacks many of the features you would expect from a modern platform, such as support for responsive layouts and new input experiences.

In 2006, as part of the release of .NET Framework 3.0, Microsoft unveiled **Windows Presentation Foundation** (**WPF**), the next evolution of the Microsoft UI platform. WPF introduced, for the first time, features that are still used today by modern UI platforms (including the Windows App SDK), such as **XAML** (which stands for **Extensible Application Markup Language**), binding, and dependency properties. WPF still supports building the UI with a designer, but it isn't as essential as it was for Windows Forms. WPF, in fact, decouples the UI from the business logic by describing the UI with XAML, an XML-based language. Additionally, WPF added support for features such as 2D/3D rendering, hardware acceleration, animations, and vector graphics. As with Windows Forms, .NET Core 3.0 welcomed WPF as a first-class citizen in the new development platform, enabling WPF developers to get access to the latest versions of runtimes, languages, and developer tools. Compared to Windows Forms, WPF is a more robust UI platform, capable of delivering more modern experiences, as you can see in the following screenshot:

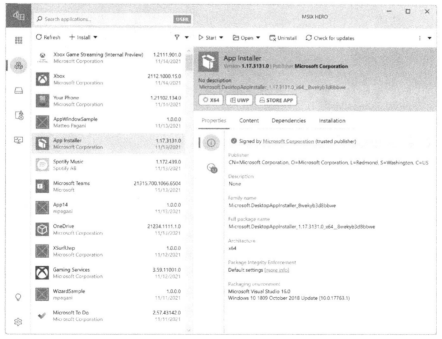

Figure 1.3 – MSIX Hero is a good example of an application that delivers a great user experience by using the WPF capabilities

However, it still has limitations, such as poor support to high **Dots-Per-Inch** (**DPI**) devices, touchscreens, digital inking experiences, and accessibility.

In 2015, with the release of Windows 10, Microsoft released UWP, which is an extension of Windows Runtime that was introduced in Windows 8. UWP is a modern development platform that enables developers to build secure and robust applications that run inside a sandbox; it gives access to all the new features added in Windows 10, such as tiles, notifications, and Windows Hello; it's based on a new UI platform called Fluent Design, which offers built-in support to responsive layout, touch and digital pen, accessibility, and so on. Many built-in Windows applications, such as **Microsoft Store**, are built with UWP and WinUI:

Figure 1.4 – Microsoft Store in Windows is a UWP application

In the first releases of UWP, the UI framework (like all the other development APIs) was built in the operating system. Over time, this approach created multiple challenges, both to Microsoft and developers:

- Despite Windows 10 adopting a much more aggressive update strategy compared to prior versions, by releasing two upgrades per year, it still forced the development team to address issues and add new UI controls and features only twice per year.

- If a developer wanted to use the new UI controls or features added to the latest version of Windows 10, all their users had to upgrade their machines to that version as well. This was a challenge, especially in enterprise environments, where the upgrade pace is slower than in the consumer world.

To overcome these challenges, in October 2018, Microsoft released the first public release of the Windows UI library, called **WinUI 2.0**. With this release, Microsoft detached most of the UI controls and features from the operating system and moved them inside a library, which is distributed as a NuGet package. The library enabled Microsoft to release more frequent updates (the current life cycle is four releases per year) and developers to get access to the latest UI enhancements without forcing their user base to upgrade to the latest Windows 10 version.

Introducing the Windows App SDK and WinUI 3.0

Now that we have learned a bit of the history of the Microsoft development platform for Windows, we can better understand how the Windows App SDK fits into the story. We've learned that, over time, Microsoft has released new UI platforms with the goal of being state of the art at the time of release. However, UWP introduced a few challenges, especially for developers building line-of-business applications. UWP is a great fit for many consumer scenarios – the sandbox enables an application to run more safely, since it has limited access to critical Windows features, such as a registry or a full filesystem; thanks to a rich capabilities system, the user is always in control of which features of the device (a webcam, a microphone, location, and so on) an application can access; and thanks to a modern life cycle, applications are more respectful of the CPU, memory, and battery life of a device. In many cases, however, enterprises require deeper control and flexibility – applications need to always run, even when they are minimized in the taskbar; they need to interact with custom hardware devices and retrieve information from the Windows Registry.

Also, Windows Runtime, the application architecture behind UWP, introduced a few challenges. This modern runtime offers a lot of benefits: it's built in C++, which means it offers great performance; it's based on asynchronous patterns, which help developers to build applications that are fast and responsive; and through a feature called projections, it enables developers to consume Windows Runtime APIs from multiple languages, such as C++, C#, and Rust. However, since it's a significantly different platform compared to the ones already on the market (such as .NET), developers who have heavily invested in C++, Windows Forms, and WPF are able to reuse little or no code when porting their applications to UWP. This requirement created a lot of friction, since in the enterprise ecosystem, it's easy to find desktop applications with decades of development that are very hard, if not impossible, to port to UWP without rewriting them from scratch. The outcome is that even if developers loved Fluent Design and the new features introduced in Windows 10, most of them weren't able to take advantage of them.

In late 2019, Microsoft announced the first milestone to overcome these challenges – WinUI 3.0. Earlier versions of WinUI already started to lift most of the UI controls and features from the operating system, but the library was still targeting only UWP. With the 3.0 release, instead, Microsoft lifted the whole UI framework from the operating system, enabling other development platforms such as .NET to start taking advantage of the new UI library.

Thanks to WinUI 3.0, developers can now build modern applications and experiences but, at the same time, leverage familiar development platforms such as .NET or C++; reuse most of the investments they did building Windows Forms, WPF, or C++ applications; and use popular NuGet packages that they had already adopted in Windows Forms and WPF applications. Additionally, since applications using WinUI 3.0 are no longer dependent on UWP, but instead run as classic desktop applications, many of the features that enterprise developers saw as a constraint (the sandbox, the life cycle optimized to reduce CPU, and memory usage) don't apply anymore.

Microsoft, during the 2020 edition of the Build conference, officially shared the next step of the journey by announcing **Project Reunion**, a new development platform with the goal to provide developers with the best of both worlds – familiar developer frameworks and languages (such as .NET, C#, and C++) and the ability to use all the Windows features. WinUI 3.0 was incorporated into this new platform, and it became its first and most important building block.

Project Reunion continued to evolve over time by gradually bringing new features that were exclusively a part of UWP, such as push notifications and activation contracts. On June 24, 2021, when Microsoft revealed to the world Windows 11, the Windows development team announced the official name of Project Reunion – **the Windows App SDK**.

The Windows App SDK and WinUI 3.0 have the ambitious goal of becoming the new reference UI platform for all Windows developers, regardless of their background. In fact, the Windows App SDK targets both C++ developers (who, before the Windows App SDK, were stuck on MFC as the UI framework offered by Microsoft) and .NET developers (who can reuse their existing skills but target a more modern UI framework than Windows Forms and WPF).

The Windows App SDK introduced the following advantages for developers:

- **Consistent support**: Windows and its development platform continue to evolve over time, and developers must rely on techniques such as adaptive coding to keep a single code base and use a specific feature only if the application is running on a version of the operating system that supports it. Over time, this approach can add complexity, so Microsoft decided to create the Windows App SDK with built-in down-level support. An application that uses the Windows App SDK can run from Windows 11 down to Windows 10, version 1809. Newer features will be automatically enabled if the application is running on the proper version of the operating system. An example of this feature is Mica, a new material introduced in Windows 11 to enrich the background of an application. If you configure your application to use it, Mica will automatically light up if the application is running on Windows 11, while it will automatically fall back to a solid color on Windows 10.

- **A faster release cadence**: Since the Windows App SDK is a library, Microsoft can ship updates faster, without being aligned to the release cycle of Windows.

- **A unified API surface**: Historically, the number of Windows features you could use in your application depended on your framework of choice. For example, many Windows 10 features were supported only by UWP apps. The Windows App SDK unifies access to Windows APIs so that, regardless of the app model you choose, you will be able to use the same set of Windows features.

The Windows App SDK and WinUI are considered the future of the Windows developer platform. UWP is now considered a stable and mature platform that will continue to receive security updates and be supported. However, all the investments for new features and scenarios will be focused on the Windows App SDK, making it the best choice to build future-proof Windows applications.

This book is dedicated to C# developers, and it will cover the usage of the Windows App SDK with classic desktop applications based on the .NET runtime. However, every concept you are going to learn can also be applied to C++ applications.

The role of the new .NET runtime

.NET Framework was introduced by Microsoft in 2002 with the goal to provide a better development experience and an easier learning curve for developers to build applications for Windows. These principles still hold true today, but the development landscape has changed significantly since 2002. New platforms have appeared, and Windows, in the server and cloud space, isn't the leading platform anymore; scenarios such as containers and microservices require a new level of optimization and performance, especially in key areas such as networking or filesystem access. Open source is now the new standard to release and evolve development platforms in collaboration with the community.

In 2014, Microsoft announced the first version of .NET Core, a new development platform based on the same principles of .NET Framework but open source, cross-platform, and lighter. The first versions of .NET Core were focused on key scenarios such as cloud services and web applications. With the release of .NET Core 3.0, Microsoft has started to invest also in the desktop ecosystem, by making Windows Forms and WPF open source and welcoming them into the new runtime.

However, the .NET ecosystem was still fragmented – if web and desktop applications could run on .NET Core, many developers were still using the full .NET Framework to build their solutions; Xamarin (the cross-platform framework to build Android and iOS apps with C#) and Blazor (the new platform to build client-side web applications using C# instead of JavaScript) were still based on Mono, the original open source implementation of .NET Framework. This fragmentation created multiple challenges over time and required the team to create solutions such as Portable Class Libraries and .NET Standard to enable developers to share code and libraries across different projects.

With .NET 5 (released in November 2020), Microsoft started an ambitious project to unify the entire .NET ecosystem. Instead of having multiple implementations of the framework for different workloads, .NET 5 started a unification journey, with the goal of supporting all the platforms and devices on the market with a single runtime and base class library. .NET 6 is going to complete this journey with a future update by abandoning Mono and bringing Xamarin (now known as .NET for Android and .NET for iOS) inside the family.

The new .NET runtime has also introduced a more predictable release plan, which makes it easier for developers to plan the adoption of one version or another:

1. .NET 5 was released in November 2020.
2. A new version will be released around the same time frame each year – for example, NET 6 was released in November 2021, and .NET 7 will be released in November 2022.

3. The releases with an even version number (.NET 6, .NET 8, and so on) are marked **Long-Term Support (LTS)**. These releases will be supported for 3 years, so they're a better fit for projects that need stability.

4. The releases with an uneven version number (.NET 5, .NET 7, and so on) are supported for 6 months until the release of the next version – for example, .NET 5 will go out of support in May 2022, 6 months after the release of .NET 6. You can check the most up-to-date support policy at `https://dotnet.microsoft.com/en-us/platform/support/policy/dotnet-core`.

By adopting a more frequent release cycle, the .NET team can be more agile and reduce the chances of introducing breaking changes between one version and the other. Even if you decide to adopt a .NET runtime that is not marked as LTS, in most cases, the transition to the next release will be smooth, with no or very few code changes needed.

Consequently, .NET Framework reached the end of the journey with the release of .NET Framework 4.8. This framework is still popular today, and as such, Microsoft doesn't have any plan of deprecating it. The current .NET Framework support policy is tied to the Windows one, so .NET Framework will continue to be supported for a long time. However, it won't receive any new features or major updates. If you're a developer building Windows apps that are continuously upgraded and evolved, the suggested path forward is to migrate your .NET Framework applications to the new .NET runtime so that you can take advantage of the constant evolution of the platform.

For all these reasons, the new .NET runtime plays a critical role in our journey. Since the Windows App SDK is the future of Windows development, its evolution is deeply connected with the new .NET runtime. For example, the first preview releases of the Windows App SDK were targeting .NET 5, but the platform is now aligned with the most recent .NET 6 release.

Additionally, all the features exposed by the Windows App SDK don't support the old .NET Framework, but they require the new .NET runtime.

Exploring the Windows App SDK

The Windows App SDK has currently reached version 1.0, and it includes the following features:

*   WinUI 3
*   Text rendering
*   Resource management
*   App life cycle

- Power state notifications
- Windowing
- Push notifications (preview)

The second section of this book will be focused on WinUI 3 and will guide you toward a modernization journey, with the goal of evolving your application from your existing UI framework (Windows Forms, WPF, or UWP) to embrace the latest innovation in Fluent Design.

This feature of the Windows App SDK is currently supported only by new apps that target WinUI 3. This means that this modernization journey will help you to reuse most of your existing code and libraries (thanks to the .NET runtime), but you won't be able to gradually move the UI of your Windows Forms or WPF application. You will have to use one of the new Visual Studio templates to create a new .NET application based on WinUI. Microsoft is working to bring a technology called XAML Islands to WinUI 3, which will instead enable developers to gradually migrate the UI of their applications by mixing controls from WinUI with controls from the existing UI frameworks (Windows Forms and WPF).

The third section, instead, will be focused on the developer platform – you will learn how to integrate the other features offered by the Windows App SDK, plus many other new APIs (such as geolocation and machine learning), which, previously, were available only to UWP applications. These features, unlike WinUI 3.0, can be easily integrated into existing desktop applications built with WPF and Windows Forms, as long as they have been migrated to at least .NET 5.

The Windows App SDK is distributed in three different channels:

- **Stable**: The releases distributed through this channel are supported in production environments. They include the latest stable and tested bits.

- **Preview**: The releases distributed through this channel include preview features that will be added in the next stable release. Being a preview, they aren't supported in production scenarios, since there might be breaking changes when the stable version gets released.

- **Experimental**: The releases distributed through this channel include experimental features, which might be discarded or completely changed in the next stable release. They aren't supported in production environments, since the included features might not even see the light.

All the features described in this book are included in the stable channel, except for push notifications, which are still in preview.

With an understanding of the Windows App SDK, our next step is to learn how to choose between an unpackaged and packaged deployment model. We will do that in the next section.

Choosing the right deployment model

We'll talk in more detail about how to deploy distributed applications that are using the Windows App SDK in *Chapter 11*, *Publishing Your Application*, but it's critical to introduce at the beginning two important concepts, since they will influence the way you create an application – packaged and unpackaged.

Packaged applications adopt MSIX as a deployment technology. MSIX is the most recent packaging format introduced in Windows, which, compared to other deployment technologies such as MSI or ClickOnce, brings many benefits:

- A cleaner install and uninstall process, thanks to the usage of a lightweight container that virtualizes many of the critical aspects of Windows, such as the filesystem and the registry

- Built-in features such as automatic updates even in unmanaged scenarios (such as a website), bandwidth optimization, and disk space optimization

- Tampering protection, which prevents an app that has been improperly changed from running and damaging an operating system

Windows applications packaged with MSIX are described by a manifest, which is an XML file that holds information such as the name, the publisher, the version number, and the dependencies. The packaged approach is the best one for applications that use the Windows App SDK, since it simplifies many of the scenarios that we're going to see in this book, such as managing the framework dependency and using Windows APIs that require an identity.

However, as a developer, you might face scenarios where a packaged app doesn't fit your requirements:

- The container provided by MSIX is very thin, but there are still situations when it might interfere with the regular execution of your application due to the isolation of the registry and the filesystem.

- To work properly, the application must deploy, during the installation, a kernel driver, or it must apply some global settings to the computer, such as creating environment variables and installing a Windows feature. The isolated nature of MSIX doesn't make these kinds of scenarios a good fit.

Because of these cases, the Windows App SDK also supports unpackaged apps, which are applications that you can deploy the way you prefer by using manual copy deployment, adopting a traditional MSI installer, building a custom setup, or using a script. The way you deploy your application is deeply connected to the way you manage the dependency that your application has with the Windows App SDK. Let's learn more in the next section.

Managing the dependency with the Windows App SDK

By detaching features and APIs from the operating system, the Windows team has gained a lot of benefits, such as the ability to add new features and fix issues outside the regular Windows update cycle. However, as developers, we cannot take for granted that the user who installs our application will have the Windows App SDK runtime installed on their machine. As such, it's our responsibility as developers to manage this dependency in the right way. This is one of the scenarios where the difference between adopting the packaged or unpackaged model is critical, since they adopt two different ways to deploy the dependency.

The Windows App SDK runtime is made up of the following components:

- **The framework package**: This package contains the binaries that are used by the applications at runtime.

- **The main package**: This package contains multiple out-of-process components that aren't included in the framework, such as background tasks, app extensions, and app services.

- **The singleton package**: This package includes a long-running process that provides access to features that are shared across all applications, such as push notifications.

- **Dynamic Dependency Lifetime Manager (DDLM)**: This package is used to control the life cycle of the Windows App SDK runtime. Thanks to this package, Windows will refrain from automatically updating the runtime if one or more unpackaged apps that use the Windows App SDK are running.

Each of these components is stored in its own MSIX package. Let's see now how to manage the deployment of the runtime based on the way you distribute your application.

Packaged apps

As previously mentioned, packaged apps have a few advantages compared to unpackaged apps. Packaged applications are described by a manifest, which includes a section called `Dependencies`. This section is already used today to manage the dependency on a specific version of Windows or the Visual C++ runtime.

The dependency with the Windows App SDK runtime can be easily declared in the same section by adding the following entry:

```
<PackageDependency Name="Microsoft.WindowsAppRuntime.1.0"
  MinVersion="0.319.455.0" Publisher="CN=Microsoft
    Corporation, O=Microsoft Corporation, L=Redmond,
      S=Washington, C=US" />
```

Thanks to this configuration, if the application is deployed to a machine that doesn't have the runtime installed, Windows will automatically download it from the Microsoft Store and install it system-wide so that future applications that you might install won't need to download it again.

In most cases, you won't have to manually include this entry. In fact, when you install the Windows App SDK NuGet package in your application, Visual Studio will add a special build task that will add this entry in the manifest for you every time you generate a new MSIX package.

However, there's still a manual requirement you have to take care of. As mentioned in the previous section, the Windows App SDK applications are dependent not just on the framework itself but, based on the features they use (for example, background tasks or push notifications), they might also need the main package and the singleton package. MSIX doesn't support having regular packages as dependencies, which means that when you deploy a packaged Windows App SDK application, the framework is automatically deployed if missing, but the other components aren't. However, these packages are included inside the framework and, as such, can be automatically deployed by using the DeploymentManager API, which belongs to the Microsoft.Windows. ApplicationModel.WindowsAppRuntime namespace. Thanks to this API, you can check whether one or more of the components are missing on the system and install them if necessary. This is an example implementation of the OnLaunched() method of a Windows App SDK application, which is executed when the application starts:

```
protected override void OnLaunched
    (Microsoft.UI.Xaml.LaunchActivatedEventArgs args)
{
    var status = DeploymentManager.GetStatus();
    if (status.Status ==
      DeploymentStatus.PackageInstallRequired)
    {
        DeploymentManager.Initialize();
    }
```

```
    m_window = new MainWindow();
    m_window.Activate();
}
```

Thanks to the GetStatus() method, we can detect the status of the various components on the system. If one or more of them is missing (which is represented by the PackageInstallRequired value of the DeploymentStatus enumerator), we can call the Initialize() method to perform the deployment.

Unpackaged apps

For unpackaged apps, Microsoft provides an installer called WindowsAppRuntimeInstall.exe, which automatically detects the CPU architecture of a system (x86, x64, or ARM64) and installs system-wide the MSIX packages that compose the runtime – framework, main, and DDLM.

The installer also supports a --quiet parameter, which enables a silent installation with no user interaction and output messages. All the technologies and tools to create setup programs (or even just a PowerShell script) support the possibility of launching an executable as part of the installation, so it's easy to configure your installer to silently launch the WindowsAppRuntimeInstall.exe executable before or after the deployment of the main application, and before the whole installation process is completed.

Compared to a packaged application, you won't need to use the deployment APIs here, since the WindowsAppSDKInstall tool will take care of installing all the required packages on the system.

Upgrading the Windows App SDK runtime

The Windows App SDK runtime will continue to evolve over time, and as such, it's important for a developer to understand how future updates will be managed.

Updates can be delivered in two ways:

- If the machine has access to Microsoft Store, they will be automatically downloaded and installed.
- If the machine doesn't have access to Microsoft Store, it will be up to the application to include a newer version of the Windows App SDK runtime installer.

To control the way updates will be installed, the Windows App SDK has adopted the Semantic Versioning rules. By reading the official website (`https://semver.org/`), we learn that the version number is defined by three numbers, MAJOR.MINOR.PATCH, which stand for the following:

- MAJOR is incremented when you make incompatible API changes.
- MINOR is incremented when you add new functionality but with backward compatibility.
- PATCH is increased when you make bug fixes that are backward-compatible.

The runtime installed on a machine will be automatically updated only if the MINOR or the PATCH revision changes. If a new release increases the MAJOR number, instead, it will be installed side by side with the existing ones. This makes sure that newer versions of the runtime that might include breaking changes won't replace the existing ones, causing your applications to stop working properly.

Thanks to the DDLM component of the runtime, Windows will keep the old instance of the runtime installed as long as there's at least one running application that is using it. Once the application is closed, the previous version of the runtime will be uninstalled, and at the next relaunch, the app will start using the updated version.

Now that we understand how to manage the Windows App SDK dependency in our application, we're ready to start creating our first application.

Creating the first Windows App SDK project

When it comes to adopting the Windows App SDK, there are two options:

- Start with a new project using one of the available Visual Studio templates.
- Integrate the Windows App SDK into an existing application. Since this book is dedicated to C# developers, we'll focus on Windows Forms and WPF.

We will discuss each possibility in the next sections.

A new packaged WinUI C# application

Once you have installed the Windows App SDK extension for Visual Studio, you'll get a few templates to create new projects. The starting point for a Windows App SDK application is called **Blank App, Packaged (WinUI 3 in Desktop)**, as shown in the next screenshot. Why WinUI? Because it's the only feature of the Windows App SDK that can't be integrated into existing applications, so you will likely start with a new project if you want to create a Windows application that uses WinUI as the UI framework:

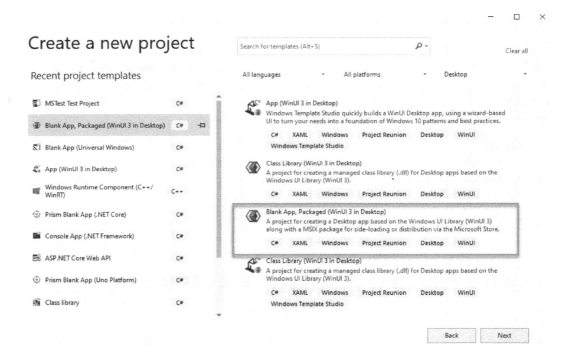

Figure 1.5 – The template to create a new packaged WinUI 3 application

The template will create a project that contains all the basic building blocks of a WinUI application. Let's take a look at the structure in more detail.

The template will create a project as shown in the following screenshot:

Figure 1.6 – A new solution for a packaged app that uses the Windows App SDK and WinUI 3

This is the project that contains the actual application, and that's where you're going to write most of your code.

One of the key differences between .NET Framework and the new .NET runtime is the adoption of a new project format, called SDK style. This format makes it quite easy to read and change the XML that describes the project. Visual Studio still supplies a more human-friendly UI (which you can see by right-clicking on the project and choosing **Properties**), but it isn't a must-have feature anymore.

This is what the project file of an application that uses the Windows App SDK looks like (you can see it by simply double-clicking on the project's name in Solution Explorer):

```
<Project Sdk="Microsoft.NET.Sdk">
  <PropertyGroup>
    <OutputType>WinExe</OutputType>
    <TargetFramework>net6.0-
      windows10.0.19041.0</TargetFramework>
    <TargetPlatformMinVersion>10.0.17763.0
      </TargetPlatformMinVersion>
    <RootNamespace>MyApplication</RootNamespace>
    <ApplicationManifest>app.manifest</ApplicationManifest>
    <Platforms>x86;x64;arm64</Platforms>
    <RuntimeIdentifiers>win10-x86;win10-x64;
     win10-arm64</RuntimeIdentifiers>
    <PublishProfile>win10-$(Platform).pubxml
      </PublishProfile>
    <UseWinUI>true</UseWinUI>
    <EnablePreviewMsixTooling>true</EnablePreview
```

```xml
        MsixTooling>
  </PropertyGroup>

  <ItemGroup>
    <AppxManifest Include="Package.appxmanifest">
      <SubType>Designer</SubType>
    </AppxManifest>
  </ItemGroup>
  <ItemGroup>
    <Content Include="Assets\SplashScreen.scale-200.png" />
    <Content Include="Assets\LockScreenLogo
      .scale-200.png" />
    <Content Include="Assets\Square150x150Logo
      .scale-200.png" />
    <Content Include="Assets\Square44x44Logo
      .scale-200.png" />
    <Content Include="Assets\Square44x44Logo
      .targetsize-24_altform-unplated.png" />
    <Content Include="Assets\StoreLogo.png" />
    <Content Include="Assets\Wide310x150Logo
      .scale-200.png" />
  </ItemGroup>

  <ItemGroup>
    <PackageReference Include="Microsoft.WindowsAppSDK"
      Version="1.0.0" />
    <PackageReference
      Include="Microsoft.Windows.SDK.BuildTools"
        Version="10.0.22000.196" />
    <Manifest Include="$(ApplicationManifest)" />
  </ItemGroup>
</Project>
```

Let's take a look at the key properties:

- `TargetFramework`: This defines which flavor of .NET we want to use. In our scenario, we need to use the specific workload that gives us access to the Windows Runtime ecosystem, which is `net6.0-windows10.0.19041.0`.

- `Platforms`: Since the Windows App SDK is a native layer of APIs, it can't be used in applications that are compiled for the generic *Any CPU* architecture. As such, our project must explicitly declare the architectures we can support, which, in the case of the Windows App SDK, are x86, x64, and ARM64.

- `RuntimeIdentifiers`: The .NET runtime supports a deployment model called self-contained, in which the runtime itself is bundled together with the application. This helps to reduce the manual steps that a user must take to satisfy all the required dependencies. By default, a packaged Windows App SDK application uses this configuration so that you can generate a standalone package, which includes everything the user needs to run it.

- `UseWinUI`: Every .NET project for a Windows application specifies which is the used UI framework so that .NET can load the proper libraries. Since we're building a WinUI application, the property to use is `UseWinUI`. In the case of a WPF application, the property would be called `UseWPF`; in the case of Windows Forms, it would be `UseWindowsForms` instead.

- Lastly, every application that uses the Windows App SDK must reference the proper NuGet packages, which contain the library itself. Without these packages, you won't be able to access any of the features exposed by the runtime. The required package is called `Microsoft.WindowsAppSDK` and it's a meta-package – it doesn't actually contain any code, but it groups together all the various packages that contain the various components of the framework.

From a project's structure perspective, a WinUI 3 application follows the same pattern as WPF or UWP. The default template contains the following components:

- `App.xaml`: This is the entry point of the application. It takes care of initializing everything that is needed to run the application. From a developer perspective, you can use it to store visual resources (such as styles) that must be shared across the whole application; you can use the code-behind class (the `App.xaml.cs` file) instead to manage the life cycle of the application and intercept various activation events, such as the application being opened via a custom protocol or a file type association.

- `MainWindow.xaml`: This is the main window of the application. If you take a look at the `App.xaml.cs` file, you will discover that there's an event handler called `OnLaunched()`, which creates a new instance of the `MainWindow` class and activates it. This means that when the user launches the application, the content of `MainWindow` is the first thing they will see.

- `Package.appxmanifest`: This is the manifest that describes the application and its identity. If you double-click on the file, Visual Studio gives you the opportunity to customize it through a visual interface. However, the designer was created during the Universal Windows App time frame, and as such, some of the features that we're going to explore in this book that are specific to the Win32 ecosystem aren't available. In these scenarios, you must manually edit the manifest file by right-clicking on it and choosing **View code**, which will give you direct access to the XML behind it.

If you analyze the XML, you will notice something peculiar in the `Capabilities` section:

```
<Capabilities>
    <rescap:Capability Name="runFullTrust" />
</Capabilities>
```

The MSIX packaging format, which was originally called AppX, was introduced for Windows Store apps (later evolved into UWP apps), and as such, it was tailored for applications that run inside a sandbox. This special capability, called `runFullTrust`, instead enables packages to also host traditional Win32 applications, such as a WinUI 3 application based on the .NET runtime.

When you use this template, you don't have to worry about managing the Windows App SDK runtime dependency. Even if you don't see it in the `Package.appxmanifest` file, Visual Studio will automatically inject the required entry into the final manifest of the package to install the runtime if it's not already available on the machine.

Using a separate packaging project

If you look at the available templates to create a new WinUI application, you will find one called **Blank App, Packaged with Windows Application Packaging Project (WinUI 3 in Desktop)**. This template was originally the only supported way to build and deploy WinUI applications, and it's made up of two different projects:

- One with the name you chose during the setup wizard (for example, `MyFirstApp`)

- One with the same name, plus the `(Package)` suffix (for example, `My FirstApp (Package)`):

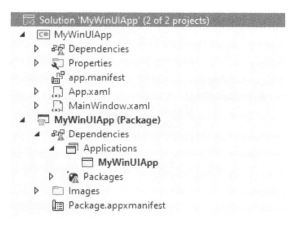

Figure 1.7 – A WinUI project that uses a separate packaging project

The first project is the main one, which contains the application. The second project, the one with the (Package) suffix, is a **Windows Application Packaging (WAP) Project**. This is a special type of project that doesn't contain actual code, but it enables you to package the main application with MSIX. If you expand the Dependencies node, you will find a section called Applications, which includes a reference to the main project.

This project isn't required anymore, thanks to a feature called single-project MSIX. When we create a new application using the **Blank App, Packaged (WinUI 3 in Desktop)** template that we have seen before, we're still using the MSIX packaging format. However, the WAP Project is now hidden and incorporated into the main project, thanks to a special property that you can see in the .csproj file:

```
<EnablePreviewMsixTooling>true</EnablePreviewMsixTooling>
```

However, there might be some scenarios where you still need to use the WAP Project. For example, if you're planning to bundle multiple applications inside the same MSIX package, you must continue to use a WAP Project.

A new unpackaged WinUI application

WinUI applications can also use the unpackaged model, which gives you the flexibility to distribute your application using the technology you prefer. The starting point for a WinUI unpackaged application is the same template we have used for the packaged version, the one called **Blank App, Packaged (WinUI 3 in Desktop)**. However, in this case, we must change the .csproj file by adding the following entry inside PropertyGroup:

```
<WindowsPackageType>None</WindowsPackageType>
```

This is how the full `.csproj` file should look:

```
<Project Sdk="Microsoft.NET.Sdk">
  <PropertyGroup>
    <OutputType>WinExe</OutputType>
    <TargetFramework>net6.0-windows10.0.19041.0</TargetFramework>
    <TargetPlatformMinVersion>10.0.17763.0</TargetPlatformMinVersion>
    <RootNamespace>App15</RootNamespace>
    <ApplicationManifest>app.manifest</ApplicationManifest>
    <Platforms>x86;x64;arm64</Platforms>
    <RuntimeIdentifiers>win10-x86;win10-x64;win10-arm64</RuntimeIdentifiers>
    <PublishProfile>win10-$(Platform).pubxml</PublishProfile>
    <UseWinUI>true</UseWinUI>
    <EnablePreviewMsixTooling>true</EnablePreviewMsixTooling>
    <WindowsPackageType>None</WindowsPackageType>
  </PropertyGroup>
```

Figure 1.8 – A WinUI project configured to support the unpackaged model

Thanks to this property, the application will automatically inject the required code to initialize the Windows App SDK runtime.

Once you make this change, you will notice that in the debugger drop-down menu, there will be a new entry with the name of your application followed by the **(Unpackaged)** suffix, as shown in the following screenshot:

Figure 1.9 – The debugger menu once you have configured a WinUI project to run unpackaged

By choosing the unpackaged version as the target and clicking on the button (or pressing *F5*), you will launch the unpackaged WinUI application with the debugger attached.

Adding support to an existing application

All the features of the Windows App SDK other than WinUI can also be used by existing applications. In this section, we're going to learn how to set up a Windows Forms or WPF application so that it can use the Windows App SDK features.

Regardless of the UI framework of your choice, remember that the Windows App SDK targets the new .NET runtime, so you will have to start with an application that targets at least .NET 5.

Here are the steps to follow:

1. Change the `TargetFramework` property to at least `net5.0–windows10.0.19041.0` or, even better, `net6.0–windows10.0.19041.0`. If you have an existing Windows Forms or WPF application, you're already using the specific .NET runtime workload for a Windows app. You can verify this by double-clicking on your project in Visual Studio. The property should be equal to `net5.0–windows` or `net6.0-windows`. However, this version of the workload isn't enough for our needs. This workload gives you access to generic Win32 APIs such as registry access or event log interaction, but it doesn't include the specific Windows 10 APIs introduced with UWP. As such, we must switch to one of the specific Windows 10 workloads. This is what the property must look like:

   ```
   <TargetFramework>net6.0-windows10.0.19041.0
     </TargetFramework>
   ```

2. Install, through NuGet, the package called `Microsoft.WindowsAppSDK`, which is a meta-package; this means that it doesn't actually include any libraries, but it will properly install all the packages that are required to use the Windows App SDK. The easiest way to install it is to right-click on your project in Visual Studio, choose **Manage NuGet Packages**, and look for the package called `Microsoft.WindowsAppSDK`.

3. Specify the runtime and platforms supported by the project by adding the following entry in the `.csproj` file:

   ```
   <Platforms>x86;x64;arm64</Platforms>
   <RuntimeIdentifiers>win10-x86;win10-x64;win10-arm64
     </RuntimeIdentifiers>
   ```

This is needed, since the Windows App SDK is a native dependency, and as such, a project that uses it can't be compiled with the generic `AnyCPU` architecture, but it must target a specific CPU platform.

> **Important Note**
>
> If you explore the content of the Windows App SDK meta-package, you might be tempted to install only a few of the sub-packages, based on the features you need. However, this isn't a supported scenario. You must always install the whole SDK, even if you aren't going to use all the available features.

The next step is different, based on the way you're going to deploy your application – packaged or unpackaged.

Distribution via a packaged app

If you want to leverage MSIX as a deployment technology, the next step is to add a WAP Project to your solution. At the time of writing, the single-project MSIX feature is supported only by WinUI apps, so you must continue using the WAP Project if you want to package an existing Windows Forms or WPF application. You can achieve this goal by right-clicking on your solution, choosing **Add | New project**, and selecting the template called **Windows Application Packaging Project**.

After you have given it a meaningful name, you must perform the following actions:

1. Select the application you want to package. To do this, right-click on the project and choose **Add | Project Reference**, and then look for your solution's project that contains the application you're looking to package.

2. Install the `Microsoft.WindowsAppSDK` package from NuGet in this project. The package will add the build steps that are required to properly generate a package – for example, thanks to this package, you won't have to manually declare in the manifest the dependency from the Windows App SDK runtime, but it will be automatically added at compile time for you.

From now on, make sure to right-click on **Windows Application Packaging Project** in **Solution Explorer** and choose **Set as startup project**. You will always need to launch and deploy this project to run and debug your application.

Distribution via an unpackaged app

Unpackaged apps use the concept of **dynamic dependencies** to dynamically take a dependency on the Windows App SDK runtime. This is achieved by using a bootstrapper API, which must be called as the first step when an application starts so that it can properly reference the runtime and start using its APIs and components. If you're creating a new WinUI application using the dedicated template, the bootstrapper API is implicitly used when you add the `WindowsPackageType` property in a project's configuration. If you want to integrate the Windows App SDK in an existing Windows Forms or WPF application, you must instead manually invoke it. This class, called `Bootstrap`, is exposed by the `Microsoft.Windows.ApplicationModel. DynamicDependency` namespace.

Thanks to this class, you will have access to two critical methods needed to reference the Windows App SDK runtime:

- `Initialize()`: This function must be called in the app's startup, passing a reference to the version of the Windows App SDK you want to use – for example, if you want to use the stable 1.0 release, you will have to pass `0x00010000` as the value. The bootstrapper will use this information to reference the version of the framework that matches the Windows App SDK version used in the application. The function also takes care of initializing the DDLM component, which makes sure that the operating system won't try to update the framework while any unpackaged app is using it.

- `Shutdown()`: This function must be called when the user quits the application or when you don't need to use any feature provided by the Windows App SDK anymore. This function cleans up the runtime reference and shuts down the DDLM so that Windows can update the framework if needed.

The way you call these functions depends on the platform you have chosen to build your Windows application. If you have a WPF application, you can take advantage of the `OnStartup()` and `OnExit()` event handlers available in the `App` class, as in the following example:

```
using Microsoft.Windows.ApplicationModel.DynamicDependency;
public partial class App : Application
{
    protected override void OnStartup(StartupEventArgs e)
    {
        Bootstrap.Initialize(0x00010000);
    }
```

```
    protected override void OnExit(ExitEventArgs e)
    {
        Bootstrap.Shutdown();
    }
}
```

In Windows Forms, you can instead leverage the `Program` class and call the `Initialize()` function before `Application.Run()`, calling `Shutdown()` immediately after:

```
static class Program
{
    /// <summary>
    ///  The main entry point for the application.
    /// </summary>
    [STAThread]
    static void Main()
    {
        Bootstrap.Initialize(0x00010000);

        Application.SetHighDpiMode
          (HighDpiMode.SystemAware);
        Application.EnableVisualStyles();
        Application.SetCompatibleTextRenderingDefault
          (false);
        Application.Run(new Form1());

        Bootstrap.Shutdown();
    }
}
```

Now that we have learned how to start a new project based on the Windows App SDK, let's see how we can enrich our project by creating new libraries and components.

Building libraries and components

A common requirement when you start to build more complex applications is to split a solution into multiple projects, with the goal of increasing reusability and having a clearer separation across various components.

When you're building a Windows application that uses the Windows App SDK, you have two options to choose from:

- A WinUI class library
- A .NET class library

Let's discuss each option in detail.

Using a WinUI class library

A WinUI class library is the best choice when you're building a library that contains code specific to the Windows App SDK and WinUI. It's a great fit when you want to store the following:

- Custom controls or user controls for WinUI
- Classes that use specific Windows 10 APIs

This type of class library is a great fit when you want to share code across multiple WinUI applications. To create a WinUI class library, you can use the template in Visual Studio called **Class Library (WinUI 3 in Desktop)**.

Using a .NET class library

A .NET class library is the best choice when you're building a library that doesn't include any code that is specific to the Windows App SDK. By using a generic .NET class library, you can share it not only with other applications based on the Windows App SDK but also with every other project type supported by .NET – web apps, cloud services, mobile apps, and so on.

When you choose the **Class library** project type in Visual Studio, there are two options to choose from:

- A specific .NET version (for example, .NET 5 or .NET 6)
- .NET Standard

The first option is the best choice when you're planning to share this library with any other project that is using the new .NET runtime. This choice will give you access to the widest surface of APIs and features, such as the latest additions to the C# language.

Alternatively, .NET Standard is the best choice when you're going to share this library with older platforms that are based on other flavors of the .NET runtime that were created before the unification journey started with .NET 5. In fact, .NET Standard was created when developers had the need to share code across different implementations of .NET, such as Mono (used by Xamarin and the first release of Blazor), .NET Core, and the full .NET Framework. .NET Standard defines a set of APIs through a contract, which must be agreed by all the various implementations of the platform. When a platform implements a specific revision of a contract, it means that it implements all the APIs that are part of the revision.

The currently most widely adopted revisions of .NET Standard are 2.0 and 2.1:

- 2.0 is the best choice when you need to share your code with .NET Framework applications.
- 2.1 is the best choice when supporting .NET Framework isn't a requirement but you need to share your code with a .NET Core or Xamarin application.

.NET Standard has played a critical role in the .NET ecosystem in the past. However, in the long-term future, as developers will gradually move their .NET projects to the new .NET runtime introduced with .NET 5, it won't be required anymore. You will just need to target the lowest version of the .NET runtime across all your projects. For example, if you need to share code between a WinUI application based on .NET 6, a web project based on .NET 7, and a mobile application based on .NET 8, it will be enough to create a class library that targets .NET 6.0 to share it across all of them.

Summary

We started this chapter by learning the fundamental concepts that will guide us through the book. We learned how the Windows App SDK works, how it compares to the other development frameworks, and how you can adopt it in your projects.

Throughout this chapter, we learned what the Windows App SDK is and how we can use it to build new projects or integrate it into existing Windows applications. Then, we reviewed why the new .NET runtime is the future for Windows developers who want to build applications using Visual Studio and C#. We also learned which deployment models you can adopt to distribute your applications.

With this, we're ready to start our modernization journey and to start using all the features that the Windows App SDK has to offer.

In the next chapter, we will learn about the differences between Windows Forms and the Windows App SDK and start with the basics.

Questions

1. Can you integrate the Windows App SDK into an existing Windows Forms application based on .NET Framework?

2. Which is the best technology to deploy a Windows application that uses the Windows App SDK?

3. Is WinUI 3 the only feature included in the Windows App SDK?

Further reading

- *Learn WinUI 3.0, Alvin Ashcraft*

- *C# 9 and .NET 5: Modern Cross-Platform Development – Fifth Edition, Mark J. Price*

Section 2: Modernization Journey

This second section takes you on a modernization journey. Based on your existing application, you can jump in from whichever starting point suits you. This allows you to take your own steps in your modernization journey.

This section contains the following chapters:

- *Chapter 2, The Windows App SDK for a Windows Forms Developer*
- *Chapter 3, The Windows App SDK for a WPF Developer*
- *Chapter 4, The Windows App SDK for a UWP Developer*
- *Chapter 5, Designing Your Application*
- *Chapter 6, Building a Future-Proof Architecture*

2

The Windows App SDK for a Windows Forms Developer

In the previous chapter, we learned the basics of the Windows App SDK. Now, it's time to put the technology into practice by learning how to use it to modernize our existing applications.

Windows Forms is, without any doubt, one of the most popular development technologies in the desktop ecosystem. It was the first desktop platform that was introduced with .NET Framework, and it has remained popular over the years, especially in the enterprise space, thanks to its easy learning curve. Owing to a powerful designer, it's relatively easy to design complex user interfaces, handle user interactions, and perform complex tasks.

Over the course of the years, however, it also started to show multiple limitations:

- Windows Forms was released almost 20 years ago, and as such, it isn't been able to handle modern scenarios, such as high-resolution screens and multiple input types.

- The importance of user experience and good design has grown significantly in the last decade. Windows Forms was designed in a different era, and it doesn't provide many ways to deeply customize the look and feel of an application.

- The **user interface** (**UI**) is deeply connected with the business logic, making it hard to apply patterns that can help to increase the testability, maintainability, and supportability of a project.

These are some of the challenges that led Microsoft to rethink the desktop development experience by introducing a different UI paradigm with **Windows Presentation Foundation** (**WPF**) in 2006. With the advent of XAML, Microsoft created a UI platform based on a markup language, which enables developers to tackle many of the challenges they faced with Windows Forms – better customization options, theming support, separation of concerns, and so on.

These are the reasons why XAML is still at the heart of many Microsoft development technologies, including WinUI. As such, this chapter is dedicated to all the Windows Forms developers or, generally speaking, developers who are coming from different platforms. We'll explore all the most important XAML concepts, from the basics to the most advanced ones, which are critical to building beautiful and robust applications and taking advantage of all the powerful features offered by WinUI.

In this chapter, we'll explore the following concepts:

- What XAML is and how it works

- XAML namespaces

- Properties and events

- Resources, styles, and templates

- Animations

- Data binding

- Data templates

- Creating user controls

In *Chapter 7, Migrating Your Windows Applications to the Windows App SDK and WinUI*, we'll learn how to put in practice many of these concepts to migrate a real Windows Forms application.

Technical requirements

As a reference, we'll use a sample Windows Forms application that you will find on GitHub at the following URL:

```
https://github.com/PacktPublishing/Modernizing-Your-Windows-
Applications-with-the-Windows-Apps-SDK-and-WinUI/tree/main/
Chapter02
```

Introducing XAML

XAML (which is the acronym of **Extensible Application Markup Language**) is a markup language based on XML to define the user interface of an application. The main difference compared to Windows Forms is that XAML enables a declarative approach to define a user interface, which makes it easier for a developer to interpret it just by looking at the code.

For example, consider the following UI:

Figure 2.1 – A simple WinUI application

This is how you describe it using XAML:

```
<Window>
    <StackPanel>
        <TextBlock>One</TextBlock>
        <TextBlock>Two</TextBlock>
        <TextBlock>Three</TextBlock>
    </StackPanel>
</Window>
```

Alternatively, this is how the same UI can be described using C#, as Windows Forms does under the hood when you're working with the designer:

```
public partial class App : Application
{
```

```
private Window m_window;
private StackPanel;
private TextBlock textBlock1;
private TextBlock textBlock2;
private TextBlock textBlock3;

protected override void OnLaunched(Microsoft.UI.Xaml.
  LaunchActivatedEventArgs args)
{
    m_window = new Window();
    stackPanel = new StackPanel();
    textBlock1 = new TextBlock();
    textBlock2 = new TextBlock();
    textBlock3 = new TextBlock();

    textBlock1.Text = "One";
    textBlock2.Text = "Two";
    textBlock3.Text = "Three";

    stackPanel.Children.Add(textBlock1);
    stackPanel.Children.Add(textBlock2);
    stackPanel.Children.Add(textBlock3);

    m_window.Content = stackPanel;

    m_window.Activate();
}
}
```

You will immediately note how the XAML version isn't just more concise; it's also easier to read and interpret. Even if you don't have experience with XAML-based technologies, the hierarchy defined by the XML structure makes it easier to understand the relationship between all the various items in the UI – we have a window, which hosts a panel, which includes three text controls. Understanding the same structure by reading the C# code instead is a tougher challenge.

Using namespaces

If you look at the top-level definition of an XAML file, such as `Window` or `Page`, you will find multiple namespaces declared as attributes, as shown in the following examples:

```
<Window
    x:Class="MyApp.MainWindow"
    xmlns="http://schemas.microsoft.com/
      winfx/2006/xaml/presentation"
    xmlns:x="http://schemas.microsoft.com/winfx/2006/xaml"
    xmlns:local="using:App14"
    xmlns:d="http://schemas.microsoft.com/
      expression/blend/2008"
    xmlns:mc="http://schemas.openxmlformats.org/markup-
      compatibility/2006"
    mc:Ignorable="d">

</Window>
```

The role of namespaces in XAML is the same as in C# – they provide a path that tells the compiler where a specific class is available. All the standard controls and properties provided by WinUI are included in the default platform namespace. This means that you can use controls such as `Button`, `TextBlock`, or `Grid` without having to specify any prefix, as shown here:

```
<TextBlock>One</TextBlock>
```

Instead, if you need to use in your XAML file a class or a control that isn't a part of the framework (for example, it's coming from a NuGet library or it's a custom control you have built), you first have to declare its namespace. For example, let's say that you want to use the following control (we'll learn how to build one later in the *Creating user controls* section):

```
namespace MyApp.Controls
{
    public sealed partial class MyUserControl : UserControl
    {
        public MyUserControl()
        {
            this.InitializeComponent();
```

```
        }
    }
}
```

As you can see, this class is declared as part of the `MyApp.Controls` namespace. As such, before using it, we have first to declare this namespace in our XAML file and assign it an alias by using the `xmlns:<alias>` syntax, as in the following snippet:

```
<Window
    x:Class="MyApp.MainWindow"
    xmlns:controls="using:MyApp.Controls">

</Window>
```

Then, to use the control inside your XAML file, you must add the alias as a prefix, as shown in the next example:

```
<Window
    x:Class="MyApp.MainWindow"
    xmlns:controls="using:MyApp.Controls">

    <controls:MyUserControl />
</Window>
```

Let's see now how we can interact with these controls.

Customizing controls with properties

Every control offers multiple properties that you can use to customize their look, feel, and behavior. For example, a `TextBlock` control offers properties to define the font size, the text color, or the font-weight. Properties are declared as XML attributes, as in the following example:

```
<TextBlock Text="Hello world" FontSize="18" />
```

Properties can be declared in an explicit way as children of the control, as shown in the following snippet:

```
<TextBlock Text="Hello world">
    <TextBlock.FontSize>18</TextBlock.FontSize>
</TextBlock>
```

For simple properties such as `FontSize`, there isn't any added value in using the extended syntax. However, there are more complex properties that can be only set explicitly. For example, let's say that you want to set the background color of a `Button` control with a gradient. In this case, a simple property isn't able to satisfy this requirement, so you'll have to configure it using the following code:

```
<Button Content="Ok">
    <Button.Background>
        <LinearGradientBrush StartPoint="0,0"
            EndPoint="1,0">
            <LinearGradientBrush.GradientStops>
                <!-- Collection type -->
                <GradientStop Offset="0.0" Color="Blue" />
                <GradientStop Offset="0.5" Color="White" />
                <GradientStop Offset="1.0" Color="Red" />
            </LinearGradientBrush.GradientStops>
        </LinearGradientBrush>
    </Button.Background>
</Button>
```

An important property exposed by every control is `x:Name`, which is used to create a reference to the control that you can leverage from the code-behind class. For example, let's add this property to a `TextBlock` control, as shown here:

```
<TextBlock x:Name="txtMessage" />
```

Now, you'll be able to change any property of the control or access any event or method directly in the code-behind class, using the following syntax:

```
txtMessage.Text = "Hello world!";
```

This approach should be familiar if you're a Windows Forms developer, since it works in a very similar way – you assign a name to the control via the designer, and then, in code, you can reference it and customize its properties.

Reacting to changes with event handlers

Most of the XAML controls offer events, which are triggered when a specific condition happens. A very simple example can be found in the `Button` control, which exposes the `Click` event that is triggered when a user clicks on the button. Events also support more complex scenarios – a `TextBox` control exposes an event to notify when text has changed, and a `Window` exposes events to notify when its position or size has changed.

All these scenarios can be managed with an event handler, which is a special method that gets invoked when the event is triggered. Thanks to a handler, you can react to the event in the most appropriate way.

The syntax to define an event handler is the same as the one used for properties. The only difference is that the value of the property must match the name of a method declared in the code-behind class. The IntelliSense feature provided by Visual Studio can generate one for you if it doesn't exist – for example, this is the code you can use to manage the Click event exposed by Button:

```
<Button Click="OnButtonClicked" />
```

This will generate the following event handler in the code-behind class:

```
private void OnButtonClicked(object sender,
    RoutedEventArgs e)
{
    //do something
}
```

As you can see, an event handler isn't a simple method, but it carries two fixed parameters:

- The sender, which is a reference to the control that raised the event. In the previous example, the sender object will contain a reference to the Button control.

- The event arguments, which contain additional information about the event context – for example, if you're working with a list, the event arguments will contain a reference to the item selected by the user.

This special signature is the reason why event handlers are deeply connected with the UI – they can be declared only in the code-behind class. You can't declare an event handler in an external class. As we're going to learn in *Chapter 6, Building a Future-Proof Architecture*, we can use different techniques to break this tight connection so that we can make our applications easier to test and maintain.

Also, in this case, event handlers should already be familiar to Windows Forms developers, except for the declaration syntax. Event handlers can also be declared in code – in this case, the syntax is the same as Windows Forms:

```
public MainWindow()
{
    this.InitializeComponent();
    btnMyButton.Click += OnButtonClicked;
```

```
}

private void OnButtonClicked(object sender,
   RoutedEventArgs e)
{
    //do something
}
```

Now that we have learned the basic XAML concepts, we can start enhancing our controls with resources.

Styling your applications with resources

Resources are one of the most powerful features of XAML, since they simplify the adoption of a unified theme across a whole application. Thanks to resources, you can define a snippet of XAML and reuse it across multiple controls in an application.

For example, let's say that you want a series of TextBlock controls to display text in red. Instead of setting the Foreground property of each control to Red, you can define a resource with this setting, as shown in the following example:

```
<StackPanel>
    <StackPanel.Resources>
        <SolidColorBrush x:Key="MyBrush" Color="Red" />
    </StackPanel.Resources>
</StackPanel>
```

Resources are defined by a special property called x:Key. It's the name we're going to use to reference this resource on our XAML page.

To apply a resource to a property, we need to introduce the concept of **markup expressions**. These are special expressions that you can use in your XAML code that, under the hood, use potentially complex logic to resolve a value. Markup expressions are set like a normal property, but they are included inside brackets. To apply a resource, you must use a markup expression called StaticResource, which will navigate through the XAML tree to find the right resource to apply, as shown in the following example:

```
<StackPanel>
    <StackPanel.Resources>
        <SolidColorBrush x:Key="MyBrush" Color="Red" />
    </Grid.Resources>
```

```
    <TextBlock Text="Hello world!"
        Foreground="{StaticResource MyBrush}" />
</StackPanel>
```

The StaticResource markup expression must be followed by a resource key; in the previous example, it was MyBrush.

The advantage of using a resource compared to setting the Foreground property directly is the reusability it gives. Let's say that the same Grid contains three different TextBlock controls. If, at any point during the life cycle of your project, you need to change the color from red to green, you just need to change the resource once instead of changing the property on every single control.

In the previous snippet, you might have noticed how the resource was defined inside a property called Resources, exposed by the Grid control. This property is exposed by every XAML control, so resources can be defined at any level. One of the most powerful features of the StaticResource markup expression is that it uses a hierarchical navigation approach to search for a resource. In the previous example, since the resource is declared at Grid level, every child control will be able to use it. For this reason, resources are typically defined at the highest possible level (for example, Page) so that every nested control can use them.

This approach also makes it possible to override specific resources if you need deeper customization. The StaticResource markup expression will resolve the resource using the closest one it finds in the hierarchy. For example, let's consider the following XAML code:

```
<Page>
    <Page.Resources>
        <SolidColorBrush x:Key="MyBrush" Color="Red" />
    </Page.Resources>

    <StackPanel>
        <StackPanel.Resources>
            <SolidColorBrush x:Key="MyBrush" Color="Green"
            />
        </StackPanel.Resources>

        <TextBlock Text="Hello world!"
```

```
        Foreground="{StaticResource MyBrush}" />
    </Grid>
</Page>
```

As you can see, we are defining two resources with the same name, which is `MyBrush`. Can you guess which will be the final output of the XAML code?

The answer is green because the `StaticResource` markup expression will use the closest resource to the `TextBlock` control, which, in this case, is the one defined at `Grid` level.

Sharing a resource with a whole application

In the previous section, I mentioned that resources are typically declared at the highest possible level, such as `Page`, so that all the nested controls can use them. But what if I want to share a resource across an entire application?

The `Application` class (defined in the `App.xaml` file), which acts as a global container, offers a `Resources` property as well, where you can declare resources that can be accessed by every `Window` or `Page` of the application, as shown in the following example:

```
<Application>
    <Application.Resources>
        <SolidColorBrush x:Key="MyBrush" Color="Red" />
    </Application.Resources>
</Application>
```

Using dictionaries to organize resources

Declaring resources directly in the `Application` class or inside `Page` can generate confusion in the long term, especially if a project grows and becomes more complex. As such, XAML supports the concept of a **resource dictionary**, which are special XAML files that can contain one or more resources. They are different from a traditional XAML page because they don't have a code-behind class, since they aren't meant to contain any logic.

Thanks to resource dictionaries, you can easily include all your resources in a dedicated file, or even create multiple files to organize your resources in a more logical way. To create a resource dictionary, right-click on your WinUI project and choose **Add | New item**. You will find a template called **Resource Dictionary (WinUI 3)** – give a name to the file and click **Ok**.

This is what a resource dictionary looks like:

```
<ResourceDictionary
    xmlns="http://schemas.microsoft.com/winfx/2006/xaml
      /presentation"
    xmlns:x="http://schemas.microsoft.com/winfx/2006/xaml"
    xmlns:local="using:MyApp">

    <SolidColorBrush x:Key="MyBrush" Color="Red" />

</ResourceDictionary>
```

The file is then registered as an application resource in the App.xaml file:

```
<Application>
    <Application.Resources>
        <ResourceDictionary>
            <ResourceDictionary.MergedDictionaries>
                <XamlControlsResources xmlns=
                  "using:Microsoft.UI.Xaml.Controls" />
                <ResourceDictionary Source=
                  "MyResources.xaml" />
            </ResourceDictionary.MergedDictionaries>
            <!—Other app resources here →
        </ResourceDictionary>
    </Application.Resources>
</Application>
```

The Resources property offers a special property called ResourceDictionary, which you can use to declare all the global resources you want across the application. All the resource dictionary files must be declared inside the MergedDictionaries property, which, at runtime, will merge all the dictionaries inside a single file. In the previous example, we have registered a single ResourceDictionary, but we could have registered as many as we need.

Now that we have learned the basics of how to use resources, let's see how we can use them for more complex scenarios.

Reusing multiple resources with styles

So far, we have explored simple resources, which are made by a single XAML element. However, XAML supports also more complex types of resources, such as styles. If you are familiar with web development, XAML styles and CSS styles are similar, since they serve the same purpose – group together a set of resources so that you can apply all of them just by referencing the style.

This is an example of what a style looks like:

```
<Style x:Key="HeaderText" TargetType="TextBlock">
    <Setter Property="Foreground" Value="Red" />
    <Setter Property="FontSize" Value="24" />
</Style>
```

As with a single resource, a style is identified by a name assigned to the x:Key property. However, we can also see a couple of important differences:

- We must specify a property called TargetType, which specifies the control type that the style is targeting.
- Since a style is used to change multiple properties at once, we have one Setter entry for each property we want to customize. Each Setter specifies the name of the property (through the Property attribute) and the value we want to assign (through the Value attribute).

The previous example defines a style called HeaderText that can be applied to TextBlock controls, which will make the text red and with a size of 24. To apply this style, we can reuse the StaticResource markup expression. This time, however, we won't assign the resource to a specific property but the Style property exposed by all the XAML controls, as shown in the following example:

```
<TextBlock Text="Hello world!" Style="{StaticResource
  HeaderText}" />
```

Styles can also be set in an implicit way by omitting the x:Key property as shown:

```
<Style TargetType="TextBlock">
    <Setter Property="Foreground" Value="Red" />
    <Setter Property="FontSize" Value="24" />
    <Setter Property="FontWeight" Value="Bold" />
</Style>
```

In this case, you won't have to manually set the `Style` property on each control, but all the controls with a type that matches the one specified in the `TargetType` property will get a style automatically applied. In the previous example, the style will be applied to all the `TextBlock` controls.

Styles also support inheritance, thanks to the `BasedOn` property. Consider the following example:

```
<Page.Resources>
    <Style x:Key="HeaderText" TargetType="TextBlock">
        <Setter Property="Foreground" Value="Red" />
        <Setter Property="FontSize" Value="24" />
    </Style>
    <Style x:Key="MainTitle" TargetType="TextBlock"
        BasedOn="{StaticResource HeaderText}">
        <Setter Property="FontWeight" Value="Bold" />
    </Style>
</Page.Resources>
```

We have defined two styles, `HeaderText` and `MainTitle`. The second one, however, extends the first one, by setting an extra property, `FontWeight`. As such, the result is that, when we apply the `MainTitle` resource, we'll display the text in red with a font size of 24 (settings are inherited from the `HeaderText` style) and in bold (which is set by the style itself).

Understanding animations

Animations are another feature that makes XAML quite powerful. They are a special type of resource that enables developers to animate any property of any control in the XAML page by creating a transition that will gradually change its value over time. What makes animations very powerful is that you don't have to manually handle timers, custom drawing, and so on. You just specify which kind of property you want to animate, which are the starting and end values, and for how long the animation must run. The XAML framework will take care of everything else for you by generating all the frames that are needed to create the animation.

Let's see an example:

```
<Page>
    <Page.Resources>
        <Storyboard x:Name="EllipseAnimation">
```

```
            <DoubleAnimation
                    From="1"
                    To="0"
                    Duration="0:0:3"
                    Storyboard.TargetName="MyEllipse"
                    Storyboard.TargetProperty="Opacity"/>
        </Storyboard>
    </Page.Resources>

    <StackPanel>
        <Ellipse Height="80" Width="80" Fill="Red"
            x:Name="MyEllipse" />
    </StackPanel>
</Page>
```

Animations are declared as resources using the `Storyboard` object, which can contain one or more animations that are triggered at the same time. There are multiple types of animations in XAML based on the type of property you want to animate. In the previous example, we're acting on the `Opacity` property, which is a number, so we use a `DoubleAnimation` object. Other examples of animations are `ColorAnimation`, `PointAnimation`, and `ObjectAnimation`.

Regardless of the type you choose, an animation is defined by the following properties:

- `From`: The value of the property when the animation starts. It's optional – if you don't specify it, the animation will use the current value as the starting point.

- `To`: The value of the property when the animation ends. This one is optional as well – in this case, the animation will use the current value as the final point.

- `Duration`: The length of the animation. The runtime will automatically generate the animation's frames based on the duration.

- `Storyboard.TargetName`: The name of the control that you want to apply the animation to.

- `Storyboard.TargetProperty`: The name of the property that you want to animate.

The previous storyboard generates a fade-out animation – the `Opacity` property of the control will change from 1 (fully visible) to 0 (hidden) in 3 seconds.

All the animations also offer a variant implementation with the `UsingKeyFrames` suffix, which puts you in control of the animation timeline. In the previous scenario, the frames of the animation were uniformly split across the duration, creating a smooth effect. In some cases, you might instead want to distribute the frames in a different way – for example, you may want the animation to be faster at the beginning and slow down toward the end.

The following snippet shows how you can achieve this goal:

```
<Page>
    <Page.Resources>
        <Storyboard x:Name="EllipseAnimation">
            <DoubleAnimationUsingKeyFrames Duration="0:0:5"
                    Storyboard.TargetName="MyEllipse"
                    Storyboard.TargetProperty="Opacity">
            <LinearDoubleKeyFrame Value="1" KeyTime="0:0:0" />
            <LinearDoubleKeyFrame Value="0.8" KeyTime="0:0:1" />
            <LinearDoubleKeyFrame Value="0.3" KeyTime="0:0:2" />
            <LinearDoubleKeyFrame Value="0.1" KeyTime="0:0:4" />
            <LinearDoubleKeyFrame Value="0" KeyTime="0:0:5" />
            </DoubleAnimationUsingKeyFrames>
        </Storyboard>
    </Page.Resources>

    <StackPanel>
        <Ellipse Height="80" Width="80" Fill="Red"
            x:Name="MyEllipse" />
    </StackPanel>
</Page>
```

We're using the `DoubleAnimationKeyFrame` object, which, in this case, doesn't need to specify the `From` and/or `To` properties, since we'll control the timeline manually. Inside the object, we add multiple `LinearDoubleKeyFrame` objects, which represent a different step in the timeline. Each of them has two properties:

- `Value`: This is the value that the property will assume in that specific step.
- `KeyTime`: This is the exact point in the timeline when the property will assume the value we have specified.

The previous example still generates a fade-out effect but this time with a different transition – the ellipse will start to fade out fast and then it will slow down at the end.

Playing an animation

You might have noticed how storyboards are a special type of resource. They are declared inside the `Resources` collection of a control or page, but they are identified by the `x:Name` property rather than the `x:Key` one.

This means that, in the code-behind, they can be referenced like a regular control. As such, all you have to do to start an animation is to invoke the `Begin()` method, as shown in the following example:

```
private void OnStartAnimation(object sender, RoutedEventArgs e)
{
    EllipseAnimation.Begin();
}
```

You can subscribe to the `Completed` event if you want to perform a task only when the animation is ended, as shown in the following example:

```
private void OnStartAnimation(object sender, RoutedEventArgs e)
{
    EllipseAnimation.Begin();
    EllipseAnimation.Completed += (obj, args) =>
        {
            //the animation is ended
        };
}
```

WinUI greatly enhanced the basic XAML animation experience by introducing new features such as transitions, composited animations, and built-in control animations. We'll learn more about them in *Chapter 5, Designing Your Application*. However, for the moment, it is important to introduce basic XAML animations, since they will play a significant role in the next section, where we'll discuss template customizations.

Customizing the template of a control

Another feature that makes XAML so powerful is the ability to customize the template of every control. Thanks to this feature, you can completely change the look, feel, or behavior of a control without creating a new one, as you would need to do in Windows Forms.

The template of a control is stored in a property called `Template`, which you can redefine to change it. Let's see, for example, how we can customize the template of a `Button` control:

```
<Button Content="Ok" Background="Red" Width="150"
  Height="50">
    <Button.Template>
        <ControlTemplate TargetType="Button">
            <Grid>
                <Ellipse Height="{TemplateBinding Height}"
                         Width="{TemplateBinding Width}"
                         Fill="{TemplateBinding
                             Background}" />
                <ContentPresenter
                  HorizontalAlignment="{TemplateBinding
                    HorizontalContentAlignment}"
                  VerticalAlignment="{TemplateBinding
                    VerticalContentAlignment}" />
            </Grid>
        </ControlTemplate>
    </Button.Template>
</Button>
```

The key of the template is the `ContentPresenter` object, which represents a placeholder where the content will be placed. You can see, for example, that we have set the `Content` property of `Button` to a label, `Ok`. When the control is rendered, the label will be displayed where we have placed `ContentPresenter` in the template.

Around `ContentPresenter`, we can define the layout we prefer to redefine the look and feel of the control. In this example, we have added `ContentPresenter` together with an `Ellipse` control inside a `Grid` container. This enables us to create a rounded button, as shown in the following screenshot:

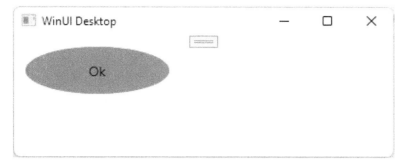

Figure 2.2 – A Button control with a custom template

Note the special `TemplateBinding` markup expression, which enables you to inherit the value from a property defined in the control itself – for example, you can see how the `Ellipse` control has a red background, since its `Fill` property is connected to the `Background` property of the `Button` control through the `TemplateBinding` expression.

If you want to customize the template of a control that is used to display a collection of data, such as `ListView`, there's a small difference – instead of `ContentPresenter`, you must use `ItemsPresenter`, which acts as a placeholder for the list of items assigned as the source. The following code shows a similar example to the previous one (using `Ellipse` as the background) but, this time, applied to a `ListView` control:

```xml
<ListView x:Name="lstPeople" Background="Red" Width="300">
    <ListView.Template>
        <ControlTemplate>
            <Grid>
                <Ellipse Height="{TemplateBinding Height}"
                         Width="{TemplateBinding Width}"
                         Fill="{TemplateBinding
                            Background}" />
                <ScrollViewer>
                    <ItemsPresenter />
                </ScrollViewer>
            </Grid>
        </ControlTemplate>
    </ListView.Template>
</ListView>
```

In the previous example, the `Template` property has been customized directly on the control itself. However, as with any other property, you can also customize it through a style and declare it as an application's resource.

If you try the previous example code in your application, you might notice how, after customizing the template, the button doesn't behave as expected anymore – for example, the hover effect when you mouse over a button is gone.

The reason is that, by redefining the `Template` property, we have completely stripped it of all the extra properties that handle the various states the control can assume (`Pressed`, `Hover`, and so on). For this reason, in a real application, when you want to change the template of a control, you typically customize the default one rather than create a completely new one from scratch.

Unfortunately, this is an area where, at the time of writing, you will face some limitations compared to other XAML-based frameworks such as WPF or Universal Windows Platform. The reason is that, for these technologies, Visual Studio provides a designer, which gives you the possibility to design the UI of your application without launching it. Additionally, the designer offers an easy way to generate a copy of a control's template.

WinUI 3, at the moment, doesn't offer a designer, so you won't be able to use it to get access to the control's default template. Let's explore a couple of workarounds.

Using a Universal Windows Platform project

The **Universal Windows Platform** (**UWP**) also uses WinUI to render the UI of an application. Unlike WinUI 3, Visual Studio supports the designer for UWP apps, so we can use an empty UWP project to copy the templates we need and import them into our Windows App SDK project.

The first step is to create a new UWP project, using the **Blank App (Universal Windows)** template that you'll find in Visual Studio. Once you have created it, double-click on the default page (the one called `MainPage.xaml`) and add the control you would like to customize in your original application – for example, let's add a `Button` control inside the page:

```
<Page>
    <Grid>
        <Button Content="Ok" />
    </Grid>
</Page>
```

At the bottom of the window, you will see a tab called **Design**, near another one called **XAML**. These two tabs can be used to switch between the **XAML** view (where you can see the code) and the **Design** view (where you can see the designer). Click on **Design** to see a preview of your application:

Figure 2.3 – The tab to access the designer in Visual Studio

Now, open the window called **Document Outline**. If you don't see it, you can find it by following the menu and going to **View | Other windows | Document Outline**. This window will show you the XAML tree of the current XAML page, which, in our case, will contain just the Button control we have added. Right-click on the button and choose **Edit Template | Edit a Copy…**:

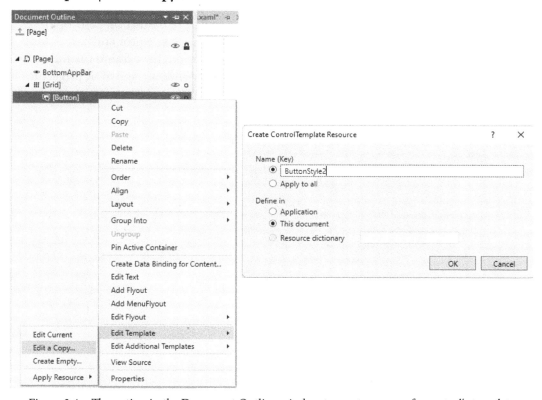

Figure 2.4 – The option in the Document Outline window to create a copy of a control's template

Give it a meaningful name (for example, `MyCustomButtonStyle`) and press **OK**. Now, go back to the **XAML** view. You will find, declared as a resource of the page, a new style with the whole template of the `Button` control, which, as you can see, is much more complex than the one we created before. Now, we just have to copy and paste this new style into our original Windows App SDK application and start customizing it.

Using the XAML source code

The Windows 10 SDK, which is installed together with Visual Studio, includes a file called `generic.xaml`, which contains all the default styles and templates for the built-in controls. You can find this file inside the `C:\Program Files (x86)\Windows Kits\10\DesignTime\CommonConfiguration\Neutral\UAP\10.0.22000.0\Generic` folder.

All the default templates are prefixed by the following comment:

```
<!–Default style for Windows.UI.Xaml.Controls.<name of the
control>
```

Let's say, for example, that you want to customize the template of a `Button` control. In this case, you open the `generic.xaml` file with your favorite text editor and search for the `Windows.UI.Xaml.Controls.Button` string. Once you have found the full template of the control, you can copy and paste it into your Windows App SDK project to start customizing it:

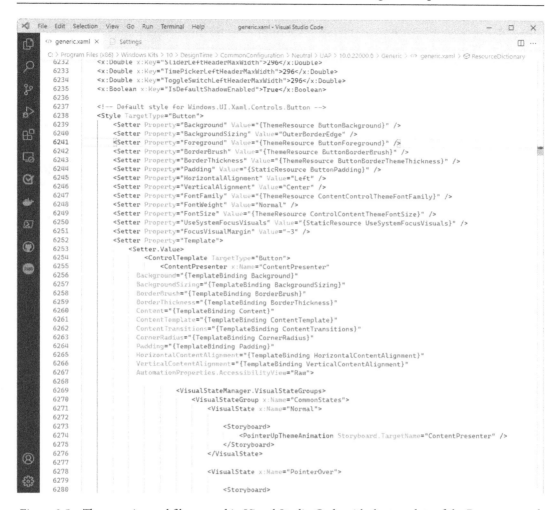

Figure 2.5 – The generic.xaml file opened in Visual Studio Code with the template of the Button control

Instead, if you want to customize the template of one of the most recent controls that were added since WinUI 2.x, you will find the template in the GitHub repository of the project (`https://github.com/microsoft/microsoft-ui-xaml/`). For example, let's say that you want to customize the template of the `InfoBar` control. In this case, you can look for the `InfoBar` folder at `https://github.com/microsoft/microsoft-ui-xaml/tree/main/dev` and open the file called `InfoBar.xaml`. Then, you can copy the template definition into your Windows App SDK project and customize it:

Figure 2.6 – A screenshot of the file on GitHub that contains the template of the InfoBar control

Regardless of your choice, if you take a deeper look at the template, you will notice that, compared to the original one we have created on our own, there's a new object we haven't met before, called `VisualStateManager`. This is the XAML feature that enables all the features that were missing in our custom template, such as the hover animation. Let's see a bit more about how it works.

Managing the states of a control

In the previous section, we saw a glimpse of a new important concept in XAML – visual states. A visual state is a way to represent the look and feel of a control in a specific situation. For example, a Button control can be pressed, disabled, or highlighted. In all these states, Button has a different aspect than the default one, as you can see in the following screenshot:

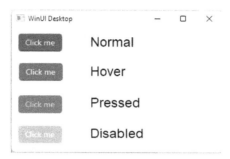

Figure 2.7 – The various states of a Button control

The power of visual states is that they simplify the management of the various states that a control can assume, since you don't have to specify the whole template each time. A visual state just specifies the differences compared to the default template of the control.

As an example, let's take a look at the VisualState definition called Pressed of a Button control:

```
<VisualState x:Name="Pressed">
    <Storyboard>
        <ObjectAnimationUsingKeyFrames
          Storyboard.TargetProperty="Fill"
          Storyboard.TargetName="BorderCircle">
            <DiscreteObjectKeyFrame KeyTime="0"
              Value="{ThemeResource SystemControlBackground
                BaseMediumLowBrush}"/>
        </ObjectAnimationUsingKeyFrames>
        <ObjectAnimationUsingKeyFrames
          Storyboard.TargetProperty="Foreground"
          Storyboard.TargetName="ContentPresenter">
            <DiscreteObjectKeyFrame KeyTime="0"
              Value="{ThemeResource SystemControl
                HighlightBaseHighBrush}"/>
```

```
        </ObjectAnimationUsingKeyFrames>
        <PointerDownThemeAnimation
            Storyboard.TargetName="RootGrid"/>
    </Storyboard>
</VisualState>
```

Instead of redefining the whole template of `Button`, the visual state is just changing a property on two controls included in the template – `BorderCircle` and `ContentPresenter`. By assigning a new color to these two properties, the visual state simulates the pressed effect.

Looking at the XAML code, it should be clear why we have previously introduced animations in this chapter. Each `VisualState` is defined with `Storyboard`, which includes one or more animations that change the value of the properties. As you can see, in most of the cases, these animations are created with the `UsingKeyFrame` variant, with `KeyTime` set to 0. This makes the animation instantaneous, leading to an immediate state change.

The transition from one state to another is controlled by a class called `VisualStateManager`, which is typically used in combination with the events exposed by the control to trigger the switch from one state to another.

For example, consider the following XAML snippet:

```
<StackPanel>
    <VisualStateManager.VisualStateGroups>
        <VisualStateGroup x:Name="ButtonStates">
            <VisualState x:Name="NormalState" />
            <VisualState x:Name="HiddenState">
                <Storyboard>
                    <DoubleAnimation
                        Storyboard.TargetName="MyButton"
                        Storyboard.TargetProperty="Opacity"
                        From="1" To="0" Duration="0:0:3">
                    </DoubleAnimation>
                </Storyboard>
            </VisualState>
        </VisualStateGroup>
    </VisualStateManager.VisualStateGroups>
```

```
<Button Content="Start animation" x:Name="MyButton"
    Click="OnStartAnimation"
            PointerEntered="MyButton_PointerEntered"
            PointerExited="MyButton_PointerExited" />
</StackPanel>
```

Inside VisualStateManager, we have defined VisualStateGroup with two visual states – NormalState and HiddenState. You can see how NormalState doesn't include any storyboard. This means that NormalState matches the default state of the control. HiddenState, instead, contains Storyboard with an animation that hides the Button control included in the page, by changing its opacity from 1 to 0 in 3 seconds.

The Button control itself is subscribed to two events – PointerEntered (which is triggered when the mouse moves over the control) and PointerExited (which is triggered when the mouse moves outside the control). Let's take a look at the two event handlers in the code-behind:

```
private void MyButton_PointerEntered(object sender,
    PointerRoutedEventArgs e)
{
    VisualStateManager.GoToState(this, "HiddenState",
        false);
}

private void MyButton_PointerExited(object sender,
    PointerRoutedEventArgs e)
{
    VisualStateManager.GoToState(this, "NormalState",
        false);
}
```

The transition to another states is triggered by the GoToState() method of the VisualStateManager class, which belongs to the Microsoft.UI.Xaml namespace. As parameters, you must pass the following:

- A reference to the control that is going to change its visual state
- The name of the state
- A boolean value to tell if you want to apply the transition with an animation

If you try this code, you will see the button disappearing when you move the mouse over it (since you will move to the `HiddenState` visual state) and reappearing when you move the mouse outside it (since you will move to the `NormalState` visual state, which is the default look and feel).

The code you've just seen is an example to demonstrate the usage of `VisualStateManager`, but it isn't very different from the actual approach used by every control in the framework to manage their different states. For example, if you look at the implementation details of the `Button` control, you will see that it uses `VisualStateManager` to move to the `Pressed` state when the `Click` event is raised or to the `PointerOver` state when the `PointerEntered` event is raised.

WinUI has introduced a new powerful way to handle visual states – triggers. Thanks to triggers, you don't have to manually subscribe to events and use the `VisualStateManager.GoToState()` API, and instead can focus only on defining the differences between one visual state and the other. We'll cover this topic in *Chapter 5, Designing Your Application*, since triggers play a key role in building responsive UIs.

Data binding

Data binding is one of the most important concepts in XAML, and it's the key to mastering many of the topics that we're going to discuss in the next chapters.

Thanks to data binding, you can create a communication channel between two different properties (either in code or in XAML) so that they get notified every time there's an update. This feature enables us to manage scenarios such as multiple controls that are connected to each other, or to create UIs that can automatically react to changes in a data layer. Data binding also enables us to define a clear separation between the UI layer and the data layer, and it's actually at the heart of the **Model-View-ViewModel** (**MVVM**) pattern, which we'll cover in *Chapter 6, Building a Future-Proof Architecture*.

Binding involves two actors – the source (which is the data layer) and the target (which is the UI control that will display the information). Let's consider the following example:

```
<StackPanel>
    <Slider x:Name="MySlider" />
    <TextBox Text="{Binding ElementName=MySlider,
        Path=Value}" />
</StackPanel>
```

We're using data binding to connect two controls – Slider and TextBox. The binding channel is created using the Binding markup expression. In this context, the binding is made by the following components:

- The Slider control is the source object. It's the one that generates the data that we want to display. We set it using the ElementName property of the binding expression.

- The TextBox control is the target object. It's the one that will display the value coming from the Slider control.

- The Value property of the Slider control is the source path. Slider, as with all the other XAML controls, has multiple properties, so we must specify which one we want to connect to the binding channel. We set it using the Path property of the binding expression.

- The Text property of the TextBox control is the target dependency property. This is the property that we want to set with the value coming from the source.

If you run this code, you will observe that the two controls are now connected – when we move the slider, TextBox will display the slider's value in real time. This is possible thanks to the fact that every property exposed by an XAML control is a dependency property. These are special properties that, other than being able to store a value, can also send a notification to the other side of the binding channel every time the value changes. Thanks to this notification, the UI control knows that it must update its state to reflect the new value.

Binding can support different modes. The default one is OneWay, which is automatically applied when you don't specify the Mode property of a binding expression (as in the previous example). In this mode, the source can update the target but not the other way around. If you test the previous code, you will see that moving the slider will change the value displayed by TextBox; however, if you manually type a new value in TextBox, the Slider control won't move. Typically, you're going to use OneWay communication when you're using, as a target, a control that is only capable of data output, such as TextBlock.

If you want to enable two-way communication, you must explicitly set Mode to TwoWay, as shown in the following example:

```
<StackPanel>
    <Slider x:Name="MySlider" />
    <TextBox Text="{Binding ElementName=MySlider,
      Path=Value, Mode=TwoWay}" />
</StackPanel>
```

You can now see that typing a new value in `TextBox` will change the position of the `Slider`. You typically use `TwoWay` binding with all the controls that enable you to input some data, such as `TextBox` or `DatePicker`.

In the end, there's also `OneTime` mode, which is helpful when you're using binding to set up the initial status of your controls but you aren't planning to have real-time updates. Binding can be expensive from a performance point of view, so it's a good choice to disable real-time notifications if you know in advance that the binding expression will never change. As we're going to learn in *Chapter 6, Building a Future-Proof Architecture*, this mode is quite helpful when you use the MVVM pattern, since you're going to use binding to connect any kind of data to the UI.

Binding expressions are evaluated at runtime, which might lead to performance issues, challenges in finding errors, and so on. To mitigate these problems, WinUI has introduced a new markup expression called `x:Bind`, which we'll explore in the next chapter.

Binding with C# objects

In the previous section, we learned about element binding, in which the source is the property of another XAML control. However, most of the time, the data you need to display in the UI is defined in code. It can be a collection retrieved from a REST API or the result of a query executed on a SQL database. In all of these scenarios, you typically have C# objects that you want to display, using one or more XAML controls.

To support this scenario with data binding, we have to introduce another important concept – `DataContext`. It's a special property, exposed by every XAML control, which defines the binding context. The XAML framework will try to resolve a binding expression by looking for the source and the source path in the binding context you have specified. In the previous example, we didn't need to explicitly set it. Thanks to the `ElementName` attribute, XAML knew where to find the `Value` property that we defined as the source path. In case you need to use C# objects instead as a data source, we must tell the runtime where it can find them.

Let's start with an example, by introducing the `Person` class:

```csharp
public class Person
{
    public string Name { get; set; }
    public string Surname { get; set; }
    public int Age { get; set; }
}
```

Now, let's connect these properties to our XAML page using binding:

```
<StackPanel x:Name="MyPanel" >
    <TextBlock Text="{Binding Path=Name}" />
    <TextBlock Text="{Binding Path=Surname}" />
    <TextBlock Text="{Binding Path=Age}" />
</StackPanel>
```

Out of the box, however, WinUI doesn't know where to find these properties. As such, we have to manually set `DataContext` in the code-behind, as shown in the following example:

```
private void Page_Loaded(object sender, RoutedEventArgs e)
{
    Person = new Person
    {
        Name = "Matteo",
        Surname = "Pagani",
        Age = 38
    };
    MyPanel.DataContext = person;
}
```

We create a new instance of the `Person` class and set it as `DataContext` of the `StackPanel` control. One of the most powerful features of the `DataContext` property is that it's hierarchical. By setting it on the `StackPanel` control, every child control will be able to access its properties. For this reason, it's common practice to set an object as `DataContext` of the entire page so that every control in it will be able to use its properties.

This example, however, has a flaw – if you try to update one of the properties of the `Person` object after setting it as `DataContext`, you won't see the new values showing up in the UI. Let's learn how to solve this problem.

Implementing the INotifyPropertyChanged interface

When we started talking about data binding in this chapter, we saw an example in which we connected two XAML controls using a binding channel. In this scenario, updates were displayed in real time, since every property exposed by XAML controls is a dependency property that can send notifications to the channel every time the value changes.

In the previous example, the `Person` class is exposing normal properties instead. Therefore, if you change one of its values after setting it as `DataContext`, you won't see them in the UI – the property isn't notifying the binding channel that its value has changed. However, dependency properties can be quite complex to declare and maintain, and as such, they aren't a good fit for plain C# objects.

To support this scenario, XAML offers an interface called `INotifyPropertyChanged` that your C# classes can implement to support sending notifications when one of their properties changes values. The following example shows how our `Person` class can be refactored to support this interface:

```csharp
public class Person: INotifyPropertyChanged
{
    private string _name;

    public string Name
    {
        get { return _name; }
        set
        {
            _name = value;
            OnPropertyChanged();
        }
    }

    public string Surname { get; set; }
    public int Age { get; set; }

    public event PropertyChangedEventHandler
        PropertyChanged;

    private void OnPropertyChanged([CallerMemberName]
        string propertyName = "")
    {
        PropertyChanged?.Invoke(this, new
            PropertyChangedEventArgs(propertyName));
    }
}
```

By implementing the `INotifyPropertyChanged` interface, we must expose an event called `PropertyChanged`, which we must invoke every time one of the properties of the class changes. Therefore, as part of the refactoring, we have also to change the way we define our properties. Instead of using automatic properties, we need to explicitly define the `get` and `set` so that when the value of the property changes, we invoke the `OnPropertyChanged()` method.

Now, our class is fully ready to act as a source for data binding. If we change the `Name` property of the `Person` class, the XAML control that acts as a target for the binding channel will update its status to display the new value.

In *Chapter 6, Building a Future-Proof Architecture*, we'll learn how, in a real application, it's unlikely that you will have to implement the `INotifyPropertyChanged` interface on your own (as shown in the previous example). Most of the toolkits and frameworks that help you implement the MVVM pattern provide a base class that does all the heavy lifting for you.

With this understanding, we'll move on to the next section to see how to display data collections.

Displaying collections of data with DataTemplates

Another powerful type of resource is `DataTemplates`, which you can use to customize the look and feel of a control. As the name says, these templates are supported by controls that you can use to display collections of data, such as `ListView` or `GridView`. Thanks to `DataTemplate`, you can define the look and feel of a single item of the collection, which will then be applied to all the items.

This is an example of what `DataTemplate` looks like:

```xml
<ListView>
    <ListView.ItemTemplate>
        <DataTemplate>
            <StackPanel>
                <TextBlock Text="{Binding Path=Name}" />
                <TextBlock Text="{Binding Path=Surname}" />
            </StackPanel>
        </DataTemplate>
    </ListView.ItemTemplate>
</ListView>
```

By looking at the code, you'll probably understand why we are talking about this type of resource only now. As you can see from the snippet, `DataTemplate` uses the `Binding` markup expression to connect the various controls inside the template with the data we want to display. Under the hood, the `DataContext` property of the template becomes a single item from the collection. In this case, we're assuming that we want to use this template to display a collection of people, which are mapped by the following class:

```
public class Person
{
    public string Name { get; set; }
    public string Surname { get; set; }
}
```

The output of the example is that, for each `Person` object included in the collection displayed by the `ListView` control, XAML will render a panel with the name and the surname of the person.

In the previous snippet, `DataTemplate` was defined inline, but it can be treated like a regular resource by assigning the `x:Key` property, as shown in the following example:

```
<StackPanel>
    <StackPanel.Resources>
        <DataTemplate x:Key="PeopleTemplate">
            <StackPanel>
                <TextBlock Text="{Binding Path=Name}" />
                <TextBlock Text="{Binding Path=Surname}" />
            </StackPanel>
        </DataTemplate>
    </StackPanel.Resources>

    <ListView ItemTemplate="{StaticResource
        PeopleTemplate}" />
</StackPanel>
```

With the standard `DataTemplate` configuration, the same template will be applied to all the items in the collection. What if, instead, you want to apply a different template based on a condition? Let's say that, in the previous example, we wanted to use a different template based on the age of the person to highlight people who are below 18 years old. We can support this scenario thanks to a special XAML feature called `DataTemplateSelector`.

The first step to implement this scenario is to create a new class that contains the logic we need to differentiate the various templates. The following example shows the class implementation for the age scenario:

```
public class PeopleTemplateSelector: DataTemplateSelector
{
    public DataTemplate YoungTemplate { get; set; }

    public DataTemplate AdultTemplate { get; set; }

    protected override DataTemplate
      SelectTemplateCore(object item)
      {
          Person = item as Person;
          if (person.Age < 18)
          {
              return YoungTemplate;
          }
          else
          {
              return AdultTemplate;
          }
      }
}
```

The first step is to inherit our class from the `DataTemplateSelector` class, which is included in the `Microsoft.UI.Xaml.Controls` namespace.

Then, we need to define a property for each template we need to support (which are defined using the `DataTemplate` class). In our scenario, we need only two of them – one for young people and one for adults.

In the end, we must override the `SelectTemplateCore` method, which gets invoked when the control that is going to use this selector needs to render an item. In this method, we can add logic that will return one `DataTemplate` object or another, based on the condition we have decided. As a parameter, we get a reference to the item of the collection that is being evaluated. Before using it in a meaningful way, however, we need to cast it first to the data type that we're actually displaying (in our scenario, it's a `Person` object). In this example, we'll return the template called `YoungTemplate` if the person's age is below 18; otherwise, we return the one called `AdultTemplate`.

Now that we have defined `DataTemplateSelector`, we can go back to XAML and define, among our resources, the two templates we need – `AdultTemplate` and `YoungTemplate`:

```xaml
<StackPanel>
    <StackPanel.Resources>
        <DataTemplate x:Key="YoungTemplate">
            <Border BorderBrush="Red" BorderThickness="3">
                <StackPanel>
                    <TextBlock Text="{Binding Path=Name}"
                        />
                    <TextBlock Text="{Binding
                        Path=Surname}" />
                </StackPanel>
            </Border>
        </DataTemplate>
        <DataTemplate x:Key="AdultTemplate">
            <Border BorderBrush="Green"
                BorderThickness="3">
                <StackPanel>
                    <TextBlock Text="{Binding Path=Name}"
                        />
                    <TextBlock Text="{Binding
                        Path=Surname}" />
                </StackPanel>
            </Border>
        </DataTemplate>
    </StackPanel.Resources>
</StackPanel>
```

As a final step, we need to declare the custom class that we have created as a resource as well, as shown in the following example:

```xaml
<StackPanel>
    <StackPanel.Resources>
        <local:PeopleTemplateSelector
            x:Key="PeopleTemplateSelector"
                AdultTemplate="{StaticResource
```

```
                    AdultTemplate}"
                YoungTemplate="{StaticResource
                YoungTemplate}" />


        <DataTemplate x:Key="YoungTemplate">
            <Border BorderBrush="Red" BorderThickness="3">
                <StackPanel>
                    <TextBlock Text="{Binding Path=Name}"
                        />
                    <TextBlock Text="{Binding
                        Path=Surname}" />
                </StackPanel>
            </Border>
        </DataTemplate>


        <DataTemplate x:Key="AdultTemplate">
            <Border BorderBrush="Green"
            BorderThickness="3">
                <StackPanel>
                    <TextBlock Text="{Binding Path=Name}"
                        />
                    <TextBlock Text="{Binding
                        Path=Surname}" />
                </StackPanel>
            </Border>
        </DataTemplate>
    </StackPanel.Resources>


    <ListView ItemTemplateSelector="{StaticResource
        PeopleTemplateSelector}"
                x:Name="lstPeopleWithSelector"
            />
</StackPanel>
```

Note how, in this case, we had to add the namespace prefix before declaring the `PeopleTemplateSelector` object; since it's a custom class we have created, it isn't part of the standard WinUI framework. As part of the definition, we connect the two `AdultTemplate` and `YoungTemplate` properties with the corresponding `DataTemplate` objects that we previously declared.

In the end, we assign the `PeopleTemplateSelector` resource to the `ItemTemplateSelector` property exposed by the `ListView` control. Note how, in this scenario, we no longer need to assign a template to the `ItemTemplate` property.

Let's populate this list now with a few values in the code-behind:

```
private void Page_Loaded(object sender, RoutedEventArgs e)
{
    List<Person> people = new List<Person>
    {
        new Person { Name = "Matteo", Surname = "Pagani",
            Age = 38},
        new Person { Name = "Marc", Surname = "Plogas",
            Age = 27},
        new Person { Name = "John", Surname = "Doe",
            Age = 15 }
    };
    PeopleList.ItemsSource = people;
}
```

This is what the final result looks like:

Figure 2.8 – A ListView control that uses an ItemTemplateSelector component

The first two items in the list are displayed using the `AdultTemplate` resource, since the `Age` property for both is higher than 18. The last item is rendered instead using the `YoungTemplate` resource, since the `Age` property is lower than 18.

In the next section, we will be seeing how to work with user controls.

Creating user controls

WinUI, like Windows Forms, supports the option to create user controls, which is a way to reuse UI components across an entire application. A user control is structured like a page – it has an XAML file, which contains existing XAML controls, and a code-behind file, which includes the logic. For example, let's say that your application offers the option to start a search. You can create a user control specific for this task, which groups together the UI (`TextBox` to write the search term, `Button` to start the search, and so on) and the logic required to perform a search across your data.

An important difference between WinUI and Windows Forms is the way you expose properties. Typically, a user control exposes one or more properties to allow customization when you add it to a specific page of your application. For example, the search box might have a different title based on the page where it's placed. As such, the title can't be hardcoded, but it must be exposed as a property so that we can customize it without being forced to create multiple copies of the same control.

In Windows Forms, you can expose them as normal properties, but in WinUI, you must expose them as dependency properties; otherwise, you won't be able to properly support binding, making the user control less flexible.

Let's start from the beginning and assume we want to build a custom control to render a search box. This control is made by `Label`, which acts as a header; `TextBox`, where the user can type the search keyword; and a `Button` control, which starts the search.

The first step is to right-click on your WinUI project in Visual Studio and choose **Add | New item**. Look for the template called **User Control (WinUI 3)**, give it a meaningful name (for example, `SearchBox`), and press **Add**.

As already mentioned, the user control is structured like a regular WinUI page. The only difference is that the base class isn't `Page` or `Window` but `UserControl`. Therefore, you can add all the elements you need to define the look and feel of your built-in and third-party XAML controls, styles, animations, and so on. For example, this is how we can create a user control in XAML:

```
<UserControl>
    <StackPanel>
```

```
        <TextBlock x:Name="txtSearchHeader" />
        <TextBox PlaceholderText="Search..."
          x:Name="txtSearchText" />
        <Button Content="Search" Click="OnSearch" />
    </StackPanel>
</UserControl>
```

The code-behind file, instead, will look like this:

```
public sealed partial class SearchBox : UserControl
{
    public SearchBox()
    {
        this.InitializeComponent();
    }

    private void OnSearch(object sender, RoutedEventArgs e)
    {
        //start the search
    }
}
```

As we would do on a regular page, we're going to use the `OnSearch()` event handler to add the logic required to start a search.

So far, there aren't many differences compared to a Windows Forms user control, except for the usage of XAML to define the UI.

Let's move on and focus on `TextBlock`, which acts as a header of the box. In this case, we aren't setting a fixed text, since we want to customize it based on the page where this user control will be placed. As such, we'll need to expose a new property called `HeaderText`, which will internally update the `Text` property of our `TextBlock` control. As mentioned, we can't expose it as a normal property, like in the Windows Forms user control; otherwise, we won't be able to use it with data binding. We have to expose it as a dependency property, as shown in the following example:

```
public string HeaderText
{
    get { return (string)GetValue(HeaderTextProperty); }
    set { SetValue(HeaderTextProperty, value); }
```

```
}
```

```
public static readonly DependencyProperty
  HeaderTextProperty =
    DependencyProperty.Register("HeaderText",
      typeof(string), typeof(SearchBox), new
        PropertyMetadata(string.Empty));
```

The `HeaderText` property is backed by a `HeaderTextProperty` object, which is the `DependencyProperty` type. These special properties are registered by calling the `DependencyProperty.Register()` method and passing the following parameters:

- The name of the property (in our case, `HeaderText`).
- The property's type (in our case, it's `string`).
- The control that exposes this property (in our case, it's our `SearchBox` user control). Typically, it's the class itself that is registering the property.
- A `PropertyMetadata` object, which holds some extra information about the property. In this first example, the only information we pass is the default value if the property isn't set, which is an empty string in our scenario.

Now, our control exposes a `HeaderText` property that fully supports binding, styles, and so on. However, we still miss a step. We have defined the property, but we aren't managing the fact that, when this property is set, our `TextBlock` control should display the value of the `HeaderText` property. We can achieve this goal by adding extra information to the `PropertyMetadata` object, which is a callback method that we want to invoke when its value changes. Here is the full implementation:

```
public string HeaderText
{
    get { return (string)GetValue(HeaderTextProperty); }
    set { SetValue(HeaderTextProperty, value); }
}

public static readonly DependencyProperty
  HeaderTextProperty =
    DependencyProperty.Register("HeaderText",
      typeof(string), typeof(SearchBox), new
        PropertyMetadata(string.Empty,
```

```
        OnHeaderTextChanged));

private static void OnHeaderTextChanged(DependencyObject d,
  DependencyPropertyChangedEventArgs e)
{
    var searchBox = d as SearchBox;
    var header = searchBox.FindName("txtSearchHeader") as
      TextBlock;
    header.Text = e.NewValue.ToString();
}
```

We have added to the `PropertyMetadata` object a reference to the `OnHeaderTextChanged()` event handler, which will be invoked every time the value of the property changes. Inside this handler, we're going to change the `Text` property of our `TextBlock` with the new value, which we can retrieve from the `NewValue` property stored in the event arguments.

However, you can see that the code to change the `Text` property isn't as straightforward as we would expect to see. The reason is that `DependencyProperty` is a static object, and as such, we can't directly access the controls declared in the XAML, which aren't static.

The workaround is to use the `DependencyObject` parameter of the event handler, which holds a reference to the object that is exposing the property (our `SearchBox` control). After having converted it to a `SearchBox` object, we can use the `FindName()` method to retrieve a reference to our `TextBlock` control by passing as a parameter its name (`txtSearchHeader`). Finally, after casting it to the control type we're expecting (`TextBlock`, in this case), we can access all of its properties, including the `Text` one we were searching for.

Now, we are ready to include our `SearchBox` control in a page and customize its header by setting the `HeaderText` property, as shown in the following example:

```
<local:SearchBox HeaderText="This is my header" />
```

Since `HeaderText` is a dependency property, we can also safely set it using a binding expression:

```
<local:SearchBox HeaderText="{Binding Path=MyHeaderText}" />
```

Remember that since the `SearchBox` control is a custom one, we need to prefix it with the proper namespace when we add it to a XAML page.

Summary

It's been quite a ride! XAML is a very powerful language and, through this chapter, we have seen some of its important features, such as resources, styles, and binding. Thanks to the knowledge you have acquired so far, you will be able to start designing complex UI interfaces and build real applications using WinUI and the Windows App SDK. If you are a Windows Forms developer, this chapter should have helped you also to understand how many of its concepts are transferrable to XAML and WinUI.

However, what we have learned so far are only the basic concepts. On top of them, WinUI has added many new exciting features, such as transitions, composite animations, and responsive layouts. We'll explore many of them in *Chapter 6*, *Building a Future-Proof Architecture*, in which we'll learn how to design great WinUI experiences. In *Chapter 7*, *Migrating Your Windows Applications to the Windows App SDK and WinUI*, we'll put into practice many of the concepts we've learned so far to migrate a real Windows Forms application.

The next chapter is dedicated to developers who are building WPF, and we'll learn how to translate some of the existing WPF concepts to WinUI. However, even if you haven't worked with WPF before, you might still learn some important concepts, so I suggest you don't skip it!

Questions

1. Which is the best approach to adopt when you want to enable data binding using your C# classes as a data source?

2. The best place to declare a style is close to the control that is going to use it – true or false?

3. By default, when you create a binding channel, both sides (source and target) are notified whenever one of the values changes – true or false?

3

The Windows App SDK for a WPF Developer

In the previous chapter, we introduced a wide range of topics – from XAML basics to binding to user controls. It was one of the biggest chapters of the book, since WinUI is significantly different from Windows Forms when it comes to building the **user interface (UI)**.

WinUI, on the other hand, has a familiar feeling for **Windows Presentation Foundation (WPF)** developers, since it shares many similarities. The most important one, without any doubt, is the UI layer – both technologies are based on XAML. This means that you'll be able to reuse all your existing knowledge when you modernize the UI of your applications.

However, what's the point of a new technology if it doesn't introduce anything new? In this chapter, we're going to learn the main differences between WPF and WinUI so that you can plan accordingly the migration phase of your application.

We'll cover the following topics:

- Importing XAML namespaces
- The new x:Bind markup expression for binding
- Using the dispatcher
- Localizing your application to support multiple languages

By the end of this chapter, you'll have a deeper knowledge of the differences between WPF and WinUI so that you can start planning the migration of your existing applications.

Technical requirements

The code for this chapter can be found at the following link:

```
https://github.com/PacktPublishing/Modernizing-Your-Windows-
Applications-with-the-Windows-Apps-SDK-and-WinUI/tree/main/
Chapter03
```

Importing XAML namespaces

One of the basic features of XAML is the ability to import namespaces so that you can use controls or classes that aren't included in a basic framework but come from a class that belongs to a custom namespace (it could be part of your project, an external class library, or a NuGet package). Namespaces are declared at the top-level container, typically the Window class or the Page class.

In WPF, you use the following syntax to declare a namespace:

```
<Window x:Class="MyApp.MainWindow"
    xmlns="http://schemas.microsoft.com/winfx/2006/
        xaml/presentation"
    xmlns:x="http://schemas.microsoft.com/winfx/2006/
        xaml"
    xmlns:d="http://schemas.microsoft.com/expression/
        blend/2008"
    xmlns:mc="http://schemas.openxmlformats.org/
        markup-compatibility/2006"
        xmlns:internal="clr-namespace:
        MyApp.MyInternalNamespace"
```

```
xmlns:external="clr-namespace:MyExternalLib
    .MyNamespace;assembly=MyExternalLib">
Title="MainWindow" Height="450" Width="800">

</Window>
```

The previous code snippet shows the declaration of two types of namespaces:

- **Internal**: This means that the namespace belongs to the same project that contains the application.

- **External**: This means that the namespace belongs to an external assembly, such as a class library or a NuGet package.

WinUI has simplified the namespace declaration by removing the differences between internal and external namespace. This is how the same namespace definition looks in WinUI:

```
<Window x:Class="MyApp.MainWindow"
        xmlns="http://schemas.microsoft.com/winfx/2006/
            xaml/presentation"
        xmlns:x="http://schemas.microsoft.com/winfx/2006/
            xaml"
        xmlns:d="http://schemas.microsoft.com/expression/
            blend/2008"
        xmlns:mc="http://schemas.openxmlformats.org/markup-
            compatibility/2006"
        xmlns:internal="using:MyApp.MyInternalNamespace"
        xmlns:external="using:MyExternalLib.MyNamespace">
        Title="MainWindow" Height="450" Width="800">

</Window>
```

The other difference is that the `clr-namespace` prefix has been replaced by the `using` one to match the same C# syntax.

There are no differences though in the way you use the namespace inside your XAML code. Let's say, for example, that your external library exposes a control called `MyCustomControl`. To use it in your XAML code, you just have to add the namespace name as a prefix:

```
<external:MyCustomControl />
```

Binding with x:Bind

Binding is, without any doubt, the most powerful feature of XAML. Thanks to binding, you can easily connect the UI with data, keeping the two layers separated. As we're going to see in *Chapter 6*, *Building a Future-Proof Architecture*, binding is at the core of the **Model-View-ViewModel** (**MVVM**) pattern, which is the most productive technique to build applications that are easy to test, maintain, and evolve.

Traditional binding in WPF, however, has a few limitations, caused by the fact that it's always evaluated at runtime. This approach introduces challenges such as the following:

- Performance issues.
- You cannot catch binding errors until an application is running.
- If you want to change the way data is displayed in a UI, you must rely on a converter, which introduces additional overhead.

With the goal of tackling these problems, Microsoft has introduced in WinUI a new markup expression called `x:Bind`. It works in a similar way to traditional binding, but it brings a huge change under the covers – the binding expression isn't evaluated anymore at runtime but at compile time. This approach has the following benefits:

- It improves performance, since bindings are now compiled together with the rest of the code of your application.
- It simplifies error management. Since the binding expression is compiled into code, errors are immediately caught during the build process.
- It opens the opportunity to create a binding channel not only with regular data but also with a function.

Let's see a basic binding sample:

```
<TextBlock Text="{Binding Path=MyText}" />
```

If we convert this expression using x:Bind, we need to use the following syntax:

```
<TextBlock Text="{x:Bind=MyText}" />
```

Looks easy! However, compared to WPF, there are a few key differences. Let's explore them!

Different context

The traditional binding markup expression is deeply connected with the DataContext property exposed by all the XAML controls. Thanks to this property, you can define which is the binding context, thus the control (and all its children) can access all the properties that are exposed by that context. If this concept isn't fully clear to you, you can read *Chapter 2, The Windows App SDK for a Windows Forms Developer* (if you skipped it), where we introduced the basic XAML concepts, including binding.

In WinUI, however, since the x:Bind expression is compiled together with the rest of the code, you can't use DataContext as the default source, but it automatically uses the page or the user control itself. This means that a binding expression created with the x:Bind expression will look for properties, fields, and methods in the code-behind class. To better understand this behavior, try to declare the following property in the code-behind of one of your pages in a WinUI application:

```
public string Name { get; set; }
```

Now, move to the XAML file and add a TextBlock control that uses the x:Bind expression:

```
<TextBlock Text="{x:Bind Name}" />
```

You will notice that, as soon as you type the x:Bind expression, the IntelliSense provided by Visual Studio will trigger and show you all the properties exposed by the code-behind class, including your new Name property.

However, in a real application, this isn't a good example. Even if you're going to change the value of the Name property during the execution, you won't see the update in the UI because Name is a simple property that doesn't implement the INotifyPropertyChanged interface. As such, it isn't able to notify the target of the binding channel (the Text property of the TextBlock control) that the value has changed.

A better example is when you adopt the MVVM pattern, which relies on binding to connect the UI controls in your page to the properties and commands exposed by your ViewModel class. How do we achieve the same with the new x:Bind expression? Using the code-behind class as the ViewModel isn't an acceptable solution, since it would defeat the benefit of using the MVVM pattern to split the UI layer from the business logic. The code-behind class is a special class that is coupled with the XAML page; you can't, for example, store it in a different class library, as you can do with a ViewModel.

The solution is to declare a ViewModel instance as a property in your class. For example, let's say that your view, called `MainPage`, is connected to a ViewModel, called `MainPageViewModel`. In this case, you would need to declare a property of the `MainPageViewModel` type in your code-behind:

```
public MainPageViewModel ViewModel { get; set; }
```

Then, in the public constructor of the code-behind class, you must add the code to initialize the ViewModel. There are multiple options to achieve this goal, based on the way you are implementing the MVVM, such as the following:

- Manually create a new instance of the ViewModel.

- If you're using **Dependency Injection (DI)**, retrieve an instance of the ViewModel from the container.

You can also keep setting the `DataContext` property in the same way as you were doing in WPF. For example, let's say that you are setting the ViewModel directly in XAML, as shown in the following example:

```
<Page x:Class="MyApp.MainPage"
        DataContext="{Binding Source={StaticResource
          ViewModelLocator}, Path=MainPageViewModel}">


        <!-- Page content -->
</Page>
```

You can set the `ViewModel` property in code-behind using the following approach:

```
public MainPageViewModel ViewModel { get; set; }
public MainWindow()
{
    this.InitializeComponent();
```

```
        ViewModel = this.DataContext as MainPageViewModel;
}
```

Once you have initialized the ViewModel property, you can use it with the x:Bind markup expression, thanks to support for property paths, which means that you can simply use a dot (.) to navigate through the properties exposed by your ViewModel. For example, let's say that ViewModel exposes a property called Name, as shown in the following example:

```
public class MainPageViewModel: ObservableObject
{
    private string _name;

    public string Name
    {
        get {   return _name; }
        set { SetProperty(ref _name, value); }
    }
}
```

If you want to connect it to a UI control, you can use the following syntax:

```
<TextBlock Text="{x:Bind Path=ViewModel.Name}" />
```

Default binding mode

In WPF, by default, binding is created in OneWay mode. As such, in the earlier example, every time the value of the Name property changes, the control will update itself to display the new value (as long as the Name property implements the INotifyPropertyChanged interface).

When you use the x:Bind expression, the default binding mode is OneTime instead, which means that the binding will be evaluated only once when the channel is created. In many cases, this new default approach helps to save memory and improve performance, since there are many scenarios where you don't need to keep the UI constantly in sync. However, the consequence is that, if the Name property changes, the UI won't be updated to reflect it. As such, if this is your scenario, it's important to specify the OneWay mode:

```
<TextBlock Text="{x:Bind=MyText, Mode=OneWay}" />
```

Using x:Bind to define a DataTemplate

A quite common scenario in which binding is used is defining DataTemplate. As mentioned in the previous chapter, DataTemplate is used to shape the look and feel of data that is displayed by a control. For example, if you're displaying a data collection using a control such as ListView or GridView, DataTemplate defines the appearance of a single item in the list.

In this context, binding is used to connect the various elements of DataTemplate with the properties of an item that you want to display. For example, let's say you want to display a collection of Person objects, which is defined as follows:

```
public class Person
{
    public string Name { get; set; }
    public string Surname { get; set; }
    public DateTime BirthDate { get; set; }
}
```

The DataTemplate object that can be used to display this collection might look like this:

```
<DataTemplate x:Key="PersonTemplate">
    <StackPanel>
        <TextBlock Text="{Binding Path=Name}" />
        <TextBlock Text="{Binding Path=Surname}" />
        <TextBlock Text="{Binding Path=BirthDate}" />
    </StackPanel>
</DataTemplate>
```

The only required step when switching to x:Bind is to specify the type of data that DataTemplate refers to. Remember that when you use x:Bind, the binding channels are compiled, so Visual Studio needs to know which type of data you're trying to visualize. This goal is achieved by setting the x:DataType template of the DataTemplate object, as shown in the following example:

```
<DataTemplate x:Key="PersonTemplate"
  x:DataType="local:Person">
    <StackPanel>
        <TextBlock Text="{x:Bind Path=Name}" />
        <TextBlock Text="{x:Bind Path=Surname}" />
        <TextBlock Text="{x:Bind Path=BirthDate}" />
```

```
    </StackPanel>
</DataTemplate>
```

Now that we have set the `DataType` property, we can change `Binding` with `x:Bind` in all the markup expressions inside `DataTemplate`.

Event binding

A new feature supported by `x:Bind` is event binding, which means that you can connect events exposed by the controls with the function you want to execute using a binding expression, instead of a traditional event handler. For example, you can implement the `Click` handler of `Button` using the following expression:

```
<Button Click="{x:Bind DoSomething}" Content=
   "Do something" />
```

Then, in the code-behind of your page or user control, you just need to declare a method called `DoSomething`, as shown in the following example:

```
public void DoSomething()
{
    Debug.WriteLine("Do something");
}
```

The method can be parameter-less (as shown in the previous example), or it can match the same signature of the original event handler. This is another example of a valid implementation:

```
public void DoSomething(object sender, RoutedEventArgs e)
{
    Debug.WriteLine("Do something");
}
```

Collection binding

Thanks to the `x:Bind` markup expression, you can also easily connect your UI controls with specific items included in a collection by specifying its index. For example, suppose you have the following collection declared in the code-behind:

```
public ObservableCollection<Person> People { get; set; }
public MainWindow()
```

```
{
    this.InitializeComponent();
    People = new ObservableCollection<Person>
    {
        new Person { Name = "John", Surname = "Doe",
            BirthDate = new DateTime(1980, 10, 1) },
        new Person { Name = "Michael", Surname = "Green",
            BirthDate = new DateTime(1975, 4, 3) }
    };
}
```

If you want to display some information about the second person in the list, you can use the x:Bind expression, as shown in the following example:

```
<TextBlock Text="{x:Bind People[1].Name}" />
```

Another type of collection supported by the x:Bind markup expression is dictionaries. In this case, you can use a similar syntax to the previous one; you just need to specify the key instead of the index. For example, let's say you have the following dictionary declared in the code-behind:

```
public Dictionary<string, Person> PeopleDictionary { get;
  set; }
public MainWindow()
{
    this.InitializeComponent();
    PeopleDictionary = new Dictionary<string, Person>
    {
        { "John", new Person { Name = "John", Surname =
            "Doe", BirthDate = new DateTime(1980, 10, 1) } },
        { "Michael", new Person { Name = "Michael",
            Surname = "Green", BirthDate = new DateTime(1975,
                4, 3) } }
    };
}
```

If you want to display some information about the second item in `Dictionary` (which is identified by the `Michael` key), you can use the following binding expression:

```
<TextBlock Text="{x:Bind PeopleDictionary['Michael']
  .Name}" />
```

Functions

One of the most powerful features exposed by `x:Bind` is function support, which enables you to create complex expressions that can manipulate data coming from the source of the binding channel before it's received by the target.

With standard binding, this feature can be achieved using converters; however, they have some downsides:

- They are evaluated at runtime, adding performance overhead.

- They don't support more than one parameter.

- They are harder to maintain, since they must be declared as XAML resources.

Thanks to functions, you can pass a method declared in the code-behind (or in a static class) as body of a binding expression. For example, let's reuse `DataTemplate` that we previously created in this section:

```
<ListView>
    <ListView.ItemTemplate>
        <DataTemplate x:DataType="local:Person">
            <StackPanel>
                <TextBlock Text="{x:Bind Path=Name}" />
                <TextBlock Text="{x:Bind Path=Surname}" />
                <TextBlock Text="{x:Bind Path=BirthDate}" />
            </StackPanel>
        </DataTemplate>
    </ListView.ItemTemplate>
<ListView>
```

When you use this template to render a collection of `Person` objects, you will get a similar outcome:

Figure 3.1 – A ListView control that displays a list of people

As you can see, we are displaying unnecessary information – since the `DateTime` property is mapped with the birth date of the person, we don't need to display the time as well. Thanks to functions, we just need to add a new method to our `Person` class to achieve the goal:

```
public class Person
{
    public string Name { get; set; }

    public string Surname { get; set; }

    public DateTime BirthDate { get; set; }

    public string ConvertToShortDate(DateTime dateTime)
    {
        return dateTime.ToShortDateString();
    }
}
```

Now, we can change our binding expression in `DataTemplate` to use this new function:

```
<DataTemplate x:DataType="local:Person">
    <StackPanel Margin="12">
```

```
        <TextBlock Text="{x:Bind Path=Name,
          Mode=OneWay}" />
        <TextBlock Text="{x:Bind Path=Surname,
          Mode=OneWay}" />
        <TextBlock Text="{x:Bind Path=ConvertToShortDate
          (BirthDate), Mode=OneWay}" />
    </StackPanel>
</DataTemplate>
```

Now, our list won't contain the time anymore when it displays the birth date of the person:

Figure 3.2 – DataTemplate now displays only the date, thanks to the x:Bind function support

In a more complex application, however, you might not appreciate this code, since it forces you to change the definition of the entity (in this case, the Person class) to add extra code that is specific to the way the data is displayed (the function to format the date). Thanks to property path support, you can easily move this code to a static class, as shown in the following example:

```
public static class DateTimeHandler
{
    public static string ConvertToShortDate(DateTime
      dateTime)
    {
        return dateTime.ToShortDateString();
    }
}
```

Then, you can use the familiar dot (.) notation to use this class in the binding expression, as shown in the following snippet:

```
<DataTemplate x:DataType="local:Person">
    <StackPanel Margin="12">
        <TextBlock Text="{x:Bind Path=Name,
          Mode=OneWay}" />
        <TextBlock Text="{x:Bind Path=Surname,
          Mode=OneWay}" />
        <TextBlock Text="{x:Bind Path=local:DateTimeHandler
          .ConvertToShortDate(BirthDate), Mode=OneWay}" />
    </StackPanel>
</DataTemplate>
```

In this case, however, since `DateTimeHandler` is a different class than `Person`, we need to specify the full namespace (`local:DateTimeHandler`).

Functions enable us to also use standard C# system functions directly in markup without having to build a custom one. For instance, we can achieve the same goal of the earlier example without building a custom `DateTimeHandler` class, directly using the `String.Format()` function included in C# instead, as shown in the following snippet:

```
<Page
    x:Class="App1.MainPage"
    xmlns:sys="using:System"
    mc:Ignorable="d">

    <Page.Resources>
        <DataTemplate x:Key="PersonTemplate"
          x:DataType="local:Person">
            <StackPanel Margin="12">
                <TextBlock Text="{x:Bind Path=Name,
                  Mode=OneWay}" />
                <TextBlock Text="{x:Bind Path=Surname,
                  Mode=OneWay}" />
                <TextBlock Text="{x:Bind
                  Path=sys:String.Format('{0:dd/MM/yyyy}',
                    BirthDate)}" />
            </StackPanel>
```

```
      </DataTemplate>
    </Page.Resources>

</Page>
```

First, you have to map the System namespace to a XAML namespace (in the previous example, we called it sys). Then, you can use the String.Format() function to format the date in the same way you would do in C#. The outcome of the previous snippet will be the same as the one with the custom function – only the date will be displayed, without the time.

Another powerful feature of functions is the ability to support multiple parameters. With converters, you are limited to supplying a single parameter through the ConverterParameter function, as shown in the following example:

```
<TextBlock Foreground="{Binding Path=BirthDate,
  Converter={StaticConverter BudgetColorConverter}
    ,ConverterParameter=20}}"
```

Additionally, since ConverterParameter isn't a dependency property, you can't use a binding expression, meaning that you can't easily pass as a parameter another property of the same binding context (for example, another property of the Person class).

Thanks to functions, both scenarios can be easily achieved, since, in the end, a function is nothing more than a C# method, so you can have as many parameters as you need and use x:Bind to assign them.

To show this feature in action, let's add two new properties to our Person class, Budget and Stocks:

```
public class Person
{
    public string Name { get; set; }
    public string Surname { get; set; }
    public DateTime BirthDate { get; set; }
    public double Budget { get; set; }
    public int Stocks { get; set; }
}
```

Now, let's assume that, in our `DataTemplate`, we want to show the item's background with a different color so that our users can quickly understand the financial status of the person. However, in our scenario, the financial status is based on the budget and the number of stocks, so we need to do some math first. Let's create a new static class that supports our requirement:

```
public static class BudgetColorHandler
{
    public static SolidColorBrush GetBudgetColor(double
      budget, double stocks)
    {
        if (budget < 100 || stocks <10 )
        {
            return new SolidColorBrush(Colors.Red);
        }
        else
        {
            return new SolidColorBrush(Colors.Green);
        }
    }
}
```

The class exposes a function called `GetBudgetColor()`, which returns a `SolidColorBrush` object based on two parameters – budget and stocks. If they are below a certain threshold, the returned color will be red; otherwise, it will be green.

Now, we can customize our `DataTemplate` to use this new function:

```
<DataTemplate x:Key="PersonTemplate"
  x:DataType="local:Person">
    <StackPanel Margin="12" Background="{x:Bind
      local:BudgetColorHandler.GetBudgetColor(Budget,
        Stocks)}">
        <TextBlock Text="{x:Bind Path=Name, Mode=
          OneWay}" />
        <TextBlock Text="{x:Bind Path=Surname,
          Mode=OneWay}" />
        <TextBlock Text="{x:Bind Path=BirthDate)}" />
```

```
    </StackPanel>
</DataTemplate>
```

As you can see, compared to the first example based on converters, thanks to `x:Bind`, we have easily been able to do the following:

- Create a binding expression calculated based on two parameters.
- Create a binding expression where the two parameters aren't static values but are actual properties of the `Person` class.

The following screenshot shows the outcome of the previous example:

Figure 3.3 – DataTemplate using a function with multiple parameters to apply a different background color

Lastly, one of the most powerful features of functions is that they are all re-evaluated every time one of the parameters changes, as long as they properly implement the `INotifyPropertyChanged` interface so that they can notify the binding channel that its value has changed. In the previous example, changing the value of the `Budget` or `Stocks` property will cause a new evaluation of the `GetBudgetColor()` function, potentially leading to a change in the background color applied to the item.

Now that we have learned how to improve the usage of binding in WinUI thanks to the `x:Bind` keyword, we can move on to learn more about the next difference between WPF and WinUI – the usage of the dispatcher.

Using the dispatcher

WinUI, like WPF, uses a single-thread model to manage the UI. This means that, in your code, you can interact with the controls included in your window only from the UI thread. Consider, for example, the following code:

```
Task.Run(() =>
{
    //do some heavy work
    myTextBlock.Text = result;
});
```

By using `Task.Run()`, we are forcing the code to be executed on a background thread. However, if you run this snippet, you will get an exception, as shown in the following screenshot:

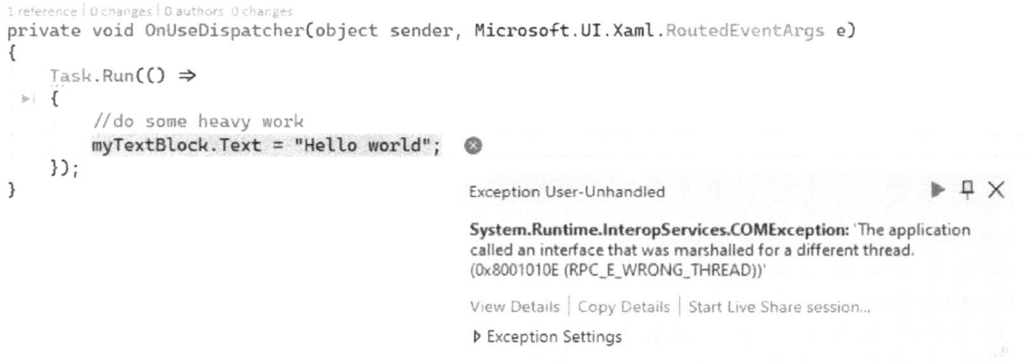

```
1 reference | 0 changes | 0 authors  0 changes
private void OnUseDispatcher(object sender, Microsoft.UI.Xaml.RoutedEventArgs e)
{
    Task.Run(() ⇒
    {
        //do some heavy work
        myTextBlock.Text = "Hello world";   ⊗
    });
}
```

Exception User-Unhandled   ▶ 🗗 ✕

System.Runtime.InteropServices.COMException: 'The application called an interface that was marshalled for a different thread. (0x8001010E (RPC_E_WRONG_THREAD))'

View Details | Copy Details | Start Live Share session...

▷ Exception Settings

Figure 3.4 – The exceptions raised when you try to access the UI from a different thread

This is expected, since we're trying to access UI control from a background thread. To manage these scenarios, all the .NET UI platforms use the concept of a dispatcher, which, as the name suggests, is able to dispatch a specific action from a different thread to the UI thread.

In WPF, the dispatcher is implemented by the `Dispatcher` class, which exposes methods such as `Invoke()` or `BeginInvoke()`.

In WinUI, it's implemented instead by the `DispatcherQueue` class, which belongs to the `Microsoft.UI.Dispatching` namespace. To execute a specific action on the UI thread, you use the `TryEnqueue()` method, as shown in the following example:

```
await Task.Run(() =>
{
    //do some heavy work
    Microsoft.UI.Dispatching.DispatcherQueue.
      TryEnqueue(() =>
    {
        myTextBlock.Text = "Hello world";
    });
});
```

This code won't raise an exception anymore, since the line of code that interacts with the UI (updating the `Text` property of a `TextBlock` control) is added to the dispatcher's queue.

The previous code example works fine if you need to use the `DispatcherQueue` class in a code-behind class, since, in this scenario, WinUI already knows which is the UI thread. If you need to use the dispatcher from a regular class, you will have to obtain first the proper reference for the current thread by using the `GetForCurrentThread()` method, as shown in the following example:

```
var dispatcher = Microsoft.UI.Dispatching.
  DispatcherQueue.GetForCurrent
  Thread();
Task.Run(() =>
{
    dispatcher.TryEnqueue(() =>
    {
        Name = "This message has been updated from a background
            thread";
    });
});
```

This usage of the dispatcher, however, could lead to a few challenges when you need to use it in combination with asynchronous APIs. Let's see how we can improve the code.

Improving dispatcher usage in asynchronous scenarios

The `DispatcherQueue` object has a downside – it doesn't provide support for asynchronous operations. This means that, if you need to execute some code only when the operation you have dispatched has been completed, you will be forced to split your flow into two parts and continue the code execution inside the dispatcher, as shown in the following example:

```
Task.Run(() =>
{
    //do some heavy work
    Microsoft.UI.Dispatching.DispatcherQueue
      .TryEnqueue(() =>
    {
        myTextBlock.Text = result;
        //continue the work
    });
});
```

Thanks to the Windows Community Toolkit (`https://docs.microsoft.com/en-us/windows/communitytoolkit/`), an open source project by Microsoft, we can improve the usage of the dispatcher in asynchronous scenarios, thanks to an extension method.

The first step is to install the `CommunityToolkit.WinUI` NuGet package in your WinUI application. Then, you must add the following namespace in your class:

```
using CommunityToolkit.WinUI;
```

Now, you can use the `EnqueueAsync()` method instead of `TryEnqueue()` to dispatch actions to the UI thread. This method implements the `Task` class, so you can use the `async` - `await` pattern to invoke it, as shown in the following example:

```
Task.Run(async () =>
{
    //do some heavy work
    await DispatcherQueue.EnqueueAsync(() =>
    {
        myTextBlock.Text = result;
    });
```

```
    //continue the work
});
```

As you can see, we don't have to continue the work inside the dispatcher's action anymore. By using the `await` keyword, we can be sure that the rest of the code will be executed only when the action we have enqueued has been completed.

You can also return a value from the dispatched action by using the `EnqueueAsync<T>()` variant of the method, as shown in the following example:

```
string result = await DispatcherQueue
    .EnqueueAsync<string>(() =>
{
    return "Hello world";
});
var output = $"Result: {result}";
```

Now that you have learned how to properly use the dispatcher, we can explore the last important difference between WPF and WinUI – localization.

Localizing your applications

Supporting multiple languages is a key requirement for applications, especially in the enterprise space. Offering an application in the native language of our users improves the user experience and also makes it easier for people who might not be familiar with the technology to use.

WinUI, as with other UI platforms, offers a way to create a UI that can be easily localized by defining placeholders that, at runtime, are replaced with strings taken from a localization file that matches the language of the user.

Localization is one of the areas that has been significantly upgraded from WPF, thanks to a new resource management system called MRT Core, which, other than supporting localization, also enables us to easily load custom resources, such as a different image based on the scaling factor of a device or based on the Windows theme that we set.

The starting point to enable localization is to add a resource file in your application, which is based on the `.resw` extension (unlike in WPF, which used files with the `.resx` extension).

These files simply contain a collection of keys with their corresponding value:

- The key is used as a placeholder, and it will be referenced in our UI and code.

- The value is the localized text that we want to display as a replacement for the placeholder.

As a developer, you won't have to manually load one resource file or another based on the language of the user. Windows will take care of loading the proper resource file. It's enough for you to add resources to your project following one of the available naming conventions. One approach is to create a folder called Strings and, inside it, create one subfolder for each language you need to support, using the language code (for example, en-US or it-IT). Then, inside each folder, include a file called Resources.resw, which will include all the localized resources for the corresponding language code. The following screenshot shows what a project that uses this naming convention looks like:

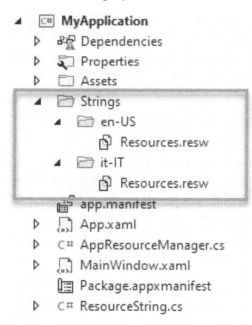

Figure 3.5 – An application that supports localization in English and Italian

This is the suggested approach, since it helps to keep the project's structure clean and well-organized. However, if you prefer, you can also use another approach, which is keeping all the resource files at the root folder of the project but adding the word lang as a suffix (after the filename), followed by the language code the file is related to. The following screenshot shows an example of this approach:

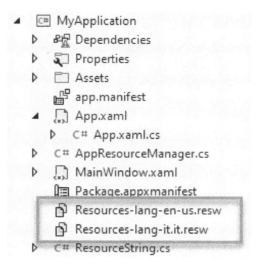

Figure 3.6 – An application that supports localization in English and Italian but using a different naming convention

Both files are placed in the root folder, but they're called `Resources-lang-en-us.resw` (which contains the strings localized in English) and `Resources-lang-it-it.resw` (which contains the strings localized in Italian).

Regardless of the approach you prefer, the way to create a new resource file is the same. You right-click on your project (or on a specific folder, in case you opt in for the approach), choose **Add | New item**, and select the template called **Resources file (.resw)**. Once you have added a resource file, Visual Studio will automatically open the resource editor, as in the following screenshot:

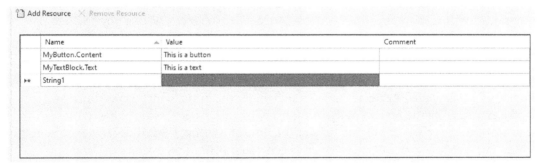

Figure 3.7 – The resource editor included in Visual Studio

As mentioned at the beginning of the section, a resource file is nothing more than a collection of names with their corresponding values. However, before starting to fill the file with one or more resources, it's important to understand how to use them.

Translating the user interface

WinUI makes it easy to use resources to localize the UI, thanks to a special property called `x:Uid`, which can be set for any XAML control. Let's say, for example, that we want to localize the label of a `Button` control, which is defined by the `Content` property. First, we need to assign the name of the resource we want to use to the `x:Uid` property, as shown in the following example:

```
<Button x:Uid="MyButton" />
```

Now, we can add the resource in our resources file, using the `<name of the resource>.<name of the property>` syntax. In our case, since we want to localize the label (which, for a `Button` control, is stored inside the `Content` property), we're going to create a resource named `MyButton.Content`. As the value, we must add the text that we want to display in the language the file refers to. Now, we have to repeat this process for every other resource file in our project. At the end, all the resources files must have a resource called `MyButton.Content` with, as the value, the translated label.

As you can see, this approach is more powerful than the one included in WPF, since it enables us to not only localize the label but also other properties based on the user's language. For example, let's say that we want the `Width` property of the `Button` control to be bigger when the application is localized in Italian, since the Italian label takes more space. We just need to add, only in the Italian resources file, a resource called `MyButton.Width` with, as the value, the size we want to set.

Handling translation in code

There are scenarios in which you need to retrieve a translated string directly in the code of the application (for example, when you need to localize the message of a popup dialog). In this scenario, things are slightly more complicated. Let's understand the reasoning a bit more.

When you enable localization using the special `x:Uid` property, WinUI generates an instance of a class called `ResourceManager` under the hood, which is responsible for properly loading the resources file and retrieving the correct resource based on the value of the property. Unfortunately, at the time of writing, WinUI doesn't expose the `ResourceManager` instance, which is generated behind the scenes, making it impossible to use it to load translations in code.

As such, we have to create and maintain our own instance of the `ResourceManager` class and use it to retrieve localized strings from a resource file. The easiest way to achieve this goal is to create a helper class, which we're going to call `AppResourceManager`, that will take care of managing the `ResourceManager` instance. Here is the full implementation:

```
public sealed class AppResourceManager
{
    private static AppResourceManager instance = null;
    private static ResourceManager _resourceManager = null;
    private static ResourceContext _resourceContext = null;

    public static AppResourceManager Instance
    {
        get
        {
            if (instance == null)
                instance = new AppResourceManager();
            return instance;
        }
    }

    private AppResourceManager()
    {
        _resourceManager = new ResourceManager();
        _resourceContext = _resourceManager
        .CreateResourceContext();
    }

    public string GetString(string name)
    {
        var result = _resourceManager.MainResourceMap
          .GetValue($"Resources/{name.Replace(".", "/")}",
            _resourceContext).ValueAsString;
        return result;
    }
}
```

Here is the most essential information to know about this class:

- To avoid having too many instances of the `ResourceManager` class, which might lead to memory leaks, we use the singleton approach to make sure that every class in our project is always going to reuse the same instance.

- We expose a method called `GetString()`, which returns the translated string based on the name of the resource that we provide. These resources are stored inside the `MainResourceMap` collection of the `ResourceManager` instance. The naming convention to retrieve a resource is *Resources/<name of the resource>/<name of the property>*. However, since in our resource file a resource is defined as *<name of the resource>.<name of the property>*, we have to apply a string replacement.

- You can see that we are also passing a `ResourceContext` object to retrieve a localized string. This object includes the whole application's context (language, scaling factor, theme, and so on), which is set automatically by the operating system. As we're going to see later, we can also use the context to override the default settings.

Once we have added this class to our project, we can simply retrieve a resource in a class using the following code:

```
var translatedLabel = AppResourceManager
  .Instance.GetString("MyButton.Content");
```

A better way to manage localization

The `x:Uid` property helps to implement localization in an easier way, compared to WPF. However, it still has a few limitations. For example, it makes it impossible to reuse the same resource for multiple controls, since each resource name includes a reference to `x:Uid` (and you can't have two controls on the same page with the same `x:Uid`). Another critical scenario is enabling users to choose their own preferred language. By default, WinUI will load the resource file based on the display language configured in the Windows settings. You can find this property in the Settings application by following the **Time & Language | Language & Region | Windows display language** path, as in the following screenshot:

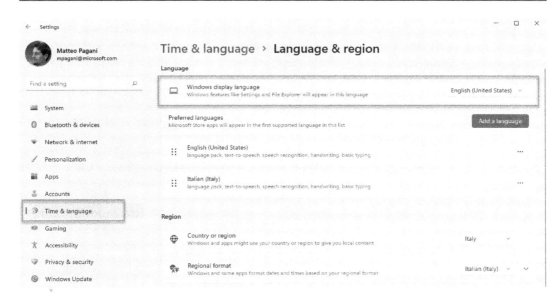

Figure 3.8 – The settings in Windows 11 to customize the Windows display language

However, there are scenarios in which you might want to allow your users to set a different language than the one used for the operating system. In this case, we can leverage the ResourceContext class to force a different context, instead of using the default one. For example, if we want to force our application to use resources in Italian, we can customize the public constructor of the AppResourceManager class that we previously created in the following way:

```
private AppResourceManager()
{
    _resourceManager = new ResourceManager();
    _resourceContext =
      _resourceManager.CreateResourceContext();
    _resourceContext.QualifierValues["Language"] = "it-IT";
}
```

By overriding the Language qualifier, our ResourceManager instance will start to retrieve resources for the specific chosen language, instead of the default one provided by the operating system.

However, there's a catch if you're using the `x:Uid` property to handle localization. Remember that, in a WinUI app, we don't have a way to retrieve the `ResourceManager` instance that is automatically used by the `x:Uid` property, so we had to create a new one inside the `AppResourceManager` class. This means that we can't set this property on the built-in instance, which will lead to a disconnection between UI and code:

- In code, since we're using the `AppResourceManager` class, we'll start to retrieve strings in the language we have forced.

- In XAML, since we're using the `x:Uid` property, WinUI will continue to use the default display language in Windows to choose the resource to load.

As such, we can enable a better way to manage localization (inspired by the following blog post from a Microsoft engineer: `https://devblogs.microsoft.com/ifdef-windows/use-a-custom-resource-markup-extension-to-succeed-at-ui-string-globalization/`) by using a custom markup expression, which we're going to use as a replacement for the `x:Uid` property. First, let's create the markup expression by adding the following class to our project:

```
[MarkupExtensionReturnType(ReturnType = typeof(string))]
public sealed class ResourceString : MarkupExtension
{
    public string Name
    {
        get; set;
    }

    protected override object ProvideValue()
    {
        string value = AppResourceManager
            .Instance.GetString(Name);
        return value;
    }
}
```

As you can see, the `ProvideValue()` method (which defines the value to return when the markup expression is used in XAML) uses the same `AppResourceManager` instance that we are using across our code, giving us the consistency we need across the entire application.

Now, we can remove the x:Uid property from our Button control in the XAML page and replace it with our markup expression, as shown in the following example:

```
<Button Content="{local:ResourceString
   Name=MyButton.Content}" />
```

An important difference, compared to x:Uid, is that we need to use the markup expression to explicitly set the property we want to change; we can't rely anymore on the *<name of the resource>.<name of the property>* naming convention. In the previous example, since we're localizing the Button label, we're using the ResourceString markup expression to set the Content property.

This approach has multiple advantages:

- We can simplify resource files, since we can use a plain string to define a resource (for example, MyButtonLabel).

- We can reuse the same resource with multiple controls.

- Since we don't have a reference to the XAML control in the resource name anymore, it's easier to do refactoring if we need to change the UI.

Support localization in a packaged app

If you're using the packaging approach to distribute your application, there are two extra steps you need to take:

- Declaring the supported languages in the application manifest

- Defining what the default language is for your application

To accomplish the first task, right-click on the Package.appxmanifest file included in your project, choose **View Code**, and locate the following entry:

```
<Resources>
  <Resource Language="x-generate"/>
</Resource
```

Delete it and replace it with a new `Resource` entry for each language you want to support and set, as the value of the `Language` attribute, the language code. For example, since in our previous example we supported English and Italian, our entry will look like this:

```
<Resources> a method declared in the code-behind (or in a
   static class)
   <Resource Language="en-US"/>
   <Resource Language="it-IT"/>
</Resource
```

The second task can be accomplished from the manifest as well. However, this time, we can use the visual editor, so it's enough to double-click on the `Package.appxmanifest` file in Visual Studio and to move to the **Application** tab. You will find a property called **Default language**, which, by default, is set to `en-US`. This is the suggested configuration – if you are using Windows in a language that you don't directly support, the application will be displayed in English. However, if your application targets a specific market (for example, it's meant to be used only in Germany), you are free to change this value to a more appropriate one.

Summary

In this chapter, we have explored the key differences between WPF and WinUI when it comes to building the UI of our applications. Compared to a Windows Forms developer, we have a significant advantage, since many of the basic concepts behind WinUI are the same ones as WPF because they are both based on XAML as a markup language. However, WinUI is a newer and more modern UI framework, and as such, we can enjoy some new powerful features, such as the new `x:Bind` expression to enable compiled bindings and the new localization approach.

Thanks to the skills acquired in this chapter, you now have a path forward to modernize your existing WPF applications by creating a new and modern UI experience, thanks to the new WinUI features.

However, when you start building an application based on WinUI, you will find other differences compared to WPF. For example, the way you enable navigation is different, there are many new controls that you can leverage to build the UI, and WinUI has a much more powerful rendering system that supports creating beautiful animations.

We didn't cover these features in this chapter though, as they aren't specific for WPF developers, but they do affect Windows Forms developers. As such, we will discuss them in *Chapter 5, Designing Your Application*, which is entirely dedicated to designing great UIs with WinUI.

The next chapter will be dedicated to developers who already have experience in building Windows applications with Universal Windows Platform and who would like to move them forward by adopting WinUI and the Windows App SDK.

Questions

1. If you need to format a string that is managed by a binding channel created with x:Bind, the only option is to create a custom function – true or false?

2. When you use the DispatcherQueue class, there's no way to wait for the action dispatched to the UI thread to be completed before performing other tasks – true or false?

3. In a WinUI application, you must detect in code what the current display language in Windows is and load the proper resource file accordingly – true or false?

4

The Windows App SDK for a UWP Developer

The **Universal Windows Platform** (**UWP**) was introduced in Windows 10 with the goal of delivering a modern development platform for Windows. Many valuable concepts that are important to provide a modern experience nowadays were included from the beginning, such as support for new input paradigms or advanced UI scaling capabilities.

As such, it shouldn't come as a surprise that the Windows App SDK can be considered a direct successor of UWP, as it reuses many of the same building blocks and fundamental UWP concepts. Some of the Windows App SDK features even come directly from UWP itself. The UI layer is, without a doubt, the most prominent one: WinUI 3 does, in fact, use the same UI layer as UWP but detached from the operating system so that it can also be consumed by Win32 apps developed with .NET or C++.

The result is that if you have experience building apps with UWP, you have a great advantage over a Windows Forms or WPF developer. You will be able to reuse most of your knowledge to build great Windows App SDK applications. However, with UWP being based on a different runtime, when you move to the Win32 ecosystem, there are some important differences around some key concepts, such as the application's life cycle or windows management.

Specifically, we'll cover the following topics in this chapter:

- Moving to a new namespace
- Working with the UI thread
- Controlling the application's life cycle
- Supporting the application's activation
- Application instancing
- Managing the application's window
- Performing operations in the background

By the end of this chapter, you'll have gained the knowledge required, from a UWP perspective, to start new projects or migrate your existing ones.

Technical requirements

The code for the chapter can be found here:

```
https://github.com/PacktPublishing/Modernizing-Your-Windows-
Applications-with-the-Windows-Apps-SDK-and-WinUI/tree/main/
Chapter04
```

Moving to a new namespace

When you are building an application with UWP, your home is the `Windows` namespace. Every API that belongs to this ecosystem is included in one of its subnamespaces. One of the most important is `Windows.UI.Xaml`, which includes all the building blocks to create the UI: controls, dependency properties, styles, and so on.

When moving the UI layer to the Windows App SDK, the team had to create a different namespace, to avoid conflicts with the one built inside the operating system. As such, all the UI building blocks for WinUI are included in the `Microsoft.UI.Xaml` namespace. Compared to a traditional UWP application, this introduces the following two main changes:

- If you want to port over some code you have created in UWP, you will have to change every reference to the `Windows.*` namespace to `Microsoft.*`.

- When you create a new WinUI application, you will find the following declaration inside the `App.xaml` file:

```
<Application>
    <Application.Resources>
```

```
        <ResourceDictionary>
            <ResourceDictionary.MergedDictionaries>
                <XamlControlsResources
                    xmlns="using:Microsoft.UI.Xaml
                    .Controls" />
            </ResourceDictionary.MergedDictionaries>

            <!—Other app resources here >
        </ResourceDictionary>
    </Application.Resources>
</Application>
```

This declaration will make sure that every UI control you use in the application (including basic ones, such as `Button` or `TextBlock`) comes from the WinUI library and not from the operating system.

If you want to know more about using namespaces in WinUI, you can refer to *Chapter 3, The Windows App SDK for a WPF Developer*.

Working with the UI thread

In *Chapter 3, The Windows App SDK for a WPF Developer*, you already met the `Dispatcher` class. It's a special class provided by the XAML framework that you can use to dispatch a task on the UI thread. Since every WinUI application has a single UI thread, `Dispatcher` becomes especially important when you are executing some code on a background thread but then, at some point, you have to interact with a UI control. Without using the `Dispatcher` class, the operation would raise an exception, since you can't update the UI from a different thread than the UI one.

UWP exposes a `Dispatcher` class on every page, which belongs to the `Windows.UI.Core.CoreDispatcher` namespace. Here is an example of how to use it:

```
await Dispatcher.RunAsync(Windows.UI.Core.
    CoreDispatcherPriority.Normal, () =>
{
    txtHeader.Text = "This is a header";
});
```

Actions on the UI thread are dispatched using the `RunAsync()` method, which is asynchronous. The Windows platform, however, offers another API called `DispatcherQueue`, which supports all types of Windows applications: Win32, UWP, and so on. If, in the UWP ecosystem, you had the option to choose between the two, when it comes to the Windows App SDK and WinUI, you must use `DispatcherQueue` since `CoreDispatcher` is built around concepts that are available only in the Windows Runtime ecosystem.

This is an example of how to use `DispatcherQueue`:

```
DispatcherQueue.TryEnqueue(() =>
{
    txtHeader.Text = "This is a header";
});
```

As mentioned earlier, the `DispatcherQueue` class (which belongs to the `Microsoft.UI.Dispatching` namespace) is already available in UWP. However, you must manually obtain a reference to it, since it isn't automatically exposed by XAML controls. In WinUI, instead, this class is directly exposed by the `Window` and `Page` classes, which means that you can use it directly in code-behind. If, instead, you need to use `Dispatcher` in a regular class, you must retrieve a reference first using the `GetForCurrentThread()` method, as shown in the following example:

```
var dispatcher = DispatcherQueue.GetForCurrentThread();
dispatcher.TryEnqueue(() =>
{
    txtHeader.Text = "This is a header";
});
```

Compared to the `Dispatcher` class in UWP, the `DispatcherQueue` API has a disadvantage: it isn't asynchronous, so you aren't able to wait until the operation is completed before moving on with other tasks. To improve the development experience, you can use the Windows Community Toolkit, as explained in *Chapter 3, The Windows App SDK for a WPF Developer*. Thanks to this library, you will have access to an extension method called `EnqueueAsync()`, which you can use as in the following sample:

```
await DispatcherQueue.EnqueueAsync(() =>
{
    myTextBlock.Text = result;
});
//continue the work
```

Thanks to this asynchronous extension method, you can replicate the behavior of the RunAsync() method exposed by the Dispatcher class available in UWP.

You can refer to *Chapter 3, The Windows App SDK for a WPF Developer*, for more information on how to integrate the Windows Community Toolkit and this extension.

Now that we know how to properly work with the UI thread, we can move on to another significant difference between UWP and Windows App SDK applications: the application's life cycle.

Controlling the application's life cycle

One of the biggest differences with the UWP model, compared to the more traditional Win32 one, is the application life cycle. Since UWP apps were originally created to run on multiple devices (including less powerful ones compared to a computer, such as mobile phones or tablets), Windows implements a series of optimizations to avoid an application consuming too much memory or CPU, making the system unstable.

Here are some of the key differences compared to the life cycle of a Win32 application:

- **Applications aren't able to run in the background**: A few seconds after an application has been minimized to the taskbar, it gets suspended and isn't able to perform any operations. If you need to perform operations in the background, you must use special features, such as background tasks.

- **When an application is suspended, it doesn't consume CPU (since it can't perform any task) but it still uses memory**: The process, in fact, is kept alive so that if the user goes back to the application, it's instantly resumed and they can continue where they left from. However, if the system is low on memory, Windows can terminate one or more suspended applications so that it can recover enough memory to launch other processes. In this scenario, it's up to the developer to save the state so that when the application is reopened, it can resume the previous status.

These features are powerful because they help to deliver applications that are respectful of the system resources. However, when you are working on enterprise apps, you might need more flexibility, such as being able to run a long-running task in the background. Windows App SDK applications are based on the Win32 model and, as such, they have a more traditional life cycle that isn't controlled by Windows. For example, they can perform operations in the background without any limitations and they never get terminated, unless an exception happens.

However, this doesn't mean that a Win32 application shouldn't be respectful of the system resources. The main difference compared to UWP is that the responsibility is moved from the operating system to the developer. For this purpose, the Windows App SDK includes a class called `PowerManager`, which belongs to the `Microsoft.Windows.System.Power` namespace and provides information about the current status of the device from a power perspective—whether it's running on battery or AC, how much battery is left, whether the user is present or not, and so on.

As a developer, you can use this information to optimize your application; for example, you can pause heavy tasks if the computer is low on battery, or you can kick out a long-running background process if the user is away. Let's see a brief example of this:

```
private void PerformTask()
{
    if (PowerManager.PowerSourceKind == PowerSourceKind.DC
      && PowerManager.BatteryStatus ==
      BatteryStatus.Discharging &&
      PowerManager.RemainingChargePercent < 25)
    {
        PauseLongRunningWork();
    }
}
```

In the preceding code snippet, you can see three different types of information exposed by the `PowerManager` API:

- `PowerSourceKind`, which tells you whether the computer is running on battery or connected to a power plug
- `BatteryStatus`, which tells you whether the battery is charging or discharging
- `RemainingChargePercent`, which tells you the current charge percentage of the battery

In the previous example, we're using these properties to determine whether it's OK to execute a long-running task—if the computer is running on battery and the charge is below 25%, we pause the execution. Another approach supported by the `PowerManager` API is event subscription. Every property exposed by the API also has an equivalent event, which is triggered when the value changes. For example, the `PowerSource` property is paired with a `PowerSourceKindChanged` event, which is triggered when the computer moves from battery to power adapter and vice versa; or the `RemainingChargePercent` property has a corresponding `RemainingChargePercentChanged`, which is triggered when the battery charge percentage changes.

Let's take a look at the following example for this:

```
public MyPage()
{
    this.InitializeComponent();
    PowerManager.PowerSourceKindChanged +=
      PowerManager_PowerSourceKindChanged;
}

private void PowerManager_PowerSourceKindChanged(object sender,
  object args)
{
    if (PowerManager.PowerSourceKind == PowerSourceKind.DC)
    {
        PauseLongRunningWork();
    }
    else
    {
        ResumeLongRunningWork();
    }
}
```

We are applying a logic similar to the one in the previous snippet, except that this time, we're proactively reacting to events. We use the PowerSourceKindChanged event to detect when the power source changes and, in the event handler, we use this information to determine whether the computer is running on battery (in this case, we pause the long-running task) or on mains power (in this case, we resume the job).

You can notice how the information about the new status isn't returned as an event argument (args is a generic object), but you have to use the PowerManager API to retrieve it.

There are other properties that you can retrieve using the PowerManager API:

- DisplayStatus tells you whether the screen is on, off, or dimmed. You can use this information, for example, to stop rendering UI graphics and start background tasks if the screen is off since it means that the user isn't interacting with the device.
- EffectivePowerMode tells you which kind of energy profile is currently applied, such as **Balanced**, **High Performance**, **Battery Saver**, or **Game Mode**.
- RemainingDischargeTime returns TimeSpan with information on how long the computer can continue to run before the battery is completely discharged.

Thanks to these properties, you can further optimize the execution of tasks that might be more impactful on battery and system performance.

Now it's time to analyze another critical difference between UWP and the Windows App SDK ecosystem: how to manage the way an application gets activated.

Supporting the application's activation

UWP offers a rich set of activation contracts, which enables the application to be launched by multiple actions—file type association, protocol association, sharing, and automatic startup are just a few examples.

To handle these scenarios, the App class in a UWP application provides multiple hooks that you can override, such as OnFileActivated (when the app is opened using a file type association), OnShareTargetActivated (when the app is opened using a share contract), or the more generic OnActivated, which can be used to handle all other scenarios for which there isn't a dedicated handler.

The UWP model offers dedicated hooks since, by default, applications are run in a single-instance mode. This means that if the application is running and it gets activated by another action (such as the user double-clicking on a file and then which type is associated with the app), the already activated instance will receive the information about the activation context. The App class also provides an event called OnLaunched, which, however, is used only to initialize the application at the first launch. It doesn't get invoked every time the application is activated, thus the need to have separate activation events.

In the Windows App SDK application model, instead, applications run as multi-instance by default. Every time the application gets launched, regardless of the activation path, a new instance will be created. This is the default model in the Win32 ecosystem and, on your machine, you can see many examples of applications that behave like this. Take, for example, the Office suite: every time you double-click on a Word file, it will be opened in a separate instance.

Due to this different model, a Windows App SDK application based on WinUI doesn't expose any more different activation events, but only a single OnLaunched event, which will be triggered every time the application gets launched.

Let's take a look at the default OnLaunched event of a WinUI application:

```
public partial class App : Application
{
    public App()
    {
```

```
            this.InitializeComponent();
    }


    protected override void OnLaunched(Microsoft.UI.Xaml
        .LaunchActivatedEventArgs args)
    {
        m_window = new MainWindow();
        m_window.Activate();
    }


    private Window m_window;
}
```

The default implementation is quite simple: it just creates a new instance of MainWindow and activates it so that it can be displayed. The Windows App SDK offers a class called AppInstance, which belongs to the Microsoft.Windows.AppLifecycle namespace, to manage application instances and you can use it to retrieve the activation context.

Let's assume that our application can be associated with one or more file types and, as such, you need to change the OnLaunched implementation to accommodate this requirement. This is an example of how you can do it:

```
protected override void OnLaunched(Microsoft.UI.Xaml.
    LaunchActivatedEventArgs args)
{
    m_window = new MainWindow();
    var eventArgs =
        AppInstance.GetCurrent().GetActivatedEventArgs();
    if (eventArgs.Kind == ExtendedActivationKind.File)
    {
        var fileActivationArguments = eventArgs.Data as
            FileActivatedEventArgs;
        m_window.FilePath =
            fileActivationArguments.Files[0].Path;
    }

    m_window.Activate();
}
```

As you can notice, we aren't using the `LaunchActivatedEventArgs` parameter of the event, since it doesn't actually contain the information we need to get the context. The right way to retrieve it is by using the `AppInstance` class—after retrieving a reference to the current instance with the `GetCurrent()` method, we can get the activation arguments by calling `GetActivatedEventArgs()`. The first lot of information that we can retrieve from the arguments is the activation type through the `Kind` property (whose type is `ExtendedActivationKind`) since our application might support different activation paths. In the previous example, our application supported only file activation, so we are checking whether the `Kind` property is equal to `ExtendedActivationKind.File`. If that's the case, it means the user has double-clicked on one or more files that we support, so we need to retrieve their path so that we can handle them in the right way. A common implementation is to redirect users to a dedicated page or window of the application, where they can see the content of the file.

To do this, however, we first have to get the selected files from the `Data` property of the arguments, which contains the activation context. This property is generic across all the various activation paths and, as such, we first have to cast it to the type we're expecting. Since in this case we're working with files, the `Data` property will contain a `FileActivatedEventArgs` object; if, for example, it were a protocol activation, the type would have been `ProtocolActivatedEventArgs`.

Now that we have converted the `Data` property to the correct type, we can use the `Files` collection to retrieve the list of all the files that the user is trying to open. In this example, we're assuming that the user can open only one file at a time, so we retrieve the path of the first item in the collection and store it in a property called `FilePath`, which we have created in the `MainWindow` class.

The window can now use this property to handle the scenario based on the requirements. For example, we can add some logic that uses the file path to load the content of the file and display it on the screen.

We have seen so far how to manage applications that can be activated in different ways, but how can we enable these activation paths? Let's check this in the next section.

Supporting multiple activation paths

If you're using the Windows App SDK packaged model (thus, you're distributing your application using MSIX), you should feel right at home if you have experience with UWP. The way you enable activation paths is the same as for a UWP application—through the manifest. If you double-click on **Package.appxmanifest** in your project and move to the **Capabilities** tab, you will find many activation paths that you can add to your application. For example, if you want to support files, you can choose **File Type Association** from the drop-down menu and configure it in the following way:

1. Under **Name**, give a meaningful name to the extension, such as the file type you want to support.

2. Under **Supported File Types**, add one or more entries for each type you want to support. The critical information to fill in is the **File type** field, which must contain the extension you want to support (such as . foo). This is shown in the following screenshot:

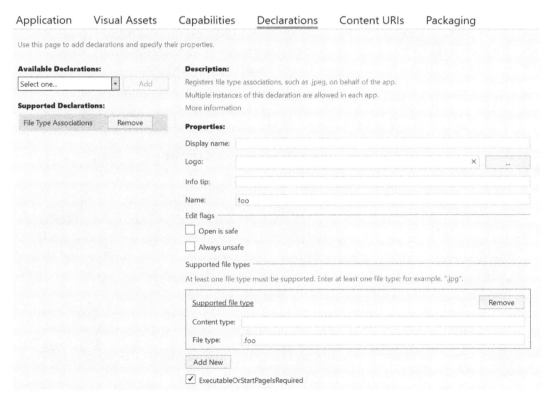

Figure 4.1 – The section of the manifest where you can set up custom
activation paths, such as file association

When the application is packaged, you can enable all the activation types of UWP, which you can find documented at https://docs.microsoft.com/en-us/uwp/api/ Windows.ApplicationModel.Activation.ActivationKind. Additionally, the packaged model provides a smoother developer experience, since activations are automatically registered when the application is deployed and unregistered when the application is removed.

Unpackaged applications, instead, at the time of writing the book, can use four specific activation types supported by the Windows App SDK:

- Launch
- File
- Protocol
- StartupTask

Since an unpackaged application behaves like a traditional Win32 application, you can register for these activations directly in the registry, as you would do with a Windows Forms, WPF, or C++ application. For example, this is how you can enable file type association: https://docs.microsoft.com/en-us/windows/win32/shell/ how-to-register-a-file-type-for-a-new-application.

However, the Windows App SDK simplifies the registration for these four paths by providing APIs you can invoke when the application is initialized. These APIs will take care, on your behalf, of adding the proper registry keys to enable the activation scenario.

These APIs are exposed by the ActivationRegistrationManager class, which belongs to the Microsoft.Windows.AppLifecycle namespace. For example, the following snippet shows how to register a file type association:

```
ActivationRegistrationManager.RegisterForFileTypeActivation(new
    string[] { ".foo", ".fee" },
    "logo.png", "File Type Sample",
    new string[] { "View" },
    Process.GetCurrentProcess().MainModule.FileName);
```

The RegisterForFileTypeActivation() method requires the following parameters:

- A collection of the file types you want to register. In this sample, we're registering to support the .foo and .fee extensions.
- Optionally, the path of the image you want to use as an icon for these files. You can also pass an empty string; in this case, the default icon of the app will be used.

- The name of the extension.

- The verbs you want to support.

- The full path of the process will manage these types of extensions. Since in this example the process that will manage the extension is the main one, we're using the `Process` APIs to get the full path of the current process. However, it might also be a different one, for example, if your application is composed of different executables and the activation is managed by a different process than the one that is registering the file type.

If you need to clean up the registration (for example, when the application is about to be uninstalled), you can use the equivalent unregister method, as in the following sample:

```
ActivationRegistrationManager.UnregisterForFileTypeActivation
(new string[] { ".foo", ".fee" }, Process.GetCurrentProcess().
MainModule.FileName);
```

The method, called `UnregisterForFileTypeActivation()`, requires the list of extensions you want to unregister and the process path.

This example was based on file type association, but `ActivationRegistrationManager` also provides the following methods:

- `RegisterForProtocolActivation()`: To register one or more protocols (such as `foo://`)

- `RegisterForStartupActivation()`: If you want your application to automatically start together with Windows

All the samples we have seen so far around activation assume that you want to use the default Win32 behavior, which is creating a new instance each time an app is activated. However, there are scenarios where you might want more control over this behavior. Let's learn more in the next section.

Application instancing

As we mentioned in the previous section, UWP apps use, by default, a single-instancing model. If you have an instance of the application running and you try to open a new one using a different activation path (for example, you have double-clicked on an associated file type in File Explorer), the already running instance will be reused. Starting from Windows 10 1803, UWP applications have also gained support for multi-instancing, but this is an opt-in feature.

In the Win32 ecosystem and, as such, in the Windows App SDK as well, the default approach is instead the opposite one: by default, different activation paths will lead to multiple instances of the application being opened.

This behavior is the most flexible one since it enhances the multitasking capabilities of your application. However, there might be scenarios in which you would like to keep using the single-instance model. This is possible thanks to the `AppInstance` class that we introduced in the previous section.

The first step is to create a new class, which will replace the standard initialization of a Windows App SDK application. Right-click on your project, choose **Add | New class**, and name it `Program.cs`. Now, before working on the class implementation, we must add a special compilation symbol to our project, which will prevent the Windows App SDK from autogenerating the app initialization code and use, instead, the `Program` class we have just created. Right-click on the project, choose **Properties**, move to the **Build** section, and add the following text to the **Conditional compilation symbols** field:

```
DISABLE_XAML_GENERATED_MAIN
```

This is how it appears on the screen:

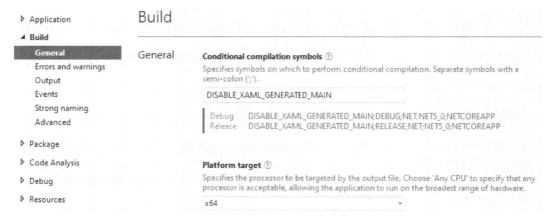

Figure 4.2 – The conditional compilation symbols configuration to enable advanced instancing scenarios

Now we can implement the `Program` class. Let's start with the `Main()` method:

```
public static class Program
{
    [STAThread]
    static async Task<int> Main(string[] args)
    {
```

```
            WinRT.ComWrappersSupport.InitializeComWrappers();

            bool isRedirect = false;

            AppActivationArguments activationArgs =
              AppInstance.GetCurrent().GetActivatedEventArgs();
            AppInstance keyInstance =
              AppInstance.FindOrRegisterForKey("Main");
            if (!keyInstance.IsCurrent)
            {
                isRedirect = true;
                await keyInstance
                  .RedirectActivationToAsync(activationArgs);
            }
            if (!isRedirect)
            {
                Microsoft.UI.Xaml.Application.Start((p) =>
                {
                    var context = new
                      DispatcherQueueSynchronizationContext(
                        DispatcherQueue.GetForCurrentThread());
                    SynchronizationContext
                      .SetSynchronizationContext(context);
                    new App();
                });
            }

            return 0;
        }
}
```

The key to managing the different instances is a method offered by the AppInstance class called FindOrRegisterForKey(). This method can be used to query Windows to see whether an instance of our application identified by a specific key already exists. In this example, we are labeling the main instance of our application with the Main key.

> **Note**
>
> Due to an issue in the Windows App SDK 1.0, the
> `FindOrRegisterForKey()` method works only if you're targeting the
> x64 architecture. The issue will be fixed in one of the future Windows App SDK
> releases, enabling you to also target x86.

The first time we launch our application, the `IsCurrent` property will be `false` since there won't be any other instance running. In this case, we use the `Microsoft.UI.Xaml.Application.Start()` method to create a new instance starting from the App class.

At every other launch of the application (regardless of which is the activation path), the `IsCurrent` property will be `true`, since we already have an instance of the application identified by the `Main` key running in the system. In this case, instead of using the `Start()` method to create a new instance, we call the `RedirectActivationToAsync()` method, which will redirect the initialization to the already existing instance. Notice how, first, we have to retrieve the activation arguments (by calling `AppInstance.GetCurrent().GetActivatedEventArgs()`) so that we can pass them to the `RedirectActivationToAsync()` method. This is important because, if we are in a file activation scenario, for example, the already running instance still needs to know which is the selected file.

If you try now to launch your application multiple times, either via the icon in the Start menu or by activating it via a file or protocol, you won't see multiple instances opening up anymore, but the behavior will be the same as the UWP apps.

However, thanks to the `AppInstance` class, we can enable even more powerful redirection scenarios, as we're going to see in the next section.

Supporting advanced redirection scenarios

As already mentioned in the previous section, Microsoft Word is a good example of an application that supports instancing in a smart way: it always launches a new instance, to maximize its multitasking capabilities, but if you double-click on a file that is already opened, it will redirect you to the already running instance where the file is open.

Thanks to the `AppInstance` class, we can enable similar redirection capabilities in our applications as well. Let's see, for example, how we can change the `Program` class to support this feature:

```
public static class Program
{
    [STAThread]
```

```
static async Task<int> Main(string[] args)
{
    WinRT.ComWrappersSupport.InitializeComWrappers();

    bool isRedirect = await DecideRedirection();

    if (!isRedirect)
    {
        Microsoft.UI.Xaml.Application.Start((p) =>
        {
            var context = new
                DispatcherQueueSynchronizationContext(
                    DispatcherQueue.GetForCurrentThread());
            SynchronizationContext
                .SetSynchronizationContext(context);
            new App();
        });
    }

    return 0;
}
}
```

The `Main()` method for this code is very similar to the previous one. However, instead of always redirecting any new instance to the existing one, we use a method called `DecideRedirection()` to understand the best approach. Let's take a look at the implementation:

```
private static async Task<bool> DecideRedirection()
{
    bool isRedirect = false;
    // Find out what kind of activation this is.
    AppActivationArguments args =
        AppInstance.GetCurrent().GetActivatedEventArgs();
    ExtendedActivationKind kind = args.Kind;
    if (kind == ExtendedActivationKind.File)
    {
```

```
    // This is a file activation: here we'll get
    // the file information,
    // and register the file name as our instance
    // key.
    if (args.Data is IFileActivatedEventArgs
      fileArgs)
    {
        IStorageItem file = fileArgs.Files[0];
        AppInstance keyInstance = AppInstance
          .FindOrRegisterForKey(file.Name);

        // If we successfully registered the file
        // name, we must be the only instance
        // running that was activated for this
        // file.
        if (keyInstance != null &&
          !keyInstance.IsCurrent)
        {
            isRedirect = true;
            await keyInstance
              .RedirectActivationToAsync(args);
        }
    }
}

    return isRedirect;
}
```

Inside the method, we check whether we are in a file activation scenario (using the same approach we learned about in the *Supporting the application's activation* section of this chapter). If that's the case, we use the AppInstance.FindOrRegisterForKey() method to retrieve a reference to the instance, which might be new or already existing. The key difference is that this time, we aren't using a fixed key (Main, in the previous example), but a dynamic one based on the filename. This means that if you are trying to open a file that was already opened, the IsCurrent property will be false, leading to redirecting the activation to the already running instance. If instead there isn't any instance with the name of the selected file, a new one will be opened.

This is just an example of a redirection scenario. Thanks to the flexibility of the `AppInstance` class, you can enable any other scenario you might need to support in your applications, leading to delivering a better user experience.

Now that we know how to properly manage instances of our application, we are ready to introduce the next topic: working with the application's window.

Managing the application's window

UWP introduced a different model to manage the application's window, compared to the more traditional one offered by the Win32 ecosystem. Since UWP was created with the goal to support multiple devices, the default experience was based on single-window applications. In this context, UWP used the `ApplicationView` and `CoreWindow` classes to host the content of a window. Thanks to these APIs, you were able to perform common tasks related to windows, such as customizing the content and changing the size.

Then, later, when Microsoft started to invest more in the desktop side of the UWP, it introduced support to multiple windows, intending to improve the multitasking story. To better support this scenario, Microsoft introduced a new class called `AppWindow`, which combined the UI thread and the window used by the application to display its content.

In Windows App SDK applications, the windowing experience has some significant differences, since it's based on the **Handle to a Window** (**HWND**) model, which is the standard approach used by the Win32 ecosystem. HWND is essentially a pointer that you can use to access the low-level window object. To make the developer's work easier, the Windows App SDK provides a class called `AppWindow`, which belongs to the `Microsoft.UI.Windowing` namespace, which abstracts many native features, making it easier to perform operations with the window.

`AppWindow` has many similarities with the equivalent API in the UWP. As such, if you are using this class in your UWP app, the migration path will be easier for you. If instead you are using `ApplicationView` and `CoreWindow`, there will be more work to do. Regardless of the class you're using, always keep in mind that the HWND model is very different from the UWP one. As such, even if you're using `AppWindow` in your UWP apps, you won't find a perfect 1:1 mapping across all the APIs and concepts.

Let's see how we can use it.

Using the AppWindow class

Out of the box, the `Window` class in WinUI doesn't give you access to the `AppWindow` abstraction, but you must manually create it starting from the original handle.

You can achieve this goal with the following code snippet:

```
private AppWindow GetAppWindowForCurrentWindow()
{
    IntPtr hWnd =
      WinRT.Interop.WindowNative.GetWindowHandle(this);
    WindowId myWndId =
      Microsoft.UI.Win32Interop.GetWindowIdFromWindow(hWnd);
    return AppWindow.GetFromWindowId(myWndId);
}
```

The first step is retrieving the HWND of the window, using the `GetWindowHandle()` helper method provided and exposed by the `WinRT.Interop.WindowNative` class. The method requires a reference to the window. If you're writing this code in the code-behind of a `Window` class, you can just pass the code itself as a reference (using the `this` keyword); if you are in another class (such as `Page`), you will have to retrieve a reference in another way (for example, by exposing the `Window` property in the `App` class as public).

The next step is to use the HWND to retrieve the identifier of the window using another helper method provided by the `Microsoft.UI.Win32Interop` class called `GetWindowIdFromWindow()`. Once we have the ID, we can finally get the corresponding `AppWindow` class by calling the static `GetFromWindowId()` method exposed by the `AppWindow` class itself.

Now that we have an `AppWindow` object, we can use it to achieve many different tasks.

Moving and resizing the window

Thanks to the `AppWindow` class, you can move the window to a specific position of the screen or resize it. Let's take a look at the following code snippet:

```
private void OnMoveWindow(object sender, RoutedEventArgs e)
{
    var appWindow = GetAppWindowForCurrentWindow();
    appWindow.Move(new Windows.Graphics.PointInt32(300,
      300));
```

```
appWindow.Resize(new Windows.Graphics.SizeInt32(800,
    600));
}
```

Once we have used the previous method to retrieve a reference to the AppWindow object, we can use the Move() method to move the window to a specific location. As a parameter, we must pass a Windows.Graphics.PointInt32 object, with the *X* and *Y* coordinates of the position.

The Resize() method works similarly, except that this time it accepts a SizeInt32 object with the width and the height we want to set.

The two methods can be used independently. However, if you want to resize and move the window at the same time, you can also use the convenient MoveAndResize() shortcut, as in the following sample:

```
private void OnMoveWindow(object sender, RoutedEventArgs e)
{
    var appWindow = GetAppWindowForCurrentWindow();
    appWindow.MoveAndResize(new Windows.Graphics.RectInt32
      (300, 300, 800, 600));
}
```

In this case, the required parameter is RectInt32 with, as values, the *X* and *Y* values of the coordinates, followed by the width and height.

Customizing the title bar

Another powerful feature supported by AppWindow is the ability to customize the title bar of the window. The most basic customization is the title, which can be customized by setting the Title property, as in the following example:

```
private void SetTitle()
{
    var appWindow = GetAppWindowForCurrentWindow();
    appWindow.Title = "This is a custom title";
}
```

A more advanced type of customization is the color scheme. The AppWindow class offers many properties to customize the colors of the title bar, including the default window buttons (minimize, maximize, and close). Let's take a look at the following code:

```
private void CustomizeTitleBar()
{
    var appWindow = GetAppWindowForCurrentWindow();
    appWindow.Title = "This is a custom title";
    appWindow.TitleBar.ForegroundColor = Colors.White;
    appWindow.TitleBar.BackgroundColor = Colors.DarkOrange;

    //Buttons
    appWindow.TitleBar.ButtonBackgroundColor =
        Colors.DarkOrange;
    appWindow.TitleBar.ButtonForegroundColor =
        Colors.White;
}
```

We are customizing the title bar with an orange background, both for the main bar and the buttons. There are many other properties you can customize, for example, the color of the title bar when the window is inactive or the color of the buttons when they are pressed. The following screenshot shows the result of the previous code:

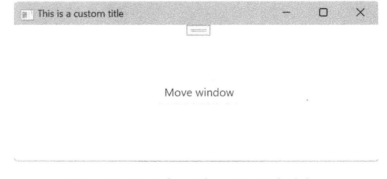

Figure 4.3 – A window with a customized title bar

If you want to have even more flexibility, you can hide the standard title bar and replace it with a XAML control, giving you the freedom of using custom fonts, adding images, and so on. This goal is achieved thanks to a special property of the AppWindow class called ExtendsContentIntoTitleBar. When it's set to true, the content area will use all the available window space, giving you the possibility of defining the title bar as part of your XAML page.

Let's see an example of a XAML page with a custom title bar:

```xml
<Window>
    <Grid>
        <Grid.RowDefinitions>
            <RowDefinition Height="Auto"/>
            <RowDefinition />
        </Grid.RowDefinitions>

        <Grid Grid.Row="0" x:Name="MyTitleBar"
            Background="LightBlue" Visibility="Collapsed">
            <Grid.ColumnDefinitions>
                <ColumnDefinition Width="*"/>
                <ColumnDefinition Width="Auto"/>
            </Grid.ColumnDefinitions>
            <Image x:Name="MyWindowIcon"
                    Source="windows.png"
                    Grid.Column="0"
                    HorizontalAlignment="Left"
                    x:FieldModifier="public"
                     Width="20" Height="20" Margin="12,0"/>
            <TextBlock
                    Text="Custom titlebar"
                    FontWeight="Bold"
                    Grid.Column="0"
                    Margin="44,8,0,0"/>
        </Grid>
        <StackPanel Grid.Row="1">
            <TextBlock Text="This is the content" />
```

```
        </StackPanel>
    </Grid>

</Window>
```

The window includes a `Grid` control with two rows. The first one contains another `Grid` called `MyTitleBar` that defines the custom title bar, which is up of an image and text. This control is hidden by default. The second row, instead, contains the actual content of the window.

Now we can use the following code, in code-behind, to replace the default title bar with the `MyTitleBar` control:

```
private void TitlebarCustomBtn_Click(object sender,
    RoutedEventArgs e)
{
    var appWindow = GetAppWindowForCurrentWindow();
    appWindow.TitleBar.ExtendsContentIntoTitleBar = true;

    // Show the custom titlebar
    MyTitleBar.Visibility = Visibility.Visible;

    //Infer titlebar height
    int titleBarHeight = appWindow.TitleBar.Height;
    this.MyTitleBar.Height = titleBarHeight;


}
```

As the first step, we set `ExtendsContentIntoTitleBar` to `true`, which will enable the content of the window to overflow in the title bar area. Then we make the `MyTitleBar` control visible and, in the end, set its height to be equal to the height of the title bar, to make sure it fits in the available space. Now our title bar will look like this:

Figure 4.4 – A custom title bar created using a XAML control

In this scenario, you might want to also use properties exposed by the AppWindow class to customize the title bar buttons so that they match the look and feel of the custom title bar.

The APIs provided by the AppWindow class are very powerful. They enable you to fully customize the title bar but, at the same time, continue to deliver a familiar user experience, by supporting features such as rounded corners in Windows 11 or buttons to minimize, maximize, and close the application.

Using custom presenters

Presenters can be used to customize the way a window is presented to the user. This feature was also supported in UWP but using different APIs and concepts. In the Windows App SDK ecosystem, presenters can easily be enabled thanks to the AppWindow class.

At the time of writing, the Windows App SDK supports the following three different presenters:

- **Default**: This is the standard way of presenting a window to the user. The user can fully resize or move the window on the screen, which stays in the foreground as long as the user doesn't move to another application.

- **Fullscreen**: With this configuration, a window is automatically maximized to use the whole screen size. The title bar is also hidden, removing the option for the user to minimize or close the application. The only way to return to the standard user experience is to change back the presenter to the default one. This mode is a good fit for games or kiosk applications.

- **Compact overlay**: With this configuration, the window is reduced to a smaller size at a 16:9 ratio and it can't be resized by the user. However, the window will always be in the foreground. Even if the user moves to another application, the window will stay on top and continue to be visible. This mode is a good fit when you want to enable **Picture-in-Picture** (**PiP**) scenarios.

To change the presenter, you just have to call the `TrySetPresenter()` method exposed by the `AppWindow` class, passing as a parameter one of the values supported by the `AppWindowPresenterKind` enumerator. Let's take a look at an example:

```
private void OnGoFullScreen(object sender, RoutedEventArgs e)
{
    var appWindow = GetAppWindowForCurrentWindow();
    if (appWindow.Presenter.Kind is not
      AppWindowPresenterKind.FullScreen)
    {
        appWindow.SetPresenter(AppWindowPresenterKind
            .FullScreen);
    }
    else
    {
        appWindow.SetPresenter(AppWindowPresenterKind
            .Default);
    }
}
```

From the preceding code, we see that first, we use the `Presenter.Kind` property exposed by `AppWindow` to check which is the current presenter. If it's the default one (`AppWindowPresenterKind.Default`), then we move to fullscreen by passing `AppWindowPresenterKind.FullScreen` to the `SetPresenter()` method. Otherwise, we return to the default mode.

If you instead want to use the compact overlay mode, you must use the `CompactOverlay` value of the `AppWindowPresenterKind` enumerator, which will enable the following result:

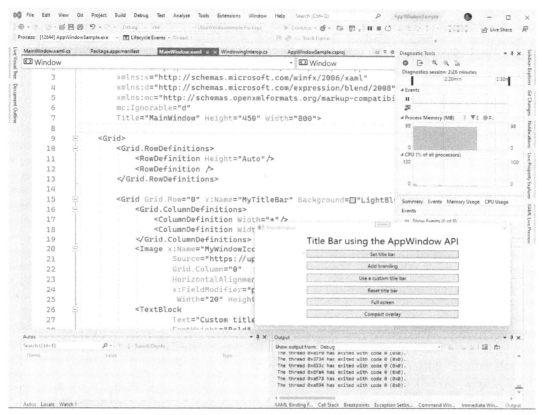

Figure 4.5 – A window using the CompactOverlay presenter

As can be seen in the preceding screenshot, despite Visual Studio being the active application, the app's window is still visible.

We have completed our journey in learning how to take advantage of the `AppWindow` class to manage the window of our application. Now we can move on to analyze another difference between the two ecosystems: background operations.

Performing operations in the background

Since one of the main goals of UWP is to enable developers to write apps that are respectful of the system resources, its background model is quite restrictive compared to the Win32 one. As we highlighted in the *Controlling the application's life cycle* section, applications aren't able to freely run in the background, but they are automatically suspended after a few seconds of inactivity. Background activities are still supported but through a unique feature called **background tasks**, which are independent snippets of code that can perform operations in the background when a specific trigger happens (every 15 minutes, when the user connects to the internet, and so on).

The Windows App SDK, being part of the Win32 ecosystem, removes most of the restrictions around background activities. The biggest one is that applications aren't suspended anymore when they are minimized, which makes it possible to continue running tasks in the background even when the user isn't actively working with the application.

However, background tasks still have a lot of potential in the Windows App SDK ecosystem, since they make it quite easy to trigger an operation as a consequence of an action or a change in the system. Additionally, background tasks are registered inside the operating system so that they can be executed even if the application that registered them is not running.

For this reason, background tasks are also fully supported by Windows App SDK apps, even if there are a few differences based on the adopted model:

- **Out-of-process**: This is the model that was originally introduced in UWP. The background task is implemented in a separate project (a Windows Runtime component) than the main application and is scheduled by Windows using a dedicated process. This model is supported by Windows App SDK apps in the same way as UWP.

- **In-process**: This model was introduced in Windows 10 1607 to simplify the creation of background tasks. Instead of having to create a different project (which also makes it harder to share code and assets with the main application), the background task is included directly in the main project. An in-process background task is executed by Windows by launching the main process and executing the `OnBackgroundActivated()` method exposed by the App class. Since, as we highlighted in the *Supporting the application's activation* section, now the App class only exposes the `OnLaunched()` event, this model is still supported but it must be implemented differently.

With this section, you have now a basic understanding of the differences in managing background operations between UWP and the Windows App SDK. We'll explore the topic in more detail in *Chapter 8, Integrating Your Application with the Windows Ecosystem*, to see how you can enhance your applications with background tasks.

Summary

UWP introduced many concepts that are still truly relevant today—delivering a modern UI experience, optimized for multiple screens and accessibility; being respectful of the system resources; supporting new technologies, such as modern authentication and artificial intelligence. These are just a few examples of features introduced by UWP that are important to build modern desktop applications.

This is the reason the Windows App SDK and WinUI were built around the same principles, but on top of the Win32 ecosystem to give developers more flexibility and make it easier for them to reuse the existing code and libraries.

Even if many of the building blocks are the same, moving to the Win32 ecosystem means that there are some differences around some key concepts, such as the application's life cycle or windowing management. In this chapter, we learned about these differences in detail as well as what the new APIs you must use to properly support these scenarios in Windows App SDK apps are.

This chapter concludes the overview of the Windows App SDK ecosystem for the different development frameworks: Windows Forms, WPF, and UWP.

In the next chapter, we'll learn how to properly design a Windows App SDK application, by adopting the right navigation pattern or implementing responsive layouts.

Questions

1. To ensure stability and performance, Windows can automatically terminate the Windows App SDK if it is consuming too many resources. True or false?

2. There is no difference between a packaged and an unpackaged application when it comes to supporting activation contracts. True or false?

3. Windows App SDK applications, unlike UWP apps, support multi-instancing by default. True or false?

5
Designing Your Application

In previous chapters, we learned all the basic WinUI and Windows App SDK building blocks, such as XAML resources, binding, dispatcher, and windows. However, these basic components aren't enough to create a real application. You also need to define the user experience and the navigation pattern you want to adopt, you must properly support themes, you need to define the UI so that it can be responsive and quickly adapt to different sizes and layouts, and much more.

In this chapter, we're going to put all the pieces we have learned together so that we can start building a real project and evolve it over time.

We're going to cover the following topics:

- Creating a responsive layout
- Supporting navigation
- Supporting Windows themes
- Creating animations
- Exploring the WinUI controls

Technical requirements

The code samples for the chapter can be found at the following URL:

https://github.com/PacktPublishing/Modernizing-Your-Windows-Applications-with-the-Windows-Apps-SDK-and-WinUI/tree/main/Chapter05

Creating a responsive layout

One of the most powerful features of WinUI, compared to other Microsoft UI frameworks, such as WPF and Windows Forms, is the built-in support for responsive layouts. Even if WinUI applications are tailored mainly for desktop devices, there are many scenarios where it's important to let the application scale properly: big screens, monitors with high resolution, applications running in a smaller window, and so on. In these situations, your application must react accordingly, so that the content presented to the user can always be meaningful.

When building the layout of your application, there are some aspects that Windows will take care of for you, and some others that must be manually tweaked by the developer.

Let's start by analyzing the first category.

Using effective pixels

One of the biggest challenges when using technologies such as WPF or Windows Forms is that all the sizes you set are translated into screen pixels. For example, if you add a Rectangle control and you set its Width property to 400, this control will be rendered as it is. The challenge is that 400 can be a very small size on a 4K screen or a big size on a 720p screen. This approach makes it complex to build UIs that can easily scale regardless of the screen size and the resolution of the device.

WinUI makes scaling easier by adopting a technology called **effective pixels**, which is based on the concept of *scaling factor*. To understand it better, open your Windows **Settings** and move to the section called **Display**. Other than the resolution of your screen, you will find a setting called **Scale**, as highlighted in the following screenshot:

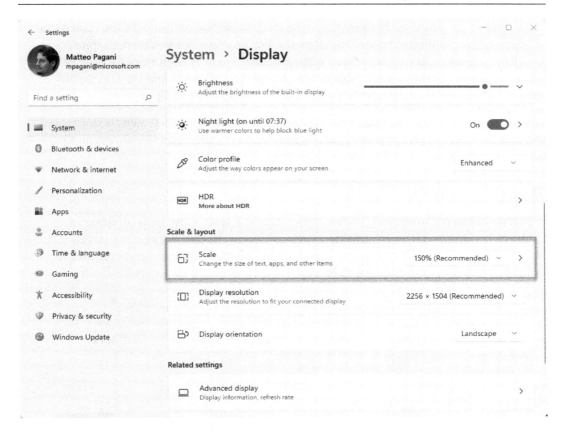

Figure 5.1 – The Scale option in the Windows 11 settings

Thanks to this setting, all the elements on the screen will be clear and easy to read even if you have a computer with a very high resolution. Without this setting, it would be impossible to effectively use the operating system on 2K or 4K monitors.

Effective pixels are based on this scaling factor. Let's take the following XAML snippet:

```
<Rectangle Width="400" />
```

In WinUI, this size will automatically be scaled based on the scaling factor set in Windows, which means that the percentage will be used as a multiplier to calculate the real pixels. For example, in the previous example (where the screen scale is set to **150%**), the real pixels that the Rectangle control takes on the screen are 400 * 1.5. On a screen with a scale of 200%, the real pixels will be 400 * 2, and so on.

Effective pixels are powerful because they take into consideration not only the screen size and the resolution but also the viewing distance. For example, a TV screen, despite it usually being considered a big screen with, potentially, a very high resolution, uses a scaling factor that helps to make things bigger and easier to view and read, since it's a screen that is used at a longer viewing distance compared to a PC monitor.

Thanks to effective pixels, you don't have to worry that items on the screen (text, shapes, and controls) might look too small or too big for the user. Windows will always apply the proper scaling factor.

Adapting the UI

When it comes to building the UI of an application, the size of the items on the screen isn't the only thing you have to consider. The position of the elements on the screen can also be impacted by the size of the screen or the window. For example, consider the following screenshot of the Microsoft Store:

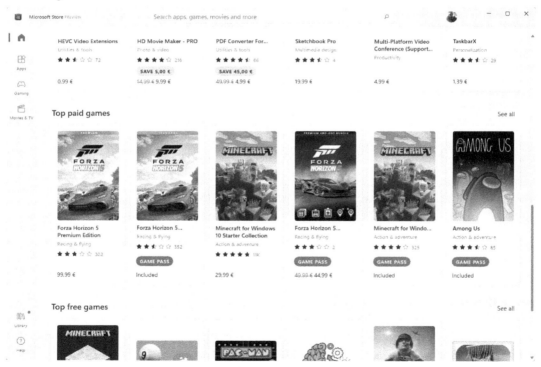

Figure 5.2 – A collection of apps and games highlighted in the Microsoft Store

As you can see, there are multiple collections of applications and games that are displayed with a UI that spans horizontally. This layout works well on a big screen or when the window is maximized, but it isn't very effective if the window is small. To mitigate these problems, you can adopt a responsive layout—the position and the layout of the elements placed on the screen can change based on the size of the window.

In *Chapter 2, The Windows App SDK for a Windows Forms Developer*, you learned about the concept of visual states and how, using VisualStateManager, you can transition the look and feel of a control from one state to another (for example, a Button control can become Pressed or Disabled and each of these states has a different aspect). This also makes this feature a good fit to build responsive UIs: we can define the different layouts as states and use VisualStateManager to transition from one to another when the size of the window changes. Let's take a look at the following code:

```xml
<Grid>
    <VisualStateManager.VisualStateGroups>
        <VisualStateGroup>
            <VisualState x:Name="SmallWindow">
                <Storyboard>
                    <ObjectAnimationUsingKeyFrames
                        Storyboard.TargetName="MyPanel"
                        Storyboard.TargetProperty=
                        "Orientation">
                        <DiscreteObjectKeyFrame KeyTime=
                        "0:0:0" Value="Vertical" />
                    </ObjectAnimationUsingKeyFrames>
                </Storyboard>
            </VisualState>
            <VisualState x:Name="LargeWindow">
                <Storyboard>
                    <ObjectAnimationUsingKeyFrames
                        Storyboard.TargetName="MyPanel"
                        Storyboard.TargetProperty=
                        "Orientation">
                        <DiscreteObjectKeyFrame
                        KeyTime="0:0:0" Value=
                        "Horizontal" />
                    </ObjectAnimationUsingKeyFrames>
```

```
                    </Storyboard>
                </VisualState>
            </VisualStateGroup>
        </VisualStateManager.VisualStateGroups>

        <StackPanel x:Name="MyPanel" HorizontalAlignment=
            "Center" VerticalAlignment="Center">
            <Button Content="Move window" Click=
                "OnMoveWindow" />
            <Button Content="Go to page 2"
                Click="OnGoToSecondPage" />
        </StackPanel>

    </Grid>
```

This code defines, using VisualStateManager, two different states: one called
SmallWindow, where the StackPanel container included in the page is set with
vertical orientation, and one called LargeWindow, where instead, the orientation
is set to horizontal.

This approach, however, is a little bit cumbersome for two reasons:

- Even if we're just changing the value of a control's property, we are forced to use
  a complex syntax that involves defining a Storyboard object despite the fact that
  we don't need a real animation; we're just changing the value of the property.

- We need to write some code to handle the transition from one state to another.
  Specifically, we have to subscribe to the SizeChanged event of the Page container
  and implement a similar event handler:

```
private void Page_SizeChanged(object sender,
    SizeChangedEventArgs e)
{
    if (e.NewSize.Width < 720)
    {
        VisualStateManager.GoToState(this,
            "SmallWindow", true);
    }
    else
    {
```

```
            VisualStateManager.GoToState(this,
                "LargeWindow", true);
        }
    }
```

WinUI makes `VisualStateManager` much easier to define, thanks to two new XAML features:

- A shorter syntax to define the value of a property in a specific state
- The `AdaptiveTrigger` class, which you can use to specify that one state should be automatically applied when the size of the window reaches a specific threshold

This is what the previous snippet looks like thanks to the WinUI implementation:

```xaml
<Grid>
    <VisualStateManager.VisualStateGroups>
        <VisualStateGroup>
            <VisualState x:Name="SmallWindow">
                <VisualState.StateTriggers>
                    <AdaptiveTrigger MinWindowWidth="0"/>
                </VisualState.StateTriggers>
                <VisualState.Setters>
                    <Setter Target="MyPanel.Orientation"
                        Value="Vertical"/>
                </VisualState.Setters>
            </VisualState>
            <VisualState x:Name="LargeWindow">
                <VisualState.StateTriggers>
                    <AdaptiveTrigger MinWindowWidth="720"/>
                </VisualState.StateTriggers>
                <VisualState.Setters>
                    <Setter Target="MyPanel.Orientation"
                        Value="Horizontal"/>
                </VisualState.Setters>
            </VisualState>
        </VisualStateGroup>
    </VisualStateManager.VisualStateGroups>
</Grid>
```

```
        </VisualStateManager.VisualStateGroups>

        <StackPanel x:Name="MyPanel" HorizontalAlignment=
          "Center" VerticalAlignment="Center">
            <Button Content="Move window" Click=
              "OnMoveWindow" />
            <Button Content="Go to page 2" Click=
              "OnGoToSecondPage" />
        </StackPanel>

    </Grid>
```

The code is now simpler to write and read:

- To change the value of a property, we can just define a `Setter` object with, for `Target`, the name of the control followed by the name of the property (separated by a dot) and, for `Value`, the new value that it must assume.

- Inside the `StateTriggers` collection of `VisualState`, we use an `AdaptiveTrigger` object specifying what the window size is that will trigger that state. In the previous example, the `SmallWindow` state will be applied when the width of the window is between 0 and 720; the `LargeWindow` state, instead, will be applied when it's bigger than 720.

What makes adaptive triggers powerful is that, as we learned in the previous section, the `MinWindowWidth` and `MinWindowHeight` properties are expressed in effective pixels and not in real pixels. As such, they will always be proportional to the size and resolution of the screen.

Thanks to the new syntax to define a visual state, we can make the previous code even simpler:

```
<Grid>
    <VisualStateManager.VisualStateGroups>
        <VisualStateGroup>
            <VisualState x:Name="SmallWindow">
                <VisualState.StateTriggers>
                    <AdaptiveTrigger MinWindowWidth="0"/>
                </VisualState.StateTriggers>
                <VisualState.Setters>
```

```
                    <Setter Target="MyPanel.Orientation"
                        Value="Vertical"/>
                </VisualState.Setters>
            </VisualState>
            <VisualState x:Name="LargeWindow">
                <VisualState.StateTriggers>
                    <AdaptiveTrigger MinWindowWidth="720"/>
                </VisualState.StateTriggers>
            </VisualState>
        </VisualStateGroup>
    </VisualStateManager.VisualStateGroups>

    <StackPanel x:Name="MyPanel" HorizontalAlignment=
        "Center" VerticalAlignment="Center">
        <Button Content="Move window" Click=
            "OnMoveWindow" />
        <Button Content="Go to page 2" Click=
            "OnGoToSecondPage" />
    </StackPanel>

</Grid>
```

Since if the size of the window is bigger than 720 we want to keep the default layout we have defined inside the StackPanel container, we can just add the AdaptiveTrigger control inside the StateTriggers collection of the LargeWindow state, but without specifying any Setter. This means that when the size of the window is bigger than 720, WinUI will revert the page to the default layout.

> **Note**
>
> At the time of writing, adaptive triggers are supported only inside a Page container. If you apply them directly to the Window object, they won't work.

Adaptive triggers are a great feature to build responsive layouts, but they aren't enough on their own. You need to make sure that you're building the layout of your page in the right way. Let's see the controls that you can use to achieve this goal.

Using the right controls to build a responsive layout

WinUI includes a wide range of controls that aren't visible on the screen, but they can be used to define the layout of the application. `Grid`, `StackPanel`, and `Canvas` are some good examples. Let's see the ones that enable you to create powerful responsive layouts.

Using the Grid control

The `Grid` control is, without any doubt, one of the most flexible layout controls available in WinUI. You can use it to create table layouts, in which you can place controls across multiple rows and columns. Let's take a look at the following code snippet:

```
<Grid>
    <Grid.RowDefinitions>
        <RowDefinition Height="100" />
        <RowDefinition Height="200" />
    </Grid.RowDefinitions>
    <Grid.ColumnDefinitions>
        <ColumnDefinition Width="300" />
        <ColumnDefinition Width="500" />
    </Grid.ColumnDefinitions>

    <TextBlock Grid.Row="0" Grid.Column="0" Text="Hello
      world!" />
    <TextBlock Grid.Row="1" Grid.Column="1" Text="Hello
      world again!" />
</Grid>
```

This XAML code creates a table with two rows and two columns, each of them with a fixed size. Controls inside the `Grid` container are placed in a specific cell by using the attached `Grid.Row` and `Grid.Column` properties, where 0 is the first cell. In the previous example, the first `TextBlock` is placed in the first cell and the second `TextBlock` in the last one.

Even if the previous example isn't fully responsive (since we have assigned a fixed size to the cells), it can still adapt well to the size of the screen since the values are expressed in effective pixels. But we can do more!

Let's see the following example:

```
<Grid>
    <Grid.RowDefinitions>
        <RowDefinition Height="Auto" />
        <RowDefinition Height="200" />
    </Grid.RowDefinitions>
    <Grid.ColumnDefinitions>
        <ColumnDefinition Width="Auto" />
        <ColumnDefinition Width="500" />
    </Grid.ColumnDefinitions>

    <TextBlock Grid.Row="0" Grid.Column="0" Text="Hello
      world!" />
    <TextBlock Grid.Row="1" Grid.Column="1" Text="Hello
      world again!" />
</Grid>
```

We have set one of the rows and one of the columns with Auto as the size. Thanks to this configuration, now the size of the cell will automatically adjust to the size of the controls that are placed within. This helps even more to build a responsive UI, since the layout can automatically adapt to the state of the controls on a page.

Last but not least, we have an even more powerful option to create responsive layouts with a Grid control: star sizing. Let's analyze the following example so that we can better understand what it means:

```
<Grid>
    <Grid.RowDefinitions>
        <RowDefinition Height="1*" />
        <RowDefinition Height="2*" />
    </Grid.RowDefinitions>
    <Grid.ColumnDefinitions>
        <ColumnDefinition Width="1*" />
        <ColumnDefinition Width="1*" />
    </Grid.ColumnDefinitions>

    <TextBlock Grid.Row="0" Grid.Column="0" Text="Hello
      world!" />
```

```
        <TextBlock Grid.Row="1" Grid.Column="1" Text="Hello
            world again!" />
    </Grid>
```

Now we have changed the size of rows and columns using a number followed by a star. This means that we are creating a proportional layout. Look at the row definitions: the current configuration means that we are ideally splitting the page into three parts; the first row will always take 1/3 of the space, while the second one will take 2/3. The columns, instead, will be evenly distributed; each of them will take 50% of the page.

As you can imagine, this approach is a perfect fit for creating responsive layouts: regardless of the size of the screen or the window, the cells in the table will always keep the same proportion.

Like with a regular table, the `Grid` control can be used to create a more complex layout by enabling cells to span across multiple columns and rows, by using the `RowSpan` and `ColumnSpan` properties, as in the following example:

```
<Grid>
    <Grid.RowDefinitions>
        <RowDefinition Height="1*" />
        <RowDefinition Height="2*" />
    </Grid.RowDefinitions>
    <Grid.ColumnDefinitions>
        <ColumnDefinition Width="1*" />
        <ColumnDefinition Width="1*" />
    </Grid.ColumnDefinitions>

    <TextBlock Grid.Row="0" Grid.Column="0"
        Grid.ColumnSpan="2" Text="Hello world!" />
    <TextBlock Grid.Row="1" Grid.Column="1" Text="Hello
        world again!" />
</Grid>
```

In this case, the first `TextBlock` will span across the entire size of the first row, since the two columns will be merged into one.

Let's move on now to see another interesting control to create responsive layouts.

Using the VariableSizeWrapGrid control

VariableSizeWrapGrid is similar to the Grid control, except that children controls are automatically placed into multiple rows and columns based on the available space.

Let's take a look at the following code:

```
<VariableSizedWrapGrid Orientation="Horizontal"
  MaximumRowsOrColumns="3" ItemHeight="200"
    ItemWidth="200">
    <Rectangle Fill="Red" />
    <Rectangle Fill="Blue" />
    <Rectangle Fill="Green" />
    <Rectangle Fill="Yellow" />
</VariableSizedWrapGrid>
```

In this specific scenario, we are setting up VariableSizedWrapGrid so that it does the following:

- The wrapping will happen at the column level since we are setting the Orientation property to Horizontal.
- The maximum number of columns that each row will have is 3. If the Orientation property had been set to Vertical, this property would have represented the maximum number of rows.
- ItemWidth and ItemHeight are used to define the size of each cell.

The following screenshot shows the output of the previous code, using two different sizes of the window:

Figure 5.3 – The VariableSizeWrapGrid control in action

In the one on the left, we have enough space to display the full content, so the `VariableSizeWrapGrid` control will create the maximum number of columns allowed (in our case, three, since it's the value we have set for the `MaximumRowsOrColumns` property). On the right, instead, the size of the window is too small to host three columns, so the control will automatically wrap the other items in a new row.

The `VariableSizeWrapGrid` control is a great fit for building responsive desktop applications, since you can create layouts that quickly react to changes in the window's size.

Let's move on now to one more control, which might be new if you're coming from other XAML-based technologies, such as WPF.

Using the RelativePanel control

`RelativePanel` is a new control that has been added to WinUI to better support the creation of responsive layouts. Instead of placing child controls in a specific position, they are arranged based on the relationship with the panel itself or with other controls. Let's take a look at the following example:

```xml
<RelativePanel>
    <StackPanel x:Name="InfoCard1" Style="{StaticResource
      InfoCardStyle}">
        <TextBlock Text="Customers" FontSize="28" />
        <controls:DataGrid x:Name="dgCustomers" />
    </StackPanel>
    <StackPanel x:Name="InfoCard2" Style="{StaticResource
      InfoCardStyle}" RelativePanel.RightOf="InfoCard1">
        <TextBlock Text="Orders" FontSize="28" />
        <controls:DataGrid x:Name="dgOrders" />
    </StackPanel>
</RelativePanel>
```

This `RelativePanel` has two children controls: two `StackPanel` layouts, with a `DataGrid` control inside. Thanks to the attached properties exposed by the `RelativePanel` control, we can specify that the two children have a relationship: the second information card must be placed at the right of the first one. The code generates the following result:

Figure 5.4 – Two controls aligned using a RelativePanel relationship

This kind of relationship works great in combination with the `AdaptiveTrigger` feature we have previously explored. Thanks to a trigger, in fact, it's very easy to change the relationship between two or more controls when the size of the window changes. For example, our page might use the following implementation of `VisualStateManager`:

```xml
<VisualStateManager.VisualStateGroups>
    <VisualStateGroup>
        <VisualState x:Name="SmallWindow">
            <VisualState.StateTriggers>
                <AdaptiveTrigger MinWindowWidth="0"/>
            </VisualState.StateTriggers>
            <VisualState.Setters>
                <Setter Target="InfoCard2.(RelativePanel
                    .Below)" Value="InfoCard1"/>
                <Setter Target="InfoCard2.(RelativePanel
                    .AlignLeftWith)" Value="InfoCard1" />
            </VisualState.Setters>
        </VisualState>
        <VisualState x:Name="LargeWindow">
            <VisualState.StateTriggers>
                <AdaptiveTrigger MinWindowWidth="720"/>
            </VisualState.StateTriggers>
        </VisualState>
    </VisualStateGroup>
</VisualStateManager.VisualStateGroups>
```

When the window is smaller than 720 effective pixels, we change the relationship between the two `StackPanel` controls: `InfoCard2`, instead of being placed to the right of `InfoCard1`, is moved below and aligned to the left, which leads to the following result:

Figure 5.5 – When the size of the window changes, the two controls are rearranged by changing their relationship

This change makes the layout a good fit for scenarios when the application is used in a small window, since it keeps all the relevant content (the two info cards) visible.

`RelativePanel` supports the following properties to set up a relationship between controls:

- `Above`
- `Below`
- `LeftOf`
- `RightOf`

`RelativePanel` can also be used to create relationships between the controls and the panel itself, which is another great way to support adaptive layouts. Consider the following example:

```
<RelativePanel>
    <StackPanel x:Name="InfoCard1" Style="{StaticResource
```

```
        InfoCardStyle}" Width="300" Height="300"
        RelativePanel.AlignHorizontalCenterWithPanel="True">
          <TextBlock Text="Customers" FontSize="28" />
          <controls:DataGrid x:Name="dgCustomers" />
     </StackPanel>
 </RelativePanel>
```

By setting the attached `RelativePanel.AlignHorizontalCenterWithPanel` property to `true`, the control will always be placed in the middle of the panel, regardless of the size of the window. Of course, you can also in this case use `AdaptiveTrigger` to change this behavior.

`RelativePanel` supports the following properties to manage the relationship of child controls with the panel:

- `AlignBottomWithPanel`
- `AlignHorizontalCenterWithPanel`
- `AlignLeftWithPanel`
- `AlignRightWithPanel`
- `AlignVerticalCenterWithPanel`

Let's end this section by taking a quick look at other types of layout controls.

Other layout controls available in WinUI

WinUI supports other controls to define the layout of a page. However, they aren't specific to building a responsive layout and, as such, we'll just briefly mention them:

- `StackPanel`: We have already seen this control in action in multiple examples. The goal of this control is to place the child controls one after the other. By default, they are aligned vertically, but you can also use it to align them horizontally by setting the `Orientation` property to `Horizontal`. This is shown in the following code:

```
    <!-- the controls are placed one below the other -->
    <StackPanel>
        <TextBlock Text="This is a text" />
        <TextBlock Text="This is another text" />
    </StackPanel>
    <!-- the controls are placed one right to
       the other -->
```

```
<StackPanel Orientation="Horizontal">
    <TextBlock Text="This is a text" />
    <TextBlock Text="This is another text" />
</StackPanel>
```

- `Canvas`: This control can be used to position content in an absolute way. Every child control can be placed inside the `Canvas` container by specifying the distance from the top-left corner. This control should only be used in specific scenarios, since it isn't a good fit to create adaptive layouts. The following code shows how you can use the attached `Canvas.Left` and `Canvas.Top` properties to specify the position of a control inside the `Canvas` container:

```
<Canvas>
    <TextBlock Text="This is a text" Canvas.Left="20"
        Canvas.Top="50" />
</Canvas>
```

Now that we know which is the right way to build a responsive user experience, it's time to understand how we can better support it by providing the proper navigation experience.

Supporting navigation

When you start to build a real application, navigation is one of the first key topics you must address. Other than a very few specific scenarios, in most cases, the application you're going to build will have multiple screens. In WinUI, there are different containers for your layouts:

- **Window**: This is the main host of your application. Theoretically, a `Window` control can directly contain content, but it has a few limitations. It doesn't expose a `DataContext`, it doesn't expose life cycle events (such as `Loaded`), and it doesn't support navigation. As such, in real-world applications, the `Window` object acts just as the main container of your application.

- **Page**: The `Page` object is the most used container for the content of your application, since it supports features such as `DataContext` and life cycle events. A real-world application typically has multiple pages, and the user can navigate from one to another by using the navigation menu or performing specific actions (such as selecting an item from a list).

- **Frame**: The `Frame` object acts as a host for pages and enables navigation between one page to another.

Applications usually adopt the `Window` | `Frame` | `Page` hierarchy, implemented in the following way:

1. `Window` hosts the navigation menu and a `Frame` object.

2. When the application starts, `Frame` loads the main page of the application.

3. When the user needs to move to a different page, `Frame` will take care of performing the navigation.

When you create a new WinUI application, you will already have a default `Window`, called `MainWindow`. Starting from there, you can create one or more pages by right-clicking on the project, choosing **Add** | **New item**, and selecting the **Blank Page (WinUI 3)** template. This is what the basic implementation of a page looks like:

```
<Page
    x:Class="MyApplication.MyPage"
    xmlns="http://schemas.microsoft.com/winfx/2006/xaml
       /presentation"
    xmlns:x="http://schemas.microsoft.com/winfx/2006/xaml"
    xmlns:d="http://schemas.microsoft.com/expression/blend
       /2008"
    xmlns:mc="http://schemas.openxmlformats.org/markup-
       compatibility/2006"
    Background="{ThemeResource Application
    PageBackgroundThemeBrush}">

    <Grid>
        <!-- the content of the page -->
    </Grid>
</Page>
```

Let's learn now how we can implement a navigation menu in our application.

Implementing the NavigationMenu control

WinUI offers a powerful control to implement your navigation experience called NavigationMenu. The default navigation experience is enabled using the approach known as a **hamburger menu** (named for the resemblance of the button used to open the menu with a hamburger)—a collapsible menu that expands from one side of the screen to display the various sections where the user can navigate to. This control is typically included in the main window of your application, along with the frame that will host the pages. Let's see a basic implementation in XAML:

```
<Window>
    <NavigationView>
        <NavigationView.MenuItems>
            <NavigationViewItem Content="Home" Icon="Home"
                Tag="Home" />
            <NavigationViewItem Content="Favorites"
                Icon="Favorite" Tag="Favorite"/>
            <NavigationViewItem Content="Messages"
                Icon="Message" Tag="Messages" />
        </NavigationView.MenuItems>

        <Frame x:Name="ShellFrame" />
    </NavigationView>
</Window>
```

The NavigationView control exposes a collection called MenuItems, which can host one or more NavigationViewItem controls. Each of them represents a section of your application the user can navigate to, which is identified by a Content property (the label) and an Icon property (which can be set using one of the built-in WinUI icons through the Microsoft.UI.Xaml.Controls.IconElement class).

One of the features that makes NavigationView so powerful is that it provides built-in support for responsive layouts. The control, in fact, will automatically be rendered in a different way based on the size of the screen, as you can see in the following screenshot:

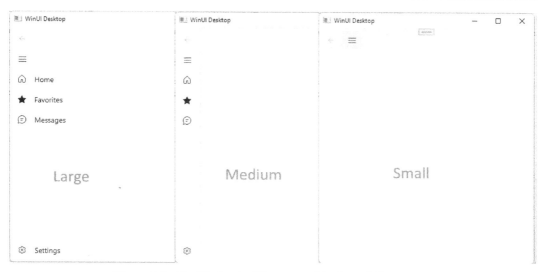

Figure 5.6 – The NavigationView control in three different sizes

When the window is wide enough, the menu will always be visible and fully expanded; when the window is smaller, the menu will continue to appear, but only the icons will be visible; when the window is small, instead, the menu will completely disappear, and the user must tap on the button at the top to see it.

Now that we have learned the basics of the control, let's see how we can handle the navigation.

Handling the navigation

The NavigationMenu control exposes an event called ItemInvoked, which is triggered whenever the user selects one of the items in the menu. Inside this event handler, through the event arguments, you can get a reference to the selected NavigationViewItem object thanks to the InvokedItemContainer property, which you can use to understand where to redirect the user. A commonly used approach is to use the Tag property exposed by every XAML control to store a string that can help you to understand which is the selected item, as in the following sample:

```
<NavigationView ItemInvoked="NavigationView_ItemInvoked">
    <NavigationView.MenuItems>
        <NavigationViewItem Content="Home" Icon="Home"
            Tag="Home" />
        <NavigationViewItem Content="Favorites"
            Icon="Favorite" Tag="Favorite"/>
```

```
        <NavigationViewItem Content="Messages"
          Icon="Message" Tag="Messages" />
    </NavigationView.MenuItems>

    <Frame x:Name="ShellFrame" />
</NavigationView>
```

Thanks to the Tag property, you can implement the ItemInvoked event in the following way:

```
private void NavigationView_ItemInvoked(NavigationView
  sender, NavigationViewItemInvokedEventArgs args)
{
    switch (args.InvokedItemContainer.Tag)
    {
        case "Home":
            //navigate to home
            break;
        case "Favorite":
            //navigate to favorites
            break;
    }
}
```

Now we have the basic infrastructure to manage the navigation. But how can we effectively trigger it? If you take a look at the previous XAML definition of the NavigationView control, you'll see that we have placed inside it a Frame control. As we learned at the beginning of the section, the Frame class is the basic building block to implement navigation, since it acts as a host for pages by providing the infrastructure required to navigate the user from one page to the other.

Let's assume that in our application we have added two new pages, HomePage and FavoritePage. This is how we can expand the previous code example to perform the navigation:

```
private void NavigationView_ItemInvoked(NavigationView
  sender, NavigationViewItemInvokedEventArgs args)
{
    switch (args.InvokedItemContainer.Tag)
```

```
    {
        case "Home":
            ShellFrame.Navigate(typeof(HomePage));
            break;
        case "Favorite":
            ShellFrame.Navigate(typeof(FavoritePage));
            break;
    }
}
```

The `Frame` class exposes a method called `Navigate()`, which accepts, as a parameter, the class type of the page we want to redirect the user to. Since we have assumed that the two pages are represented by the `HomePage` and `FavoritePage` classes, we use them in combination with the `typeof()` function to retrieve their type.

Implementing the navigation in our `MainWindow` class is easy since it's the one that is hosting both the `NavigationView` and `Frame` controls. As such, we can directly reference the `Frame` object to trigger the navigation when one of the menu items is selected. What if, instead, we need to trigger the navigation from a page? For example, one of your pages displays a list of items and when you select one of them, you want to redirect the user to another page with some details about the selected item. In this case, you can't directly use the `Frame` class, because it's defined in a different class (`MainWindow`).

If you have experience with the **Universal Windows Platform** (**UWP**), this is one of the areas where the Windows App SDK implementation requires a bit more work. In UWP applications, in fact, since there's a unique window shared across the whole application, the `Page` class offers a `Frame` property, which gives you easy access to the application's frame. As such, in the code-behind of every `Page` class, you can simply call `Frame.Navigate()` to redirect the user to a different page.

In WinUI desktop applications, instead, the `Page` class still has a `Frame` property, but it's not set (you will get a `null` reference if you try to use it). As such, we need to store somewhere a reference to the `Frame` object included in the `MainWindow` class, so that we can reuse it from every page of our application. The best place to do this is the `App` class, since it's a global singleton that can be accessed from every class of the application. Let's see the changes to make. As the first step, we must define in the `App` class a static property to host our `Frame` object. In the following example, we have called it `ShellFrame`:

```
public partial class App : Application
{
    public Window m_window { get; set; }
```

```
public static Frame ShellFrame { get; set; }

public App()
{
    this.InitializeComponent();
}
protected override void OnLaunched
    (Microsoft.UI.Xaml.LaunchActivatedEventArgs args)
{
    m_window = new MainWindow();
    m_window.Activate();
}
}
```

Now we can move to the MainWindow class and, in the constructor, store inside the ShellFrame property that we have just created a reference to the Frame control included in MainWindow:

```
public sealed partial class MainWindow : Window
{
    public MainWindow()
    {
        this.InitializeComponent();
        App.ShellFrame = ShellFrame;
        ShellFrame.Content = new HomePage();
    }
}
```

We use the static property we have just defined in the App class to store a reference to our ShellFrame control. At the end, we trigger the navigation to the main page of our application (in the example, it's called HomePage).

Now that we have a reference to the frame stored in the App class, we can recall it on every page of our application to trigger a navigation. For example, let's say that our HomePage has a ListView control and when the user selects one item from the list, we trigger the navigation to another page. This is what the SelectionChanged event handler of the ListView control would look like:

```
private void lstPeople_ItemClick(object sender,
    SelectionChangedEventArgs e)
{
    App.ShellFrame.Navigate(typeof(DetailPage));
}
```

We use the same code as before to retrieve a reference to the App class and then, through the ShellFrame property, we invoke the Navigate() method.

Now that we have discovered how to enable the navigation from one page to another, a new question might arise: how can we manage the life cycle of a page through navigation? Let's learn more in the next section.

Supporting the page's life cycle

The Page class offers three methods that you can override to manage the life cycle of a page:

- OnNavigatedTo(), which is triggered when the user has just landed on the current page after a navigation. This is a great place to load the content you need to display on the page since, unlike the page's constructor, it supports performing asynchronous tasks.

- OnNavigatedFrom(), which is triggered when the user is navigating away from the current page to another page, immediately after the page has been unloaded. This is a great place to clean up content that you don't need to keep in memory anymore.

- OnNavigatingFrom(), which is also triggered when the user is navigating away from the current page to another page, but immediately before the page has been loaded.

The `OnNavigatedTo()` method is also important for another scenario. Let's consider again the example we saw in the previous section, in which the user selects an item from a list and is redirected to a detail page. That sample was missing an important implementation detail: we were redirecting the user to a detail page, but without any reference to which was the selected item. To support this scenario, the `Navigate()` method of the `Frame` class supports a second parameter to store the information we want to carry over to the destination page. Let's adjust the previous sample to cover this scenario:

```
private void lstPeople_ItemClick(object sender,
    SelectionChangedEventArgs e)
{

    Person selectedPerson = e.ClickedItem as Person;
App.ShellFrame.Navigate(typeof(DetailPage), selectedPerson);
        (typeof(DetailPage), e.AddedItems[0]);
}
```

Other than the page where we want to redirect the user (`DetailPage`), we are adding as information the selected item from the `ListView` control, which is stored inside the `AddedItems` collection of the event arguments.

This information can be retrieved in the destination page thanks to the `OnNavigatedTo()` method, which includes a property called `Parameter` in the event arguments:

```
public sealed partial class MyPage : Page
{

    protected override void OnNavigatedTo
        (NavigationEventArgs e)
    {

        Person selectedPerson = e.Parameter as Person;
        //do something with the selectedPerson object

    }
}
```

Now we know exactly which item was selected by the user in the previous page and we can populate the content of the page in the proper way. The `Navigate()` method also supports another feature: page transitions. Let's learn more in the next section.

Supporting page transitions

The Navigate() method also supports a third parameter, which you can use to enable page transitions. These are special animations that you can use to create a smooth transition when the user moves from one page to another. WinUI supports the following transitions:

- **Page refresh**: This is a combination of a slide-up animation of the current content, followed by a fade-in animation for the incoming content. It's represented by the EntranceNavigationTransitionInfo class. However, this transition is applied by default, so you don't have to specify it if it's the one you want to implement.

- **Drill**: This is used mainly in master-detail scenarios. The user clicks on an item in a gallery and the animation helps to create a connection with the same item on the detail page. It's represented by the DrillInNavigationTransitionInfo class.

- **None**: In some scenarios, you might not want an animation at all. Since, by default, the page refresh animation is applied during navigation, you must explicitly disable it by using the SuppressNavigationTransitionInfo class.

To apply one of these transitions, you just need to pass a new instance of one of the preceding classes as the third parameter of the Navigate() method. For example, if you want to use a drill transition, you must use the following code:

```
private void OnGoToSecondPage(object sender,
    RoutedEventArgs e)
{
    App.ShellFrame.Navigate
      (typeof(MySecondPage), null, new
        DrillInNavigationTransitionInfo());
}
```

In the previous example, we didn't have the need to pass any parameter to the second page, so you can just use a null object as a second parameter.

Now we have almost completed the implementation of a full navigation system; however, we're still missing one important feature: backward navigation. So, let's check that out.

Managing backward navigation

Whenever you navigate from one page to another, Windows keeps track of the movement using a stack. This approach enables developers to easily redirect users to the page of your application they had previously visited.

Backward navigation can be easily implemented using the GoBack() method exposed by the Frame class, as in the following sample:

```
private void NavigationView_BackRequested(NavigationView
    sender, NavigationViewBackRequestedEventArgs args)
{
    if (ShellFrame.CanGoBack)
    {
        ShellFrame.GoBack();
    }
}
```

The method can be paired, as in the previous sample, with the CanGoBack property, which tells you whether there is any page in the stack. This way, we are sure that we trigger backward navigation only if there's an actual page to redirect the user to.

The NavigationView control offers built-in support for backward navigation, by providing a back button integrated inside the UI. As a developer, you can intercept the selection of this button by subscribing to the BackRequested event handler and using the GoBack() method to redirect the user to the previous page. The NavigationView control also offers a property called IsBackEnabled, which you can use to enable or disable the built-in back button. By connecting it to the CanGoBack property of the Frame class, you can easily automatically disable the button if there are no pages in the stack, which helps to improve the experience since the user can immediately realize when they're not allowed to navigate back. You can achieve this goal easily thanks to binding, as in the following example:

```
<NavigationView
            BackRequested="NavigationView_BackRequested"
            ItemInvoked="NavigationShell_ItemInvoked"
            IsBackEnabled="{x:Bind
                ShellFrame.CanGoBack, Mode=OneWay}">
    <NavigationView.MenuItems>
        <NavigationViewItem Content="Home" Icon="Home"
```

```
            Tag="Home" />
        <NavigationViewItem Content="Favorites"
            Icon="Favorite" Tag="Favorite"/>
        <NavigationViewItem Content="Messages"
            Icon="Message" Tag="Messages" />
    </NavigationView.MenuItems>


    <Frame x:Name="ShellFrame" />
</NavigationView>
```

Now that we have completed the implementation of the navigation experience, we can see a few more features that the `NavigationView` control enables to further customize it.

Adding sections to the footer

The `NavigationView` control supports adding menu items to the footer rather than at the top. It's a great fit for sections that shouldn't be the primary focus of the user, but they're important enough to be quickly accessible. These sections are defined in the same way as the primary ones, with the only difference being that they're added to a collection called `FooterMenuItems`, as in the following example:

```
<NavigationView>
    <NavigationView.MenuItems>
        <NavigationViewItem Content="Home" Icon="Home"
            Tag="Home" />
        <NavigationViewItem Content="Favorites"
            Icon="Favorite" Tag="Favorite"/>
        <NavigationViewItem Content="Messages"
            Icon="Message" Tag="Messages" />
    </NavigationView.MenuItems>
    <NavigationView.FooterMenuItems>
        <NavigationViewItem Content="Support" Icon="Help"
            Tag="Support" />
        <NavigationViewItem Content="Account"
            Icon="Account" Tag="Account" />
    </NavigationView.FooterMenuItems>
```

```
    <Frame x:Name="ShellFrame" />

</NavigationView>
```

These sections are managed like the primary ones, so you will keep receiving the information about the selected item through the `ItemInvoked` event.

The menu items support having nested menu items as well, in case some sections of the application have a more complex navigation architecture. To support this feature, the `NavigationViewItem` control can host a `MenuItems` collection on its own, where you can store other `NavigationViewItem` controls, as in the following sample:

```
<NavigationView>
    <NavigationView.MenuItems>
        <NavigationViewItem Content="Home" Icon="Home"
            Tag="Home" />
        <NavigationViewItem Content="Favorites"
            Icon="Favorite" Tag="Favorite"/>
        <NavigationViewItem Content="Messages"
            Icon="Message" Tag="Messages">
            <NavigationViewItem.MenuItems>
                <NavigationViewItem Content="Inbox"
                    Icon="Mail" Tag="Inbox" />
                <NavigationViewItem Content="Sent"
                    Icon="MailReply" Tag="Sent" />
            </NavigationViewItem.MenuItems>
        </NavigationViewItem>
    </NavigationView.MenuItems>

    <Frame x:Name="ShellFrame" />
</NavigationView>
```

Also, in this case, the way you manage the selection of one of these sections doesn't change. You use the `ItemInvoked` event and, through the `ItemInvokedContainer` property of the event arguments, use the `Tag` property to understand where to redirect the user.

Lastly, `NavigationMenu` offers built-in support for a section that is very common among applications: settings. To enable it, you just need to set the `IsSettingsVisible` property to `true`, which will automatically enable a settings menu item in the footer:

```
<NavigationView IsSettingsVisible="True">
    <!—navigationview implementation à
</NavigationView>
```

To understand whether the user has clicked on the **Settings** menu item, you can keep using the `ItemInvoked` event. However, this time, you must use the `IsSettingsInvoked` property exposed by the event argument to determine whether that's the case, as in the following example:

```
private void NavigationShell_ItemInvoked(NavigationView
    sender, NavigationViewItemInvokedEventArgs args)
{
    if (args.IsSettingsInvoked)
    {
        ShellFrame.Navigate(typeof(SettingsPage));
    }
}
```

The following screenshot shows all the advanced features we have discussed so far (footer menu items, nested menu items, and settings) enabled in the `NavigationView` control:

Figure 5.7 – A NavigationView control with footer menu items, nested menu items, and settings enabled

As a developer, however, you might prefer other ways to organize the menu of your application. Let's see them.

Displaying a menu at the top

In some cases, you might prefer to have a more traditional menu at the top of the screen, rather than using the hamburger menu approach we have adopted so far. The `NavigationView` control supports this approach thanks to the `PaneDisplayMode` property, which you can use to customize the way the panel is displayed. By setting it to `Top`, you can change its look and feel, as in the following screenshot:

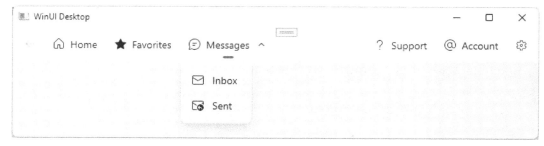

Figure 5.8 – A NavigationView control with PaneDisplayMode set to Top

The `PaneDisplayMode` property can also assume other values, such as `Left`, `LeftCompact`, or `Minimal`. These values can be used to force one of the display modes that, by default, is used to support adaptive layouts. For example, if you set it to `Minimal`, the menu will always be hidden, regardless of the window's size.

If you want to create a more traditional desktop experience, there's also another approach: using the `MenuBar` control, which enables the creation of top-level menus, with multiple submenus and options. Let's take a look at the following code:

```
<Window>
    <Grid>
        <Grid.RowDefinitions>
            <RowDefinition Height="50" />
            <RowDefinition Height="*" />
        </Grid.RowDefinitions>

        <MenuBar Grid.Row="0">
            <MenuBar.Items>
                <MenuBarItem Title="File">
                    <MenuBarItem.Items>
```

```xml
                    <MenuFlyoutSubItem Text="New">
                        <MenuFlyoutItem Text="New
                        document" />
                        <MenuFlyoutItem Text="New
                        project" />
                    </MenuFlyoutSubItem>

                    <MenuFlyoutItem Text="Save" />
                    <MenuFlyoutItem Text="Close" />
                    <MenuFlyoutSeparator />
                    <ToggleMenuFlyoutItem Text="Auto
                    save" />
                </MenuBarItem.Items>
            </MenuBarItem>
        </MenuBar.Items>

    </MenuBar>

    <Frame x:Name="ShellFrame" Grid.Row="1" />
  </Grid>

</Window>
```

The MenuBar control has an Items collection, which can host one or more MenuBarItem controls. Each of them represents a top-level menu. Each MenuBarItem control has an Items collection as well, which can host several types of controls, such as the following:

- MenuFlyoutItem, which represents a single entry in the menu
- MenuFlyoutSubItem, which can host multiple MenuFlyoutItem controls that will be displayed as a nested menu
- ToggleMenuFlyoutItem, which represents an option that you can enable or disable
- RadioMenuFlyoutItem, which represents a radio button that is included inside the menu
- MenuFlyoutSeparator, which you can use to separate different sections inside the same menu

The following screenshot shows how the previous XAML code is rendered in an application:

Figure 5.9 – A MenuBar control with various menus and options

To handle the selection, each MenuFlyoutItem control exposes a Click event that you can manage in code-behind. If you don't want to implement a different event handler for each menu item, you can use a generic event handler and, through the Tag property (like we did with the NavigationView control), identify which item was selected and act accordingly. For example, let's say that you configure the Tag properties of your MenuFlyoutItem controls in the following way:

```
<MenuBarItem Title="File">
    <MenuBarItem.Items>
        <MenuFlyoutItem Text="Save" Tag="Save"
          Click="OnMenuItemSelected" />
        <MenuFlyoutItem Text="Close" Tag="Close"
          Click="OnMenuItemSelected"/>
    </MenuBarItem.Items>
</MenuBarItem>
```

You can use this property in the `OnMenuItemSelected` event handler to detect which item was selected, as in the following code:

```
private void OnMenuItemSelected(object sender,
    RoutedEventArgs e)
{
    if (sender is MenuFlyoutItem menuItem)
    {
        switch (menuItem.Tag)
        {
            case "Save":
                //code to save the file
                break;
            case "Close":
                //code to close the file
                break;
        }
    }
}
```

With this, we have completed our journey into the WinUI navigation system. Now we can move on to another important topic to build beautiful applications: themes.

Supporting Windows themes

In the modern tech ecosystem, one of the most common features included in every app or device is support for light and dark themes. This feature has become increasingly important not only to accommodate personal preferences but also to better support accessibility scenarios. Additionally, when you're using applications on battery-powered devices (such as a tablet or a phone), dark themes have a less significant impact on battery life, since they reduce the number of pixels that the screen must render.

Windows provides built-in support for themes and the user can set their preferences through the **Settings** app. Additionally, Windows provides a high-contrast theme, which is dedicated to people who have visual impairments.

By default, WinUI applications automatically follow the Windows theme, thanks to the built-in support provided by all the WinUI controls. If you add any standard control to a page, you will notice how its look and feel will automatically change when you switch from one theme to another, as you can observe in the following screenshot:

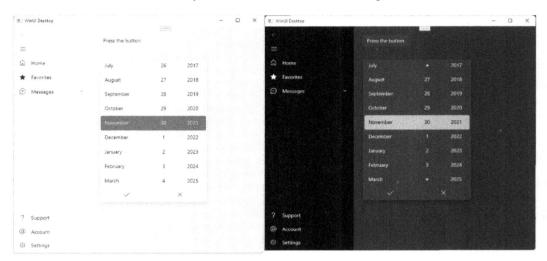

Figure 5.10 – The built-in WinUI controls automatically adapt to the
Windows theme selected by the user

However, it's very common when you're building an application to not just rely on the default Windows look and feel, but to add your own branding and use different colors and styles. In these scenarios, there might be situations where a page looks perfectly fine with the light theme, but the content becomes unreadable when you switch to the dark theme (or the other way around). To manage these cases, WinUI supports the possibility to load a specific set of resources based on the current theme, thanks to the ThemedDictionaries collection exposed by the Resource dictionary class. Let's take a look at the following code, declared in the App.xaml file:

```
<Application>
    <Application.Resources>
        <ResourceDictionary>
            <ResourceDictionary.ThemeDictionaries>
                <ResourceDictionary x:Key="Light">
                    <Style x:Key="CustomButton"
                        TargetType="Button">
                        <Setter Property="Foreground"
                            Value="Black" />
```

```xml
                    <Setter Property="Background"
                        Value="Yellow"/>
                </Style>
            </ResourceDictionary>
            <ResourceDictionary x:Key="Dark">
                <Style x:Key="CustomButton"
                    TargetType="Button">
                    <Setter Property="Foreground"
                        Value="White" />
                    <Setter Property="Background"
                        Value="Green"/>
                </Style>
            </ResourceDictionary>
        </ResourceDictionary.ThemeDictionaries>
      </ResourceDictionary>
    </Application.Resources>
</Application>
```

Inside the `ThemeDictionaries` collection, we have created two different dictionaries, identified by two different keys: `Light` and `Dark`. Inside them, we are defining a style that targets the `Button` control with the same name, `CustomButton`. However, the style configuration is different: in the case of the light theme, the button will have a yellow background with black text; in the case of the dark theme, the background will be green and the text white.

Now, we just need to apply our style to any `Button` control in the standard way:

```xml
<Button Content="Press the button" Style="{StaticResource
    CustomButton}" />
```

WinUI will apply the proper `CustomButton` style based on the current theme, as you can see in the following screenshot:

Figure 5.11 – WinUI will pick the correct style based on the Windows theme

As we saw in *Chapter 2, The Windows App SDK for a Windows Forms Developer*, resources can be better organized thanks to **Resource Dictionary** files. You can use them in combination with themes simply by setting the `Source` property with the file path, as in the following sample:

```
<Application>
    <Application.Resources>
        <ResourceDictionary>
            <ResourceDictionary.ThemeDictionaries>
                <ResourceDictionary x:Key="Light"
                  Source="Resources/LightTheme.xaml" />
                <ResourceDictionary x:Key="Dark"
                  Source="Resources/DarkTheme.xaml" />
            </ResourceDictionary.ThemeDictionaries>
        </ResourceDictionary>
    </Application.Resources>
</Application>
```

There might be scenarios, however, in which you don't want to just adapt to the user's theme, but to force a specific one (either controlled by you as a developer or by the user). We'll learn about this opportunity in the next section.

Forcing a specific theme

If your application or a company has unique branding, it might happen that the color palette can't fit both dark and light themes well. In this case, you can force your application to always use a specific theme, regardless of the user's settings, by setting the `RequestedTheme` property in the `App` class, as in the following example:

```
<Application
    x:Class="MyApplication.App"
    xmlns="http://schemas.microsoft.com/winfx/2006/xaml
      /presentation"
    xmlns:x="http://schemas.microsoft.com/winfx/2006/xaml"
    RequestedTheme="Dark">
</Application>
```

In the previous example, WinUI will always apply the dark theme to all the built-in controls, even if the user has applied the light theme in Windows.

Another option is to give users the choice of the theme at the app level, so that they can set a different theme than the Windows one. Also, in this case, you can use the `RequestedTheme` property, however not on the `App` class but on the main window of the application. Dynamically changing the `RequestedTheme` property at the `App` level, in fact, isn't supported.

To achieve this goal, an approach that is easy to implement is to change the code of the `App` class so that the `MainWindow` object is exposed as a static property, as in the following example:

```
public partial class App : Application
{
    public static Window AppWindow { get; set; }
    public App()
    {
        this.InitializeComponent();
    }

    protected override void OnLaunched(Microsoft.UI.Xaml
      .LaunchActivatedEventArgs args)
    {
        MainWindow = new MainWindow();

        MainWindow.Activate();
    }
}
```

Now, in any page or class of your application, you can call the following code to apply a different theme (in this example, the dark one):

```
private void ChangeTheme()
{
    if (App.MainWindow.Content is FrameworkElement
      rootElement)
    {
        rootElement.RequestedTheme = ElementTheme.Dark;
    }
}
```

If you want to provide such a feature in your application, you might want to use another markup expression provided by XAML called `ThemeResource` to apply styles and resources, instead of the usual `StaticResource` one. `ThemeResource` works like `StaticResource` (so it follows the same resource lookup principles), but it can dynamically react to changes. A resource assigned with the `StaticResource` markup expression, instead, is evaluated only once when the control gets rendered; if the resource changes at runtime, the update won't be visible until you restart the application.

For instance, the `Button` control we previously used in our examples will look like this now:

```
<Button Content="Press the button" Style="{ThemeResource
  CustomButton}" />
```

Thanks to this change, now if you use the previous code to switch the theme at runtime, `Button` will automatically be rendered again, this time using the `CustomButton` style defined for the dark theme.

> **Note**
>
> This feature is supported only when it's used inside a `Page` container. If you try to change the theme at runtime directly inside a `Window` object, all the themed resources will change, but the background of the application will stay the same.

Now that we have added support for themes, let's see how we can improve the UI of our application even further by integrating animations.

Creating animations

Animations are one of the most powerful ways to create a sense of smoothness in your applications. Animations shouldn't be abused, and you should use them carefully, to avoid overloading your applications with effects that, in the end, will just slow down the user workflow. However, when they are used in the right way, they can really enhance the user experience and make your application stand out from the others.

In *Chapter 2, The Windows App SDK for a Windows Forms Developer*, we have already seen a way to implement animations using storyboards. WinUI still supports storyboards, but it also includes a new and powerful animation engine, called **Composition APIs**. These APIs interact with a special layer called the **visual layer**, which is placed between the XAML layer and the native DirectX one, and gives you much more power compared to traditional animations. Animations executed on this visual layer, in fact, don't run on the UI thread, but on an independent thread at 60 FPS. Moreover, they give you the opportunity to apply effects and motions that aren't available with the traditional Storyboard approach. At the heart of this feature, there's the concept of Visual, which enables the interoperability between XAML and the visual layer.

Composition APIs can be quite complex to master. Even if they abstract many of the concepts, you're still working with the UI layer closer to the native one. For this reason, Microsoft has included inside the Windows Community Toolkit a set of helpers that can help you implement powerful animations. As such, the first step is to install the toolkit inside your WinUI application. Right-click on your project, choose **Manage NuGet Packages**, and look and install the package with the **CommunityToolkit.WinUI. UI.Animations** identifier. Now we can start exploring a few of these utilities.

Creating animations in C#

The Windows Community Toolkit includes a powerful class called AnimationBuilder, which supports creating complex animations directly in C#. One of its key features is the ability to completely abstract the type of animation: with AnimationBuilder, you can define a single schedule that can include a mix of animations that run on the XAML layer and the composition layer.

Let's start with an example, which will help you to understand the basic structure of the AnimationBuilder class:

```
AnimationBuilder.Create()
    .Opacity(from: 0, to: 1)
    .RotationInDegrees(from: 30, to: 0)
    .Start(MyButton);
```

You start by creating a new `AnimationBuilder()` object thanks to the `Create()` method. Then, using fluent syntax, you can define one or more animations that you want to execute among the available ones. These animations can affect a control's size, scale, position, opacity, rotation, and many more. In the previous example, you can see two animations: the first one changes the opacity, while the second one applies a rotation. In both cases, we specify two properties (`from` and `to`) to define the starting and ending values of the property. At the end, we call the `Start()` method passing a reference to the control that we want to animate.

Similar to the `Storyboard` control, by default the XAML framework will automatically generate all the frames of the animations with the default settings (the standard duration is 400 milliseconds). However, if you want, you can use more properties to further customize the animation, as in the following example:

```
await AnimationBuilder.Create()
    .Opacity(from: 0,
             to: 1,
             duration: TimeSpan.FromSeconds(3),
             repeat: RepeatOption.Forever)
    .RotationInDegrees(from: 30, to: 0)
    .StartAsync(MyButton);
```

In this case, other than setting the starting and ending values, we also define the duration (3 seconds) and that we want to repeat the animation in a loop. This example also gives us the opportunity to showcase another powerful feature of the `AnimationBuilder` class: you can execute the animation asynchronously, by using the `StartAsync()` method to trigger it.

Exactly like with the `Storyboard` object, you can also use `AnimationBuilder` to create animations with a custom timeline, by manually specifying the keyframes, as in the following sample:

```
AnimationBuilder.Create()
    .Opacity().NormalizedKeyFrames(x =>
                x.KeyFrame(0.0, 0)
                 .KeyFrame(0.5, 0.5)
                 .KeyFrame(1.0, 1))
    .Start(MyButton);
```

The goal is achieved with the `NormalizedKeyFrames()` method, in which you can use a lambda expression to specify one or more `KeyFrame()` methods. For each of them, you can specify the timing of the frame and the value that the property must assume at that time.

Creating animations in XAML

There are many reasons to prefer creating the animations directly in XAML. For example, if you're adopting the MVVM pattern, defining the animations in XAML helps to keep the UI layer clearly separated from the business logic.

The Windows Community Toolkit supports the `AnimationSet` control, which you can use to express `AnimationBuilder` animations in XAML rather than in C#. Let's take a look at the following sample:

```xml
<Button>
    <animations:Explicit.Animations>
        <animations:AnimationSet x:Name="FadeInAnimation">
            <animations:OpacityAnimation From="0" To="1" />
            <animations:RotationAnimation From="30"
                To="0" />
        </animations:AnimationSet>
    </animations:Explicit.Animations>
</Button>
```

As a first step, we must declare the `animations` namespace in our `Page`, as in the following example:

```xml
<Page
xmlns:animations="using:CommunityToolkit.WinUI
    .UI.Animations">
<!-- page content -->
</Page>
```

Inside the control, we have defined an `AnimationSet` object inside the `Explicit.Animations` collection, where we can store one or more animations we want to apply. In the previous code snippet, we recreated the same motion that we have previously seen in C#: we applied to the `Button` control two animations, the first one changing the opacity from 0 to 1, and the other one the rotation from 30 to 0.

Like in the case of `AnimationBuilder`, also in this case, the animation will run with the default settings, with a duration of `400` milliseconds. If we want to customize its behavior, we can wrap the animations inside an `AnimationScope` object, as in the following example:

```
<Button>
    <animations:Explicit.Animations>
        <animations:AnimationSet x:Name="FadeInAnimation">
            <animations:AnimationScope Duration="0:0:3"
              Delay="0:0:1">
                <animations:OpacityAnimation From="0"
                  To="1" />
                <animations:RotationAnimation From="30"
                  To="0" />
            </animations:AnimationScope>
        </animations:AnimationSet>
    </animations:Explicit.Animations>
</Button>
```

Both the `OpacityAnimation` and `RotationAnimation` objects are declared as children of the `AnimationScope` object, which will make them run for 3 seconds, with an initial delay of 1 second.

As with `AnimationBuilder`, with `AnimationSet` we can also manually define the pace of the animation by manually setting the keyframes, as in the following example:

```
<Button>
    <animations:Explicit.Animations>
        <animations:AnimationSet
          x:Name="KeyframeFadeInAnimation">
            <animations:AnimationScope Duration="0:0:3"
              Delay="0:0:1" EasingMode="EaseIn">
                <animations:OpacityAnimation>
                    <animations:ScalarKeyFrame Key="0"
                      Value="0" />
                    <animations:ScalarKeyFrame Key="0.3"
                      Value="0.2" />
                    <animations:ScalarKeyFrame Key="0.8"
                      Value="0.5" />
```

```
                    <animations:ScalarKeyFrame Key="1"
                        Value="1" />
                    </animations:OpacityAnimation>
                    <animations:RotationAnimation From="30"
                        To="0" />
                </animations:AnimationScope>
            </animations:AnimationSet>
        </animations:Explicit.Animations>
    </Button>
```

The `CommunityToolkit.WinUI.UI.Animations` namespace offers multiple types of keyframe objects, based on the property type. In this case, since we're changing the `Opacity` value (which is a number), we use the `ScalarKeyFrame` object.

Now that we have defined an `AnimationSet` object, how can we run it? The simplest way is by assigning a name to the animation through the `x:Name` property, so that we can trigger it in code using the `Start()` or `StartAsync()` methods, as in the following sample:

```
private void OnStartAnimation(object sender,
    RoutedEventArgs e)
{
    MyAnimationSet.Start();
}
```

However, there's a better way! The previous approach, in fact, partially ruins our goal of keeping the whole animation definition in XAML since, at the end, we have to write some code in the code-behind class to trigger it. Thanks to a XAML feature called **Behaviors**, we can implement the same approach in XAML. Behaviors, in fact, is a special XAML object that you can use to perform actions in XAML that otherwise would need to be declared in code.

Before seeing it in action, however, we need to install another package from NuGet, called **CommunityToolkit.WinUI.UI.Behaviors**.

Once you have installed it, you must add a few additional namespaces:

```
<Page
    xmlns:controls="using:CommunityToolkit
        .WinUI.UI.Controls"
    xmlns:animations="using:CommunityToolkit
```

```
    .WinUI.UI.Animations"
  xmlns:interactivity="using:Microsoft
    .Xaml.Interactivity"
  xmlns:core="using:Microsoft.Xaml
    .Interactions.Core"
  xmlns:behaviors="using:CommunityToolkit
    .WinUI.UI.Behaviors">

<!-- page content -->

</Page>
```

Now we can add the following code to our `Button` definition:

```
<Button x:Name="MyButton" Content="Press the button"
    Click="OnGoToSecondPage">
    <interactivity:Interaction.Behaviors>
        <core:EventTriggerBehavior EventName="Click">
            <behaviors:StartAnimationAction
              Animation="{x:Bind StandardFadeInAnimation}" />
        </core:EventTriggerBehavior>
    </interactivity:Interaction.Behaviors>
    <animations:Explicit.Animations>
        <animations:AnimationSet x:Name="FadeInAnimation">
            <animations:OpacityAnimation From="0" To="1" />
            <animations:RotationAnimation From="30"
              To="0" />
        </animations:AnimationSet>
    </animations:Explicit.Animations>
</Button>
```

When we define a behavior, we specify two things:

- The trigger, which is the event that will execute the behavior
- The action that we want to perform when the behavior is executed

In this example, the trigger is implemented by the `EventTriggerBehavior` class, which we can use to manage events directly in XAML. We are using the `EventName` property to subscribe to the `Click` event of the button. The action, instead, is implemented by the `StartAnimationAction` class, which, as the name says, we can use to start an animation. Through binding, we connect the `Animation` property with the `AnimationSet` object we have defined just below the behavior.

Now, when you click on the button, the animation will start, recreating the same behavior we have seen before, but without needing to write any C# code.

Composition APIs include not only animations but also special effects. In the next section, we're going to learn how to apply and animate them.

Applying effects and animating them

The composition layer also gives you access to many powerful effects that you can apply to any control on the screen. They are all rendered using Composition APIs, maximizing performance and quality. However, the code required to set up one of these effects is a bit complicated and it must be declared in the code-behind. When we are working with the UI layer, instead, we have seen with Behaviors how it's much more convenient to keep everything in XAML.

Thanks to the Windows Community Toolkit, we have access to a series of controls that we can use to apply these effects easily through the concept of Visual Factory, which enables us to apply an effect directly to any `UIElement` control (which is the base class all the WinUI controls derive from).

To enable these effects, we first have to install another NuGet package into our project, identified by the name `CommunityToolkit.WinUI.UI.Media`. Now we can declare the following namespace in `Page`:

```
<Page xmlns:media="using:CommunityToolkit.WinUI.UI.Media">
    <!-- page content -->
</Page>
```

Thanks to this namespace, we can set up a pipeline to apply one or more effects directly to a control, as in the following example:

```
<Image Width="400" Source="Assets/MicrosoftLogo.png">
    <media:UIElementExtensions.VisualFactory>
        <media:PipelineVisualFactory
          Source="{media:BackdropSource}">
            <media:BlurEffect x:Name="ImageBlurEffect"
```

```
            Amount="32" IsAnimatable="True"/>
        </media:PipelineVisualFactory>
    </media:UIElementExtensions.VisualFactory>
</Image>
```

First, we declare a `UIElementExtensions.VisualFactory` property as a child of the control we want to apply the effect to, in this case, an `Image` one that displays the Microsoft logo. Inside the `VisualFactory` object, we add a `PipelineVisualFactory` object, which we can use to create a pipeline of effects, empowering us to apply multiple effects at once. In the previous example, we're just assigning a single one, a `BlurEffect` object. Using the `Amount` property, we can customize the effect by setting the blur intensity.

In the following screenshot, you can see the result with and without the effect:

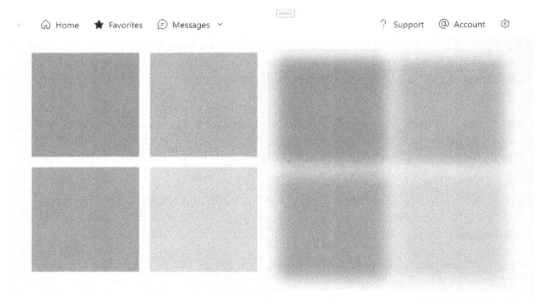

Figure 5.12 – A blur effect applied to an Image control

As already mentioned, the `PipelineVisualFactory` collection can support more than one effect. For example, the following code shows how, other than the blur effect, you can also apply a sepia and a grayscale filter:

```
<Image Width="400" Source="Assets/MicrosoftLogo.png">
    <media:UIElementExtensions.VisualFactory>
        <media:PipelineVisualFactory
```

```
        Source="{media:BackdropSource}">
          <media:BlurEffect x:Name="ImageBlurEffect"
            Amount="12" IsAnimatable="True"/>
          <media:SepiaEffect Intensity="15" />
          <media:GrayscaleEffect />
        </media:PipelineVisualFactory>
      </media:UIElementExtensions.VisualFactory>
</Image>
```

Inside the `CommunityToolkit.WinUI.UI.Media` namespace, you will find many other available effects, such as `Invert`, `Opacity`, `Saturation`, and `Tint`.

You might have noticed how `BlurEffect` has a property called `IsAnimatable` set to `True`. This is another one of the powerful features supported by the Windows Community Toolkit. These effects, coming from the `CommunityToolkit.WinUI.UI.Media` namespace, in fact, can be animated as well by using an `AnimationSet` object. Let's take a look at the following sample, which puts together a few of the concepts we have learned so far:

```
<Image Width="600" Source="Assets/MicrosoftLogo.png">
    <interactivity:Interaction.Behaviors>
        <core:EventTriggerBehavior EventName="Tapped">
            <behaviors:StartAnimationAction
              Animation="{x:Bind MyAnimationSet}" />
        </core:EventTriggerBehavior>
    </interactivity:Interaction.Behaviors>

    <media:UIElementExtensions.VisualFactory>
        <media:PipelineVisualFactory
          Source="{media:BackdropSource}">
          <media:BlurEffect x:Name="ImageBlurEffect"
            Amount="32" IsAnimatable="True"/>
          <media:SepiaEffect Intensity="15" />
          <media:GrayscaleEffect />

        </media:PipelineVisualFactory>
    </media:UIElementExtensions.VisualFactory>
```

```
<animations:Explicit.Animations>
    <animations:AnimationSet x:Name="MyAnimationSet">
        <animations:ScaleAnimation From="2" To="1"/>
        <animations:BlurEffectAnimation From="32"
            To="0" Target="{x:Bind ImageBlurEffect}"/>
    </animations:AnimationSet>
</animations:Explicit.Animations>

</Image>
```

From the preceding code, we observe the following:

- We have a behavior like the one we've already seen in a previous sample, which we can use to start an animation when an event is triggered (in this case, `Tapped`, which means that the user has clicked on the image).

- The second part should be familiar as well: it's a `PipelineVisualFactory`, object, which applies a series of effects to our image control.

- The last part is new, but still familiar: we have created an `AnimationSet`, object, which is going to animate a few properties of our control. In this case, we're applying a traditional animation (`ScaleAnimation`, to change the scale of the control from 2 to 1) and an effect animation (`BlurEffectAnimation`, which is going to change the `Amount` property of our `ImageBlurEffect` object from 32 to 0) at the same time.

- The final effect is that when the user clicks on our image, the blur effect will fade away and the image will return clearly visible.

Always remember to set the `IsAnimatable` property on the effect you want to animate; otherwise, the animation won't work.

Connected animations

Connected animations are another powerful feature provided by the composition layer that helps to stay focused on the selected item while transitioning from one page to another. Connected animations start with one element on one page and they finish on the same element, but on another page. Think, for example, of a photo gallery, with a series of thumbnails. When you click on one of the photos, you can use a connected animation to transition from the thumbnail on the gallery page to the bigger photo on the detail page, helping the user to focus on what's important (the photo, in this case).

This is another feature that requires quite a bit of code to be implemented but, thanks to the Windows Community Toolkit, we can easily set it up directly in XAML. Let's see an example of our first page:

```
<Page xmlns:animations="using:CommunityToolkit.WinUI.
  UI.Animations">
    <StackPanel>
        <Image Height="100" Width="100"
            Source="Assets/MicrosoftLogo.png"
            VerticalAlignment="Center" HorizontalAlignment=
            "Center" animations:Connected.Key="item" />

        <Button Content="Go to page 2"
            Click="OnGoToSecondPage" />
    </StackPanel>

</Page>
```

The page contains an `Image` control and a `Button` control, with an event handler that triggers the navigation to another page of the application (using the APIs we have learned how to use in this chapter). However, the `Image` control has something special: an attached property called `Connected.Key`, which is part of the `CommunityToolkit.WinUI.UI.Animations` namespace. Take note of the value we have assigned to this property, which is `item`.

Let's now see the content of the second page:

```
<Page xmlns:animations="using:CommunityToolkit.
  WinUI.UI.Animations">
    <StackPanel Orientation="Horizontal">
        <Image x:Name="HeroElement" Height="300"
            Width="300" Source="Assets/MicrosoftLogo.png"
            animations:Connected.Key="item" VerticalAlignment=
            "Top" />

        <StackPanel x:Name="HeroDetailsElement"
            Margin="20,0"
            VerticalAlignment="Top" MaxWidth="500"
            animations:Connected.AnchorElement="{x:Bind
```

```
        HeroElement}">
            <TextBlock Text="This is the title"
                FontSize="50" /</TextBlock>
            <TextBlock Text="This is the description" />
        </StackPanel>

    </StackPanel>
</Page>
```

The first thing to highlight is that this second page has the exact same control as we have seen on the first page: an `Image` control that renders the Microsoft logo, only at a bigger size. Notice also how we are using the same `Connected.Key` property, with the same value we have assigned on the first page, `item`. This is the key feature to enable connected animations; since we have set the `Connected.Key` property with the same value in both controls, WinUI will apply a transition animation between the two during the navigation.

We can also anchor other elements to the animation so that they will be included in the transition as well. In the previous snippet, you can see how the second page, other than the bigger image, also has a `StackPanel` container called `HeroDetailsElement`, which includes some information about the image. By binding the `Connected.AnchorElement` property with `Image`, the two sections of the page will be part of the same connected animation.

One of the most common scenarios for using connected animations is a master-detail scenario: the user selects an item from a collection and moves to a detail page. Let's see how to implement it.

Using a connected animation in a master-detail scenario

Connected animations are especially effective when they are used in a master-detail scenario, since they can help the user to focus on the selected item during the transition. The Windows Community Toolkit offers a few attached properties that we can use to easily implement this scenario. Let's see an example of the master page:

```
<GridView x:Name="lstPeople"
            SelectionMode="None"
        animations:Connected.ListItemElementName="ItemPhoto"
            animations:Connected.ListItemKey="listItem"
            IsItemClickEnabled="True"
```

```
                        ItemClick="lstPeople_ItemClick">
    <GridView.ItemTemplate>
        <DataTemplate x:DataType="local:Person">
            <StackPanel>
                <Image x:Name="ItemPhoto" Height="100"
                  Width="100" Source="{x:Bind Photo}" />
                <TextBlock Text="{x:Bind Name}" />
                <TextBlock Text="{x:Bind Surname}"/>
            </StackPanel>
        </DataTemplate>
    </GridView.ItemTemplate>
</GridView>
```

We are using a `GridView` control to display a collection of people. Each of them is represented with a `DataTemplate` object that displays the photo, the name, and the surname. Compared to a traditional collection, we have added two extra properties:

- `Connected.ListItemElementName`, which contains the name of the control in the template that we want to use as main driver of the animation; in this case, it's `ItemPhoto`. The same element will also be displayed on the detail page, where the transition will happen.

- `Connected.ListItemKey`, which contains a unique key that we're going to reuse on the second page to connect the elements.

In this `GridView` control, we have subscribed to the `ItemClick` event, which is triggered when one of the items is selected. The event handler simply takes care of redirecting the user to the detail page, using the navigation approach we have learned about in this chapter:

```
private void lstPeople_ItemClick(object sender,
  ItemClickEventArgs e)
{
    App.ShellFrame.Navigate
{
    Person person = e.ClickedItem as Person;
      (typeof(DetailPage), Person, new
        SuppressNavigationTransitionInfo());
}
```

As an extra parameter for the `Navigate()` method (other than the type of the detail page), we are passing the selected item (stored in the `ClickedItem` property of the event arguments) and a `SuppressNavigationTransitionInfo` object to disable the transition (since we're already using a connected animation).

Let's now take a look at the XAML layout of the detail page:

```xml
<Page xmlns:animations="using:CommunityToolkit
  .WinUI.UI.Animations">
    <StackPanel Orientation="Horizontal">
        <StackPanel x:Name="HeroDetailsElement"
          Margin="20,0"
        VerticalAlignment="Top" MaxWidth="500"
        animations:Connected.AnchorElement="{x:Bind
          HeroElement}">
            <TextBlock Text="{x:Bind Person.Name}"
              FontSize="50" />
            <TextBlock Text="{x:Bind Person.Age}" />
        </StackPanel>

        <Image x:Name="HeroElement" Height="300"
              Width="300" Source="{x:Bind Person.Photo}"
              animations:Connected.Key="listItem"
              VerticalAlignment="Top" />
    </StackPanel>
</Page>
```

The code here is the same as the first scenario we explored around connected animations. By using the `Connected.Key` property, we assign to the destination control the same key we have defined on the master page (in this example, it's `listItem`). Also, in this case, we can use `Connected.AnchorElement` to anchor other controls to the main animation.

Let's now explore one last feature to enable animations, animated icons.

Using animated icons

One of the newest features introduced in WinUI is the `AnimatedIcon` control, which you can set as content of other controls (such as a `Button` control) to render an icon that is automatically animated when the user interacts with it. For example, this is how you can define a back button that animates the back arrow when the user clicks on it:

```
<Button>
    <AnimatedIcon>
        <AnimatedIcon.Source>
            <animatedvisuals:AnimatedBackVisualSource/>
        </AnimatedIcon.Source>
    </AnimatedIcon>
</Button>
```

The `AnimatedIcon` control requires a `Source` property, which is the animation to display. WinUI provides many built-in animations inside the `Microsoft.UI.Xaml.Controls.AnimatedVisuals` namespace, which you must declare in your XAML page:

```
<Page
  xmlns:animatedvisuals="using:Microsoft.UI.Xaml.
    Controls.AnimatedVisuals">
```

The one we have used, called `AnimatedBackVisualSource`, will automatically render an animated back button. Inside the namespace (which is documented at https://docs.microsoft.com/en-us/windows/winui/api/microsoft.ui.xaml.controls.animatedvisuals?view=winui-3.0), you can also find classes to render a search icon (`AnimatedFindVisualSource`), a settings icon (`AnimatedSettingsVisualSource`), or a navigation menu (`AnimatedGlobalNavigationButtonVisualSource`).

Based on the Windows version or on the configuration of the computer, animations might not be supported. In this case, you can specify a property called `FallbackIconSource`, which you can use to display a static alternative. Let's take a look at the following example:

```
<Button>
    <AnimatedIcon>
        <AnimatedIcon.Source>
            <animatedvisuals:AnimatedBackVisualSource/>
        </AnimatedIcon.Source>
```

```
        <AnimatedIcon.FallbackIconSource>
            <SymbolIconSource Symbol="Back"/>
        </AnimatedIcon.FallbackIconSource>
    </AnimatedIcon>
</Button>
```

We are using a `SymbolIconSource` object as a fallback. If animations aren't supported, the `Button` control will be rendered with a static back icon.

Lastly, if you want to build your own custom animation, you can use **LottieGen**, which is a tool that is able to generate a WinUI component starting from an animation created with Adobe After Effects. You can learn more at `https://docs.microsoft.com/en-us/windows/apps/design/controls/animated-icon#use-lottie-to-create-animated-content-for-an-animatedicon`.

In this chapter, we have explored a few of the available options, but the Windows Community Toolkit includes many other helpers to create animations and make your application even more beautiful. Here are a few examples:

- **Implicit animations**, which you can use to create animations that are automatically triggered when a specific event happens (such as the control being rendered or a control's property changing). These animations are a great companion of responsive layouts, since you can animate controls on the screen as they are reorganized on the screen when the size of the window changes. You can learn more at `https://docs.microsoft.com/en-us/windows/communitytoolkit/animations/implicitanimationset`.

- **Expression builder**, which gives you the option to create complex animations, backed up by mathematical expressions, in a type-safe mode with C#. You can learn more at `https://docs.microsoft.com/en-us/windows/communitytoolkit/animations/expressions`.

- `FadeHeader`, which you can use to implement a page header that automatically fades away when you scroll down the content of the page. You can learn more at `https://docs.microsoft.com/en-us/windows/communitytoolkit/animations/fadeheader`.

- **Lottie**, which is a library to render Adobe After Effects animations in your application. You can learn more at `https://docs.microsoft.com/en-us/windows/communitytoolkit/animations/lottie`.

We have completed our overview of the powerful animation system offered by WinUI. Let's wrap up the chapter by taking a quick peek at the controls we can use to build our applications.

Exploring the WinUI controls

WinUI comes with a wide range of built-in controls to support a lot of scenarios: data input, animations, user interaction, and so on. Before starting to create complex applications, it's important to know what WinUI has to offer so that you can always choose the best control for the scenario you're looking to implement.

A great starting point to explore all the controls and features included in WinUI is the **XAML Controls Gallery**, which you can download from the Microsoft Store at `https://www.microsoft.com/store/productId/9MSVH128X2ZT`:

Figure 5.13 – The XAML Controls Gallery application

This application is a great companion for developers: other than highlighting all the available controls, you can also find many code examples and links to the documentation. Additionally, the whole source code of the project is available at `https://github.com/microsoft/xaml-controls-gallery/`, so you can recreate all the examples in your application.

If you need even more flexibility, however, there are many other libraries that you can use to enhance your Windows application. We have already mentioned the Windows Community Toolkit multiple times in this chapter to introduce more helpers and services. However, the toolkit also provides a set of additional controls, including a really important one when it comes to building enterprise applications: **DataGrid**. The DataGrid control is a highly effective way to display data through tables that can be deeply customized, both by the developer and the user. You can reorder columns, apply filters, provide inline editing capabilities, and do many more things, as you can see in the following screenshot taken from the Windows Community Toolkit sample app:

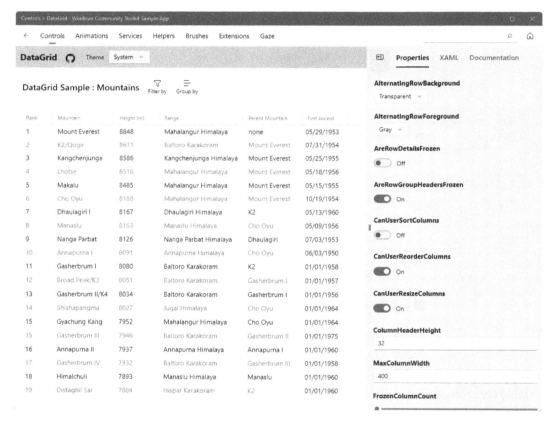

Figure 5.14 – The Windows Community Toolkit sample application

Similar to the XAML Controls Gallery application, you can use the Windows Community Toolkit sample application to explore all the controls, helpers, and services provided by the library. This application can be downloaded from the Microsoft Store as well, using this link: https://www.microsoft.com/store/productId/9NBLGGH4TLCQ.

In the end, you can also rely on partners to enrich your development experiences. Companies such as Syncfusion, Telerik, DevExpress, and GrapeCity offer commercial libraries with hundreds of extra controls to create forms, barcode readers, and charts.

There are no limits to the UI features that you can enable in your applications thanks to the WinUI extensibility!

Summary

In the initial chapters of this book, we learned about many key concepts that are essential to building an application: from XAML properties to data binding, managing the application's life cycle to customizing windows. However, in most cases, the basic concepts aren't enough to build real-world applications. In this chapter, we started to learn about a few of the topics that are critical to turning your application from a simple prototype to a real one. We first learned how to build a responsive layout, which is essential in today's world where applications can run on a wide range of devices, with different resolutions, screen sizes, and formats. Then we explored the navigation concept, which is critical to building complex applications that are made up of multiple sections and contents.

Later, we moved on to theming support, which is another important requirement in the modern world: enabling users to choose a light or dark theme is not only a matter of preference but also a way to better support accessibility and optimize power consumption.

At the end, we explored the complex but powerful world of animations enabled by the XAML visual layer and the Composition APIs. Thanks to the Windows Community Toolkit, we can integrate motion and effects in an easier way than relying on the native APIs included in WinUI.

In the next chapter, we'll continue our journey in learning how to build real-world applications by adopting techniques such as the MVVM pattern and dependency injection, which we can use to improve testability, reliability, and maintainability.

Questions

1. Canvas is a great control to use when you're building responsive layouts. True or false?

2. When you implement a NavigationView control, you must manually manage its look and feel when the size of the window changes. True or false?

3. What's the name of the layer that enables us to run powerful animations at 60 FPS?

6
Building a Future-Proof Architecture

When you start to build real-world enterprise applications, creating a beautiful UI and delivering a good user experience are important, but not sufficient to create the fundamentals for a successful project. Enterprise applications are usually developed over a span of multiple years by teams made of multiple developers; as the application grows and becomes successful, you will need to onboard new developers into the team to keep up with the pace of business expansion.

For all these reasons, it's critically important to build applications that can easily be tested in an automated way so that as the application grows you can rest assured that none of the changes you're going to introduce will break other features and can be easily maintained. This will mean that when a problem does occur, you won't need 3 days just to understand where the issue might be coming from; this you can move forward in your application adding new features without rewriting it from scratch; and that, when you onboard new developers into the team, they won't need 6 months just to understand how the architecture works.

Over time, many architectural patterns have emerged in the developer ecosystem to help teams to create projects to be sustainable over long periods of time by helping developers to separate the various concerns of the application, including the business logic, UI, and user interaction. In the Windows ecosystem, the **Model-View-ViewModel** (**MVVM**) pattern is one of the most popular ways to achieve this goal, especially if you're building your applications using XAML-based platforms such as WPF, WinUI, and Xamarin Forms. MVVM, in fact, leverages many of the building blocks of XAML (such as binding) to provide a better way for developers to build complex applications.

In this chapter, we'll explore how to implement the MVVM pattern through the following topics:

- Learning the basic concepts of the MVVM pattern
- Exploring frameworks and libraries
- Supporting actions with commands
- Making your application easier to evolve with dependency injection
- Exchanging information between different classes
- Managing navigation with the MVVM pattern

By adopting all these principles, we'll achieve the goal of building future-proof applications that can handle change (be it in the customer's requirements, the technical implementation, or the adoption of new technology) in a smart way. As you can see, we have a lot of content to cover, so let's get started!

Technical requirements

The code for the chapter can be found here:

```
https://github.com/PacktPublishing/Modernizing-Your-Windows-
Applications-with-the-Windows-Apps-SDK-and-WinUI/tree/main/
Chapter06
```

Learning the basic concepts of the MVVM pattern

Before starting to dig into the MVVM pattern, let's try to better understand the problem we're trying to solve. Let's assume we're building an e-commerce application that enables users to add products to a cart and submit an order. The application will have a summary page with a recap of the order and a button to submit the order. This is how the event handler connected to the Button control might look like:

```
private void OnBtnOrderClick(object sender, Routed
```

```
        EventArgs e)
{
    var itemsInBasket = _basketListBox.Items.Cast
        <IOrderItem>().ToArray();
    if (itemsInBasket.Length <= 0)
    {
        return;
    }
    // Validate Shopping Basket
    if (itemsInBasket.Any(item => item.Quantity < 0 ||
        item.Price <= (decimal)0.0))
    {
        throw new Exception("Shopping card manipulation
            detected");
    }
    // Get discount and calculate total
    var discount = _customerDao.Load(_currentCustomer)
        .Discount;
    var total = itemsInBasket.Sum(item => item.Price *
        item.Quantity * discount);
    MessageBox.Show($"Are you sure you want to order? Your
        total is {total}.", "Confirmation",
            MessageBoxButton.YesNo, MessageBoxImage.Question);
    _orderSystem.PlaceOrder(itemsInBasket);
    _basketListBox.Items.Clear();
}
```

Of course, this code is a simplification of what would happen in a real application, but it's a good starting point to highlight its problems. The main one is that this code is a mix of tasks that belong to different domains:

- We're interacting with the UI layer by accessing the `Items` property of the `_basketListBox` control or by calling `MessageBox.Show()` to display an alert.

- We are declaring some business logic that determines whether the cart is in a valid state before moving on to the purchase process.

- We're interacting with the data layer by calling the `_orderSystem.PlaceOrder()` method, which will likely store the order in a database.

This approach makes it really complex to do the following:

- **Test the code**: Testing is difficult since all the domains are interconnected. We can't test whether the business logic to validate the cart is correct without raising the event handler or interacting with the database.

- **Maintain the code**: Let's say that the order submission fails. Where is the problem? In the business logic? In the data layer? In the UI layer?

- **Evolve the code**: Let's assume we have a new requirement, and we have to adjust the way discounts are applied. The task will be executed by a developer who just joined the team and is still ramping up on the project. How can he quickly identify where the change must be made if the code that calculates the discount is mixed with other domains?

Let's read into this in more detail.

Moving a bit deeper into the MVVM pattern

Here, we will go into further detail about how the MVVM pattern is split and how it is implemented.

The goal of the MVVM pattern is to ideally split the application into three different layers:

- **The Model**: This domain includes all the entities and services that work with the application's data. For example, in the previous example of the e-commerce application, the Model would contain the `Order` and the `Product` classes and a data service that reads and writes data to the database. The model should be completely agnostic to the UI platform. You should be able to take your model layer and reuse it without changes in any other project based on the .NET ecosystem, such as a Blazor website in Blazor or a mobile application built with MAUI.

- **The View**: This is the UI layer, which contains the definitions of pages, animations, and so on. This is the only layer that is specific to a given platform. In the case of WinUI applications, it's made by all the XAML files and their corresponding code-behind classes.

- **The ViewModel**: This is the glue between the Model and the View. The model retrieves the data to be displayed in the UI but, first, it's processed by the ViewModel so that it can be presented in the right way. On the other side, when the user interacts with the UI, the data gets collected and processed by the ViewModel before sending it back to the model. In a WinUI application, most of the tasks that are performed in a normal application by the code-behind class (such as managing the user interaction) are now moved to the ViewModel layer. The key difference is that code-behind classes have a tight relationship with the UI: you can't, for example, store your code-behind class in a different project than the one containing the XAML files. A ViewModel, instead, is considered a **Plain Old CLR Object** (**POCO**): it doesn't have any specific dependency and, as such, it can be easily stored in a class library, tested, and so on.

The MVVM pattern has become popular in the XAML ecosystem since it relies heavily on the features of the XAML language, as you can understand from the following diagram:

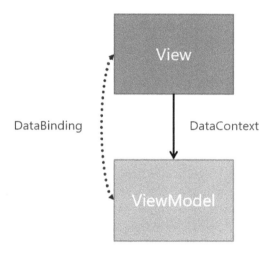

Figure 6.1 – The connection between View and ViewModel is established thanks to binding

In a traditional application, you can reference the controls in the UI simply by setting up a name using the x:Name property. When you switch to the MVVM pattern, instead, you don't have this direct connection anymore, so you replace it using data binding. Every page or component in your application is backed by a ViewModel class, which is set as DataContext of the whole page. This enables your XAML pages to access to all the properties exposed by the ViewModel so that you can display the data coming from the Model layer.

Let's understand this concept better with a concrete example. Let's assume that you're building a form to collect the name and surname of a user. In a traditional application, the XAML would look as follows:

```xml
<StackPanel>
    <TextBlock Text="Name" />
    <TextBox x:Name="txtName" />
    <TextBlock Text="Surname" />
    <TextBox x:Name="txtSurname" />
    <Button Content="Add" Click="OnAddPerson" />
</StackPanel>
```

Since in the code-behind we have direct access to the UI elements, we can directly access the `Text` property of the `TextBox` controls to retrieve the name and surname inserted by the user:

```csharp
private void OnAddPerson(object sender,
    Microsoft.UI.Xaml.RoutedEventArgs e)
{
    string name = txtFirstname.Text;
    string surname = txtLastname.Text;
    //save the user in the database
}
```

This approach, however, creates a tight connection between the UI and the business logic. We can't move this logic to another class, it must stay in the code behind. Let's translate it into a ViewModel. We have mentioned how a ViewModel is just a simple class, so right-click on your project, go to **Add | New class**, and give it a meaningful name (such as `MainViewModel`). Since the View can access the ViewModel data through properties, we need first to expose the two bits of information we need to collect (name and surname) in the ViewModel:

```csharp
public class MainViewModel
{
    public string Name { get; set; }
    public string Surname { get; set; }
}
```

Before using these properties in the UI, we need first to connect the ViewModel class to the View. In other XAML technologies such as WPF, the way to do this was by creating a new instance of the ViewModel and setting it as `DataContext` of the page, like in the following sample:

```
public partial class MainView : Page
{
    public MainView()
    {
        InitializeComponent();
        this.DataContext = new MainViewModel();
    }
}
```

As we learned in *Chapter 3, The Windows App SDK for a WPF Developer*, however, WinUI supports a more powerful binding expression called `x:Bind` which, instead of being evaluated at runtime, is compiled together with the application. This makes binding faster, more efficient, and type safe as you can find binding errors before running the application.

One of the key differences between traditional binding and `x:Bind` is that, if the former uses the `DataContext` property exposed by every XAML control to resolve the binding expression, the latter looks for properties declared in code-behind. As such, the most common approach adopted in WinUI applications to connect a View with the ViewModel is to expose the ViewModel as a property in the code-behind, like in the following sample:

```
public sealed partial class MainPage : Page
{
    public MainViewModel ViewModel { get; set; }

    public MainPage()
    {
        this.InitializeComponent();
        ViewModel = new MainViewModel();
    }
}
```

Now you can use the `ViewModel` property together with the `x:Bind` markup expression to connect these properties to the UI:

```xml
<StackPanel>
    <TextBlock Text="Name" />
    <TextBox Text="{x:Bind ViewModel.Name, Mode=TwoWay}" />
    <TextBlock Text="Surname" />
    <TextBox Text="{x:Bind ViewModel.Surname,
      Mode=TwoWay}" />
    <Button Content="Add" Click="OnAddPerson" />
</StackPanel>
```

We no longer need to set the `x:Name` property to get access to the name and surname entered by the user. Thanks to binding, we can retrieve this information directly through the `Name` and `Surname` properties of our ViewModel. Since we have set the `Mode` property to `TwoWay`, it means that not only will the UI be updated when the two properties change, but also that we'll be able to access the values entered by the user in the ViewModel.

However, if you remember what we learned about binding in *Chapter 2, The Windows App SDK for a Windows Forms Developer*, you should realize that the example we have created so far has a flaw: `Name` and `Surname` in the `MainViewModel` class are two simple properties; whenever one of them changes, the UI isn't notified and, as such, the controls won't update. `MainViewModel` is a simple class, which means that the way to solve this problem is by implementing the `INotifyPropertyChanged` interface, which gives us a way to dispatch notifications to the UI layer when one of the properties change.

This is how a proper implementation of the `MainViewModel` class looks:

```csharp
public class MainViewModel: INotifyPropertyChanged
{
    private string _name;
    public string Name
    {
        get { return _name; }
        set
        {
            NotifyPropertyChanged();
            _name = value;
        }
```

```
    }
    private string _surname;
    public string Surname
    {
        get { return _surname; }
        set
        {
            NotifyPropertyChanged();
            _surname = value;
        }
    }

    public event PropertyChangedEventHandler
        PropertyChanged;

    private void NotifyPropertyChanged([CallerMemberName]
        string propertyName = "")
    {
        PropertyChanged?.Invoke(this, new
            PropertyChangedEventArgs(propertyName));
    }
}
```

Our ViewModel now inherits from the INotifyPropertyChanged interface, which requires us to implement the PropertyChanged event. To make the implementation easier to use, we define a method called NotifyPropertyChanged(), which will invoke the event for us, dispatching the notification to the UI layer. Then, as the last step, we must change the implementation of the Name and Surname properties. Instead of using the standard initializer, we have to add some logic to the setter: when the property changes, other than storing its value, we must call NotifyPropertyChanged() to dispatch the notification to the binding channel.

This code works perfectly fine, but there's space for improvement. What if our application has 10-15 ViewModels? We would need to repeat the PropertyChanged implementation for each of them. This is one of the areas where libraries can help you in speeding up the implementation of the MVVM pattern.

Exploring frameworks and libraries

MVVM is an architectural pattern, not a library or a framework. However, when you start adopting the MVVM pattern in a real project, you need to have a series of building blocks that can help you speed up the implementation, by avoiding the need to reinvent the wheel at each step. In the previous section, we have seen one of these scenarios: each ViewModel class must implement the `INotifyPropertyChanged` interface, otherwise, the connection between the View and the ViewModel would be broken.

This is where libraries and frameworks become helpful: they give you a set of ready-to-use building blocks that you can use to quickly start building all the components of your application. The library we're going to leverage in the examples of this chapter is called the **MVVM Toolkit** and is a part of the Windows Community Toolkit from Microsoft. It's the spiritual successor of the MVVM Light Toolkit by Laurent Bugnion, one of the most popular MVVM libraries used in the past decade to support the development of XAML-based technologies, since it follows the same principles. The MVVM Toolkit is a very lightweight and simple framework. This means that you might need to do some extra work if you need to manage more complex scenarios; however, thanks to its flexibility, it's very easy to extend and learn. The MVVM Toolkit also makes it easy to onboard new developers into the team: since all its features are based on basic XAML concepts, it's very easy to learn it even if you have never worked with it before.

The first step to use the MVVM Toolkit is to right-click on your project, click on **Manage NuGet packages**, and look for the package with **CommunityToolkit.Mvvm** as an identifier. Once you have installed it, let's see how we can refactor our `MainViewModel` class:

```
public class MainViewModel : ObservableObject
{
    private string _name;
    public string Name
    {
        get { return _name; }
        set { SetProperty(ref _name, value); }
    }

    private string _surname;
    public string Surname
    {
        get { return _surname; }
        set { SetProperty(ref _surname, value); }
```

```
        }

}
```

Inside the `CommunityToolkit.Mvvm.ComponentModel` namespace, we find a class called `ObservableObject` that we can use as the base class for our ViewModel. Thanks to this class, we don't need to manually implement the `PropertyChanged` event anymore. We can simply use the `SetProperty()` method in the setter of our property, which takes care of storing the value and dispatching the notification to the binding channel.

The MVVM Toolkit also has a feature currently in preview that makes the code even less verbose, thanks to **source generators**. It's a C# feature introduced in .NET 5 that enables developers to generate code on the fly that gets added during the compilation phase. Thanks to this feature, you can use attributes to decorate your properties, which will generate all the boilerplate code you need to implement the `INotifyPropertyChanged` interface. This is how our `MainViewModel` can be changed using this feature:

```
public partial class MainViewModel : ObservableObject
{
    [ObservableProperty]
    private string name;

    [ObservableProperty]
    private string surname;
}
```

Neat, isn't it? All we need is to declare our `MainViewModel` as a partial class (since the source generator will create, under the hood, another declaration of the class with the code required to support the `INotifyPropertyChanged` interface) and decorate two simple private properties with the `[ObservableProperty]` attribute. That's it. Now our `MainViewModel` class will automatically expose two public properties, called `Name` and `Surname`, which are able to dispatch notifications to the UI layer.

There's another popular library to implement the MVVM pattern in a XAML-based application—**Prism**. It was born as a project created by a Microsoft division called **Patterns and Practices**; now it's an open source project maintained by two Microsoft MVPs, Brian Lagunas and Dan Siegel, with the support of the community. Unlike the MVVM Toolkit, Prism is a complete framework that, other than giving you the basic components, also supports many advanced features including the following:

- Creating custom services to support dialogs
- Supporting navigation from a ViewModel

- Enabling the splitting of the UI into multiple components

- Supporting modules, which are a way to split an application into multiple sub-libraries and components that can be dynamically loaded when needed

Prism is more powerful than the MVVM Toolkit, but it has a tougher learning curve. Unlike the MVVM Toolkit, onboarding new developers on a project built with Prism is more challenging, since they must first learn all the key components of the framework.

They're both great choices when it comes to supporting the MVVM pattern in your application: you should evaluate which one works best for you based on your requirements.

> **Note**
>
> You can learn more about Prism on the official website at `https://prismlibrary.com/`.

Let's continue our journey of implementing the MVVM by learning another key concept, commands.

Supporting actions with commands

Let's go back to the starting point of our journey into the MVVM pattern: a page where users can fill in their names and surnames. By adding properties and implementing the `INotifyPropertyChanged` interface in the ViewModel, we have been able to remove the tight connection between the UI layer and code-behind class when it comes to storing and retrieving the data. But what about the action? Our previous example has a `Button` control with an event handler, which includes the code to save the data. However, this is another XAML scenario that creates a deep connection between the code-behind class and the UI layer: an event handler can only be declared in the code-behind. We can't declare it in our ViewModel, where the data we need (the name and surname) is actually stored.

It's time to introduce commands, which are a way to express actions not with an event handler, but through a regular property. This scenario is enabled by a built-in XAML interface called `ICommand`, which requires you to implement two properties:

- The first describes the code to execute when the command is invoked.

- Optionally, a condition that must be satisfied for the command to be invoked.

This is another one of the scenarios where the MVVM Toolkit greatly helps. Without it, in fact, you would need to create a dedicated class that implements the ICommand interface for each action you need to handle in your application. The MVVM Toolkit instead offers a generic class to implement commands, called RelayCommand. Let's see how we can implement it in our MainViewModel class:

```
public class MainViewModel : ObservableObject
{
    public MainViewModel()
    {
        SaveCommand = new RelayCommand(SaveAction);
    }

    public RelayCommand SaveCommand { get; set; }

    private void SaveAction()
    {
        Debug.WriteLine($"{Name} {Surname}");
    }

    private string _name;
    public string Name
    {
        get => _name;
        set => SetProperty(ref _name, value);
    }

    private string _surname;
    public string Surname
    {
        get => _surname;
        set => SetProperty(ref _surname, value);
    }
}
```

We expose a new `RelayCommand` property, called `SaveCommand`. In the ViewModel's constructor, we initialize it by passing a method, called `SaveAction()`, that we want to execute when the command is invoked. In this example, it just logs the name and surname of the user in the console; in a real-world application, you would use this action to store the user data in a database.

Thanks to the `RelayCommand` class, we can expose our action with a simple property instead of an event handler. Now we can connect it to the `Command` property exposed by all the XAML controls that handle user interaction, such as the `Button` control. Here is how our XAML page gets updated:

```
<Page>
    <StackPanel>
        <TextBlock Text="Name" />
        <TextBox Text="{x:Bind ViewModel.Name,
            Mode=TwoWay}" />
        <TextBlock Text="Surname" />
        <TextBox Text="{x:Bind ViewModel.Surname,
            Mode=TwoWay}" />
        <Button Content="Add" Command="{x:Bind
            ViewModel.SaveCommand}"/>
    </StackPanel>
</Page>
```

As you can see, we can treat the command like any other property exposed by the ViewModel, which means that we can simply use `x:Bind` to connect it to the `Button` control.

Commands can be implemented using the new preview feature of the MVVM Toolkit based on source generators. This is how the previous code can be simplified:

```
[ICommand]
private void Save()
{
    Debug.WriteLine($"{Name} {Surname}");
}
```

Instead of manually creating a new `RelayCommand` object, we just define the action we want to execute, and we decorate it with the `[ICommand]` attribute. The toolkit will automatically generate a command object named like the method followed by the `Command` suffix. In the previous example, the name of the generated command will be `SaveCommand`, since the method name is `Save()`. This feature helps you to make your ViewModel even easier to read.

There's also another implementation of the `RelayCommand` class, which supports passing a parameter from the View to the ViewModel via the `CommandParameter` property. For example, this is an alternative way to pass the name of the user to the command:

```
<Button Content="Add" Command="{x:Bind
    ViewModel.SaveCommand}"CommandParameter="{x:Bind
        ViewModel.Name, Mode=OneWay}"/>
```

To support this approach, you must use the `RelayCommand<T>` implementation, where `T` is the type of data you have set in the `CommandParameter` property. In the previous example, since `Name` is a string, this is how the ViewModel implementation changes:

```
public MainViewModel()
{
    SaveCommand = new RelayCommand<string>(SaveAction);
}


public RelayCommand<string> SaveCommand { get; set; }


private void SaveAction(string name)
{
    Console.WriteLine(name);
}
```

Thanks to this implementation, the parameter is directly passed as an argument of the `SaveAction()` method.

A common scenario in modern application development is to use asynchronous APIs, which are implemented using the `async` and `await` pattern in C#. If your command must perform one or more asynchronous calls, you can use the `AsyncRelayCommand` class (or the `AsyncRelayCommand<T>` one in case you want to pass a parameter). Let's take a look at the following example:

```
public MainViewModel()
{
    SaveCommand = new AsyncRelayCommand(SaveActionAsync);
}


public AsyncRelayCommand SaveCommand { get; set; }


private async Task SaveActionAsync()
{
    await dataService.SaveAsync(Name, Surname);
}
```

`SaveActionAsync()` is an asynchronous method since inside it we're calling an asynchronous API with the `await` prefix. As such, it's masked with the `async` keyword, and it returns `Task` instead of `void`. Being asynchronous, we must wrap it using an `AsyncRelayCommand` object, which will enable us to perform the operation on a background thread, leaving the UI thread free.

However, commands aren't just a way to expose actions through properties. They give you more powerful options compared to a normal event handler. Let's take a look.

Enabling or disabling a command

As I mentioned at the beginning of this section, the `ICommand` interface also supports defining a condition to enable or disable the command. This feature is very powerful because it's automatically reflected on the visual layer of the control, which is connected to the command via binding. For example, if a command is connected to a `Button` control and the command is disabled, the `Button` control will also be disabled, blocking the user from clicking it.

Let's change our code so that the `Button` control to save a person to the database is enabled only if the `Name` and `Surname` properties actually contain values:

```
public class MainViewModel : ObservableObject
{
```

```
    public MainViewModel()
    {
        SaveCommand = new RelayCommand(SaveAction,
            () => !string.IsNullOrWhiteSpace(Name) &&
            !string.IsNullOrWhiteSpace(Surname));
    }

    public RelayCommand SaveCommand { get; set; }

    private void SaveAction()
    {
        Console.WriteLine($"{Name} {Surname}");
    }
}
```

Now we are passing a second parameter when we initialize the `SaveCommand` object: a `boolean` function, which must return `true` when the command should be enabled, and `false` when it should be disabled instead. In our scenario, this condition is based on the values of the `Name` and `Surname` properties: if one of them is empty, we disable the button, since we don't have all the information we need to save the user in the database.

However, this code isn't enough. If you try the application, you will notice that the `Button` control will indeed be disabled at startup, but it won't be enabled again once you have filled the two `TextBox` controls with some values. The reason is that, by default, the `boolean` condition we have set in the `SaveCommand` object is evaluated only the first time the command gets initialized. We must evaluate the condition, instead, every time the value of the `Name` or `Surname` properties changes, since they are the two ones that can influence the status of the command. To support this requirement, the `RelayCommand` class exposes a method called `NotifyCanExecuteChanged()` that we must call every time one of our properties changes. This is how our `ViewModel` looks after we make this change:

```
public class MainViewModel : ObservableObject
{
    public MainViewModel()
    {
        SaveCommand = new RelayCommand(SaveAction,
            () => !string.IsNullOrWhiteSpace(Name) &&
            !string.IsNullOrWhiteSpace(Surname));
```

```
    }

    public RelayCommand SaveCommand { get; set; }

    private void SaveAction()
    {
        Console.WriteLine($"{Name} {Surname}");
    }

    private string _name;
    public string Name
    {
        get => _name;
        set
        {
            SetProperty(ref _name, value);
            SaveCommand.NotifyCanExecuteChanged();
        }
    }

    private string _surname;
    public string Surname
    {
        get => _surname;
        set
        {
            SetProperty(ref _surname, value);
            SaveCommand.NotifyCanExecuteChanged();
        }
    }
}
```

The Name and Surname properties, other than calling the SetProperty() method in the setter, also invoke the NotifyCanExecuteChanged() method exposed by the SaveCommand object. Thanks to this change, our application will now behave as expected. As soon as you fill in some values both in the Name and Surname fields, the Button control will be enabled, allowing the user to complete the action.

Now that we have learned the basic concepts of the MVVM pattern (binding and commands), let's see how we can make it even more effective by adopting a pattern called dependency injection.

Making your application easier to evolve

When you build applications with the MVVM pattern, one of the consequences is that you start to move all the data layers into separate classes, maybe even into separate class libraries. As mentioned at the beginning of this chapter, in fact, ideally the model should be completely platform-agnostic and you should be able to reuse the same classes also in any other .NET project, such as a Blazor web app or an ASP.NET Web API.

Let's come back to our example of a page that contains a form to add a new user. If we follow the principle of isolating the model, it means that your ViewModel should never contain code like this:

```csharp
public MainViewModel()
{
    SaveCommand = new RelayCommand(SaveAction,
        () => !string.IsNullOrEmpty(Name) &&
        string.IsNullOrEmpty(Surname));
}


public RelayCommand SaveCommand { get; set; }

private void SaveAction()
{
    using (var context = new UserContext())
    {
        var person = new Person()
        {
            Name = Name,
            Surname = Surname
        };
        context.People.Add(person);
        context.SaveChanges();
    }
}
```

This implementation of the `SaveAction()` method (even if simplified, since it lacks many other elements) is using the Entity Framework Core library to add a new user to the database. However, the ViewModel isn't the right place to contain such data logic.

Instead, you should have a similar implementation, where the data logic is wrapped in a dedicated class (`DatabaseService` in the following example), which is used by the ViewModel:

```
private void SaveAction()
{
    DatabaseService dbService = new DatabaseService();
    Person person = new Person
    {
        Name = Name,
        Surname = Surname
    };
    db.SavePerson(person);
}
```

This means that the ViewModel now has a dependency from the `DatabaseService` class since, without it, it wouldn't be able to work properly. In the previous example, we have created a new instance of the `DatabaseService` class directly in the ViewModel which, however, can lead to many problems as the application grows:

- One of the best practices to adopt in software development when it comes to building complex projects is **unit testing**. This book isn't the right place to discuss such a deep topic in detail, so you can refer to the official Microsoft documentation at `https://docs.microsoft.com/en-us/visualstudio/test/unit-test-basics` as a starting point. The goal of unit testing is to help you evolve your application more safely, by making sure that when you add new code or make changes you don't break existing features.

  A unit test should be as small as possible and must focus on testing a specific logic. However, the previous code isn't a good fit for a unit test, because we are using a `DatabaseService` object, which performs operations on a real database. Let's say now that our unit test of the `SaveAction()` method fails. Did it happen because the logic is incorrect? Or because there was a problem with the connection to the database? To properly execute the unit tests, you must abstract the `DatabaseService` class, but the current implementation makes it impossible.

- The application becomes hard to evolve and adapt to requirement changes. Let's assume that, at some point, your application needs to scale more efficiently, so you plan to switch your data layer from a relational database based on SQL Server to a document database based on Azure Cosmos DB. You would need to go into each ViewModel of your application and manually replace the initialization of the data layer with another service to replace the `DatabaseService` class.

To solve these challenges, the MVVM pattern is often paired with another pattern called **Inversion of Control**. When you adopt this approach, you change the perspective around object initialization. Instead of you as the developer manually creating new instances of a class, it's the application itself that takes care of initializing the objects whenever it's needed.

The basic building block to implement this pattern is a container, which you can think of like a box inside which you can store all the classes you're going to need in your application. Whenever any actor in your application needs one of these classes, you won't manually create a new instance, but you will ask the container to give it one for you.

Let's see step by step how to implement this approach. The first requirement is to create an interface to describe our service. An interface describes only the properties or methods exposed by a class, without defining the implementation. This is how the interface of our `DatabaseService` class would look:

```
public interface IDatabaseService
{
    public void SavePerson(Person person);
}
```

By providing an interface, we make it easier to address the two challenges we have previously outlined:

- If I need to write unit tests for a ViewModel, I can provide a fake implementation (called a **mock**) of the `DatabaseService` that doesn't require a real connection with the database. There are many mocking libraries in the .NET ecosystem (the most popular one is `Moq`) that enable you to create a fake implementation of an object and return fake data whenever a method is called.

- If you need to swap the `DatabaseService` implementation to switch from SQL Server to Azure Cosmos DB, as long as your new class implements the same `IDatabaseService` interface, you don't need to change the ViewModel implementation.

Now we need a `DatabaseService` class that implements the `IDatabaseService` interface. In our case, it will contain our logic to add a new item to the database (the following snippet is a simplification since it lacks all the code required to initialize Entity Framework Core):

```
public class DatabaseService : IDatabaseService
{
    public void SavePerson(Person person)
    {
        using (var context = new UserContext())
        {
            context.People.Add(person);

            context.SaveChanges();
        }
    }
}
```

Now we have everything we need to effectively manage the `DatabaseService` dependency. However, we are still missing something important—the container. There are many libraries that give you a container ready to be used: some of the most popular ones are Ninject, Unity, and Castle Winsdor. In this book, we're going to use the Microsoft implementation provided for the .NET ecosystem, which is available through the NuGet package called `Microsoft.Extensions.DependencyInjection`. This is a perfect companion for the MVVM Toolkit since it integrates with the Inversion-of-Control support it offers through a series of classes included in the `CommunityToolkit.Mvvm.DependencyInjection` namespace.

Thanks to these libraries, we can use the `OnLaunched` event of the `App` class to set up our container and store inside it all the building blocks we need for our application:

```
protected override void OnLaunched(Microsoft.UI.Xaml
    .LaunchActivatedEventArgs args)
{
    m_window = new MainWindow();

    Ioc.Default.ConfigureServices(new ServiceCollection()
        .AddSingleton<IDatabaseService, DatabaseService>()
        .BuildServiceProvider()
```

```
        );

    m_window.Activate();
}
```

The container is exposed as a singleton (so a static instance is shared across all the applications) through the Default property of the IoC class. We use the ConfigureServices() method to set up the container by passing a ServiceCollection() object that registers all the dependencies we need to use in our application. In this example, we are creating a connection between the IDatabaseService interface and the DatabaseService class. Whenever we need an IDatabaseService implementation in our ViewModels, the container will return to us a concrete DatabaseService object. By referencing only interfaces in our ViewModel instead of concrete classes, we make sure that we can easily swap the implementation if we need. The connection is created with the AddSingleton() method, which means that every ViewModel will receive the same instance of the DatabaseService class.

Now we have a container that contains everything we need to bootstrap our application. How do we retrieve the classes that we need? One way could be to use the Ioc singleton in the SavePerson() method of our ViewModel to retrieve the DatabaseService, instead of manually creating a new instance, as in the following example:

```
private void SaveAction()
{
    IDatabaseService dbService =
      Ioc.Default.GetService<IDatabaseService>();
    Person person = new Person
    {
        Name = Name,
        Surname = Surname
    };
    dbService.SavePerson(person);
}
```

However, this approach isn't the best one, since we still need to manually retrieve a reference to each of the services we need to use in our ViewModel.

The best approach is to adopt a pattern called **dependency injection**, which means creating a dependency between the ViewModel and the services that it needs to work properly. Then, instead of asking the container for an instance of each service, we directly ask for an instance of the ViewModel. If all the dependencies are properly registered, the container will give us a ViewModel instance with all the services we need already injected.

But how do we create such a tight dependency? It's easy in C#: we just need to add our services as constructor parameters of the ViewModel. For example, this is how we create a dependency between the `MainViewModel` class and an `IDatabaseService` object:

```
public class MainViewModel : ObservableObject
{
    private IDatabaseService _databaseService;

    public MainViewModel(IDatabaseService databaseService)
    {
        this._databaseService = databaseService;
    }
}
```

After this change, we won't be able to create a new instance of the `MainViewModel` class anymore without passing, as a parameter, an object that implements the `IDatabaseService` interface. To enable dependency injection, however, we need to make a change to the code in the `App` class that initializes the container:

```
protected override void OnLaunched(Microsoft.UI.Xaml
    .LaunchActivatedEventArgs args)
{
    m_window = new MainWindow();

    Ioc.Default.ConfigureServices(new ServiceCollection()
        .AddSingleton<IDatabaseService, DatabaseService>()
        .AddTransient<MainViewModel>()
        .BuildServiceProvider()
    );

    m_window.Activate();
}
```

Other than the `IDatabaseService` interface, we must register the ViewModel inside the container as well. In this case, it's registered with the `AddTransient()` method, which means that we'll get a new instance of the class every time someone requests it. You can see how, in this case, we aren't creating a connection with an interface, which means that the container will simply create a new instance of the `MainViewModel` class. In our scenario, it's a valid solution because we are forecasting that at some point, we might need to change the `DatabaseService` implementation, while the `MainViewModel` will always stay the same. However, for more complex applications, it's good practice to describe ViewModels with an interface as well.

The last step we are missing is to supply the ViewModel to the page from the container instead of manually creating a new instance. We'll have to switch the initializer in the `MainPage` class to the following one:

```
public sealed partial class MainPage : Page
{
    public MainViewModel ViewModel { get; set; }

    public MainPage()
    {
        this.InitializeComponent();
        ViewModel = Ioc.Default.GetService
            <MainViewModel>();
    }
}
```

We aren't manually creating a new instance of the `MainViewModel` class anymore for the `ViewModel` property (which we use to connect the controls in the UI through binding) – instead, we're getting it from our container. Since the `MainViewModel` class has a dependency from the `IDatabaseService` interface, it means that the container will check whether we have registered any class that implements it. In our case, the answer will be affirmative, so our `MainPage` will be instantiated without errors. If by any chance, we had forgotten to register the `DatabaseService` class in the container, we would have received an error since the container doesn't know how to resolve the dependency.

Thanks to Inversion of Control and dependency injection, we have solved all the challenges that we outlined at the beginning of the section:

- Since the `MainViewModel` class has a dependency from an interface (`IDatabaseService`) instead of a concrete class (`DatabaseService`), it means that, when we write unit tests, we can simply provide a fake implementation of the interface that, instead of working with the real database, returns mock data. This is an example of this approach:

```
[TestMethod]
public void SaveCommand_CannotExecute
  _WithInvalidProperties()
{
    // Arrange.
    var dataServiceMock = new
      Mock<IDatabaseService>();
    var vm = new MainViewModel
      (dataServiceMock.Object);
    //Act
    vm.Name = string.Empty;
    vm.Surname = string.Empty;
    // Assert.
    Assert.IsFalse(vm.SaveCommand.CanExecute(null));
}
```

This unit test verifies that the `SaveCommand` command can't be executed if the `Name` and `Surname` properties of the ViewModel are empty. Thanks to the usage of dependency injection, we can use the `Moq` library to create a mock implementation of the `IDatabaseService` interface. This way, we can test whether the `MainViewModel` behavior is correct without having to establish a connection with a real database.

- If we need to switch our implementation from a classic SQL Server database to Azure Cosmos DB, we just need to create a new class that implements the `IDatabaseService` and, in the `App` class, register it in the container as a replacement of the existing one. For example, let's say we have created a new class called `CosmosDbService` with the following implementation:

```
public class CosmosDbService : IDatabaseService
{
```

```
    public void SavePerson(Person person)
    {
        this.cosmosClient = new
          CosmosClient(EndpointUri, PrimaryKey);
        this.database = await this.cosmosClient
          .CreateDatabaseIfNotExistsAsync();
        this.container = await this.database
          .CreateContainerIfNotExistsAsync
            (containerId, "/Surname");
        await this.container.CreateItemAsync
          <Family>(person, new PartitionKey
            (person.Surname));
    }
}
```

This implementation is simplified of course, but it's using the .NET SDK for Azure Cosmos DB to implement the `SavePerson()` method. Even if the implementation is different, the signature is still the same, which means that we don't have to change anything in the ViewModel to use it.

We just need to go to the `App` class and change the initialization code of the container to be like the following sample:

```
protected override void OnLaunched(Microsoft.UI.Xaml
  .LaunchActivatedEventArgs args)
{
    m_window = new MainWindow();

    Ioc.Default.ConfigureServices(new ServiceCollection()
        .AddSingleton<IDatabaseService, CosmosDbService>()
        .AddTransient<MainViewModel>()
        .BuildServiceProvider()
    );

    m_window.Activate();
}
```

Instead of registering the `DatabaseService` class for the `IDatabaseService` interface, we replace it with the `CosmosDbService` class. Now, regardless of the number of ViewModels that our application has, if any of them have a dependency on the `IdatabaseService` interface, they will start to use the `CosmosDbService` class right away.

Dependency injection can help to solve many problems in real-world application development, but it doesn't solve one of the challenges you will face when you start to adopt the MVVM pattern: how to exchange data between different classes. Let's learn more in the next section!

Exchanging messages between different classes

Splitting the application into different domains independent of each other provides many great advantages, but it also creates a few challenges. For example, there are scenarios where you need to exchange data between two ViewModels; or you need the ViewModel to communicate with the View to trigger an operation on the UI layer, such as starting an animation.

Creating a dependency between multiple elements would break the MVVM pattern, so we must find another solution. One of the most adopted solutions to support these scenarios is the **Publisher-Subscriber pattern**, which enables every class in the application to exchange messages. What makes this pattern a great fit for applications built with the MVVM patterns is that these messages aren't tightly coupled: a publisher can send a message without knowing who the subscribers are, and subscribers can receive messages without knowing who the publisher is.

Let's see a concrete example in the following scenario, based on the previous example: when the `SaveCommand` is executed, we want to trigger an animation defined on the page with a `Storyboard`. If you remember what we have learned in *Chapter 2, The Windows App SDK for a Windows Forms Developer*, animations can be triggered in the code-behind by calling the `Begin()` method exposed by the `Storyboard` class. This is an example of code that, even if you adopt the MVVM pattern, should stay in the code-behind rather than the ViewModel. We're talking, in fact, about code that deeply interacts with the UI layer. However, due to the separation of concerns, we have a disconnection: the `SaveCommand` task is handled in the ViewModel, while the animation must be triggered in the code-behind. We're going to use messages to handle the communication between these two layers, without creating a tight dependency.

Let's start by creating our animation in XAML. We're going to hide the button once the user has clicked on it:

```
<Page>
    <Page.Resources>
        <Storyboard x:Key="AddButtonAnimation">
            <DoubleAnimation
                Storyboard.TargetName="AddButton"
                Storyboard.TargetProperty="Opacity"
                From="1.0" To="0.0" Duration="0:0:3"
                />
        </Storyboard>
    </Page.Resources>

    <StackPanel>
        <TextBlock Text="Name" />
        <TextBox Text="{x:Bind ViewModel.Name,
            Mode=TwoWay}" />
        <TextBlock Text="Surname" />
        <TextBox Text="{x:Bind ViewModel.Surname,
            Mode=TwoWay}" />
        <Button x:Name="AddButton" Content="Add"
            Command="{x:Bind ViewModel.SaveCommand}" />
    </StackPanel>
</Page>
```

Now we need a message that the ViewModel will send to the View when the animation must be triggered. A message is nothing more than a C# class: it can be a plain one (if the message should act just as a notification that an event has occurred) or it can contain one or more properties if you need to also exchange some data. Our use case is the first scenario (we must notify the UI layer when the animation should start), so this is how our message looks:

```
public class StartAnimationMessage
{
    public StartAnimationMessage() { }
}
```

Now we are ready to dispatch this message. We do it inside the `SaveAction()` method of our `MainViewModel` class, which is invoked when the user presses the `Save` button through the `SaveCommand` object:

```
public void SaveAction()
{
        _databaseService.SavePerson(new Person { Name =
            Name, Surname = Surname });
        WeakReferenceMessenger.Default.Send(new
            StartAnimationMessage());
}
```

`WeakReferenceMessenger` is a class offered by the MVVM Toolkit and is included in the `CommunityToolkit.Mvvm.Messaging` namespace. As the name says, it uses a weak approach to register messages, which means that unused recipients are still eligible for garbage collection, reducing the chances of memory leaks. The class exposes a singleton through the `Default` object since we need a central point of registration for messages if we want all the classes to be able to exchange them. To send the message, we simply call the `Send()` method, passing as a parameter a new instance of our `StartAnimationMessage` class. As you can see, we don't have any reference to the recipient. Our ViewModel continues to be independent of the View.

Let's move now to the View – let's subscribe to receive this message in the code-behind class of `MainPage`:

```
public partial class MainPage : Page,
    IRecipient<StartAnimationMessage>
{
    public MainPage()
    {
        this.InitializeComponent();
        WeakReferenceMessenger.Default.
            Register<StartAnimationMessage>(this);
    }

    public void Receive(StartAnimationMessage message)
    {
        Storyboard sb = this.Resources
            ["AddButtonAnimation"] as Storyboard;
```

```
            sb.Begin();
    }
}
```

As a first step, in the page constructor, we again use the WeakReferenceMessenger class and its singleton, this time to register for incoming messages using the Register<T>() method, where T is the type of message want to receive (in our case, it's StartAnimationMessage). As a parameter, we must pass the class that will handle the message: we use the this keyword, since we want the code-behind class itself to manage it.

As a second step, we let the page implement the IRecipient<T> interface, where T is the type of message we're subscribing to. By adding this interface, we are required to implement the Receive() method, which is triggered when there's an incoming message. As a parameter, we get a reference to the message itself, so we can retrieve data from its properties if needed. In our scenario, we're using the message just as a notification, so all we do is retrieve a reference to the Storyboard object we have created in XAML and invoke the Begin() method to run the animation. Also, in this case, note how we don't have a direct reference to the MainViewModel class that sent the message.

If you prefer not to implement an interface, there's also another way to subscribe for messages:

```
public partial class MainPage : Page
{
    public MainPage()
    {
        InitializeComponent();
        WeakReferenceMessenger.Default.Register
            <StartAnimationMessage>(this,
                (sender, message) =>
            {
                Storyboard sb = this.Resources
                    ["AddButtonAnimation"] as Storyboard;
                sb.Begin();
            });
    }
}
```

When you call the `Register()` method, you can pass as a second parameter a function with the code you want to execute when the message is received.

Thanks to the Publisher-Subscriber pattern, we have achieved our goal: when the user clicks on the `Save` button on the page, it will disappear after a few seconds thanks to the animation. Most of all, we have implemented this without creating a tight relationship between the View and ViewModel, which are still decoupled.

Let's see now that last topic of our journey of building a future-proof architecture with the MVVM pattern: managing the navigation.

Managing navigation with the MVVM pattern

We explored navigation in *Chapter 5, Designing Your Application*, and we have learned that it's a critical component of a Windows application. Especially if you're building enterprise applications, it's very likely that you will have multiple pages and the various interactions will lead the user to move from one to another. However, it's also one of the most challenging features to implement when you start adopting the MVVM pattern. The reason is that the goal of the pattern is to decouple the UI from the business logic; however, this creates a fracture in this case. The actions performed by the user are managed by the ViewModel, while navigation is managed in the View layer by using the `Frame` class to trigger the navigation and by subscribing to events such as `OnNavigatedTo()` and `OnNavigateFrom()` to manage the life cycle of a page.

There are multiple solutions to this challenge, but the most common one is based on the implementation of a class called `NavigationService`, which can wrap the access to the underline `Frame` of the application so that you can trigger the navigation from one page to another directly to a ViewModel. Many advanced MVVM frameworks, such as Prism, provide an implementation of this class that, by using dependency injection, you can inject into your ViewModels.

The MVVM Toolkit doesn't provide it, so we're going to reuse the implementation included in Windows Template Studio, which is available at `https://github.com/microsoft/WindowsTemplateStudio`. It's a Visual Studio extension that you can use to accelerate the bootstrap of a new Windows application by giving you many building blocks that makes it easier to create a complex architecture by supporting the MVVM pattern and by providing services to enable navigation, storage access, and much more. At the time of writing, Windows Template Studio doesn't support Visual Studio 2022 and Windows App SDK 1.0 yet and as such, we won't cover it in this book. However, we're going to reuse the same `NavigationService` implementation so that we can adopt it in our application.

Specifically, we're going to reuse the `NavigationService` class and the `INavigationService` and `INavigationAware` interfaces from the WinUI project created by the template.

In this section, we won't focus on how `NavigationService` is implemented or how it works under the hood. You can take a look at the implementation in the example application at `https://github.com/PacktPublishing/Modernizing-Your-Windows-Applications-with-the-Windows-Apps-SDK-and-WinUI/tree/main/Chapter06/05-Navigation`.

We'll focus, instead, on how we can use it to handle the navigation. Let's start!

Navigating to another page

The first step is to create a second page in our application with a dedicated ViewModel. In our scenario, we have called it `DetailPage`, which is backed by the `DetailPageViewModel` class.

Now we must register `NavigationService` in our container so that we can inject it into our ViewModels. The `NavigationService` comes with an interface that describes it called `INavigationService`, so we can just add it to our container using the approach we learned in this chapter. We also register the new `DetailPageViewModel` we have just created. We achieve this goal by adding the following code in the `OnLaunched` event of the `App` class:

```
Ioc.Default.ConfigureServices(new ServiceCollection()
    .AddSingleton<INavigationService, NavigationService>()
    .AddSingleton<IDatabaseService, DatabaseService>()
    .AddTransient<MainPageViewModel>()
    .AddTransient<DetailPageViewModel>()
    .BuildServiceProvider());
```

The `NavigationService` is registered as a singleton so that the same instance can be reused across every ViewModel.

The main window of the application is configured in the same way that we learned in *Chapter 5, Designing Your Application*. The core is a `Frame` control, which we're going to use to load the pages and handle the navigation. This is a basic implementation:

```
<Window>
    <Frame x:Name="ShellFrame" />
</Window>
```

In the code-behind, we use the `Frame` reference to load the default page of the application, the one called `MainPage`:

```csharp
public sealed partial class MainWindow : Window
{
    public MainWindow()
    {
        this.InitializeComponent();
        ShellFrame.Content = new MainPage();
    }
}
```

Now we are ready to start using the `NavigationService` class in our ViewModel. If you remember the previous implementation we saw of the `MainPageViewModel` class, we have a `RelayCommand` called `SaveCommand`, which triggers a method called `SaveAction()`. Let's change it to trigger navigation to the `DetailPage` application we previously created:

```csharp
public class MainPageViewModel : ObservableObject
{
    private readonly INavigationService _navigationService;

    public MainPageViewModel(INavigationService
      navigationService)
    {
        SaveCommand = new RelayCommand(SaveAction,
            () => !string.IsNullOrWhiteSpace(Name) &&
              !string.IsNullOrWhiteSpace(Surname));
        _navigationService = navigationService;
    }

    public RelayCommand SaveCommand { get; set; }

    private void SaveAction()
    {
        _navigationService.NavigateTo(typeof(DetailPage));
    }
}
```

We simply need to add an `INavigationService` parameter to the constructor of the ViewModel. Thanks to dependency injection, an instance of the `NavigationService` class will be automatically injected inside the ViewModel. Now we can trigger the navigation by calling the `NavigateTo()` method exposed by the service, passing the page type as a parameter.

> **Note**
>
> If you want to implement a cleaner approach, you can change the `NavigationService` implementation to use a key (such as the page's name) instead of the page's type as a parameter of the `NavigateTo()` method. The page's type, in fact, creates a relationship between the ViewModel and the View, so it might create challenges when you want to reuse the same ViewModel in other types of projects.

The command will now redirect the user to the `DetailPage` of the application. However, what if we want to pass some data to it? Let's explore a solution.

Passing parameters from a ViewModel to another

To handle this scenario, the `NavigationService` automatically subscribes to the navigation event exposed by the `Page` class, such as `OnNavigatedTo()` and `OnNavigatedFrom()`. We can easily access them through another helper that we have imported from Windows Template Studio: the `INavigationAware` interface. By implementing this interface, we'll be able to manage the navigation events directly in the ViewModels.

Let's start by passing a parameter to the navigation action, which works in the same way that we learned in *Chapter 5*, *Designing Your Application*:

```
private void SaveAction()
{
    Person person = new() { Name = Name, Surname = Surname };
    _navigationService.NavigateTo(typeof(DetailPage),
        person);
}
```

We have created a new `Person` object and we have passed it as the second parameter of the `NavigateTo()` method exposed by `NavigationService`.

Now we can implement the INavigationAware interface in our DetailPageViewModel, which gives us the option to receive the Person object:

```
public class DetailPageViewModel: ObservableObject,
  INavigationAware
{
    private string _name;

    public string Name
    {
        get { return _name; }
        set { SetProperty(ref _name, value); }
    }

    private string _surname;

    public string Surname
    {
        get { return _surname; }
        set { SetProperty(ref _surname, value); }
    }

    public void OnNavigatedTo(object parameter)
    {
        if (parameter != null)
        {
            Person person = parameter as Person;
            Name = person.Name;
            Surname = person.Surname;
        }
    }

    public void OnNavigatedFrom()
    {
    }
}
```

Thanks to this new interface, we can use the `OnNavigatedTo()` method in our ViewModel to get notified when the navigation to the page connected to this ViewModel has happened. Thanks to the parameter, we can get access to the `Person` object that we previously passed to the `NavigateTo()` method of `NavigationService`. And also, because of this object, we can fill two properties (`Name` and `Surname`) whose values are displayed in the View thanks to the binding.

This topic concludes our journey into the MVVM pattern. In the next chapter, we're going to put into practice many of these concepts by migrating a Windows application to Windows App SDK and WinUI.

Summary

One of the biggest challenges in software development is building projects that can survive the test of time. Of course, we can't control every aspect: there are many external factors outside our control, like the advent of new and more powerful technologies. However, by choosing the right architecture, we can protect our application from aging too fast. The MVVM pattern is a great example of how you can define the architecture of your Windows applications in a way that makes them easier to maintain, evolve, and test over time. By decoupling the UI from the business logic and the data layer, it becomes easier to create modular applications, in which you can replace one component or another when a new requirement arrives, rather than having to rewrite the application from scratch.

At the same time, however, you must remember that architectural patterns are best considered as guidance to help you build better applications, not a set of rules that you have to follow blindly. You must always find the right balance between good architecture and simplicity, otherwise you risk falling into the opposite problem: over-engineering.

The knowledge we acquired in this chapter will be very valuable in the next one: we're going to migrate a real application to WinUI and the Windows App SDK, and as part of the journey, we're going also to evolve it from the traditional code-behind approach to the MVVM pattern.

Questions

1. Using a dedicated library or framework is a critical requirement to implement the MVVM pattern. True or false?

2. What's the main advantage of using messages and the Publisher-Subscriber pattern in an MVVM architecture?

3. It's fine to implement the `INotifyPropertyChanged` interface in classes that are considered part of the Model domain. True or false?

Section 3: Integrating Your App with the Windows Ecosystem

This section will explain how you can take your existing application and use the Windows App SDK to enhance and integrate it with the latest Windows features.

This section contains the following chapters:

- *Chapter 7, Migrating Your Windows Applications to the Windows App SDK and WinUI*
- *Chapter 8, Integrating Your Application with the Windows Ecosystem*
- *Chapter 9, Implementing Notifications*
- *Chapter 10, Infusing Your Apps with Machine Learning Using WinML*

7

Migrating Your Windows Applications to the Windows App SDK and WinUI

In the previous chapters, we learned about the main features of the Windows App SDK and WinUI and how they are differentiated from the existing features of other popular Windows development platforms, such as Windows Forms, WPF, and **Universal Windows Platform (UWP)**. We have learned a lot of concepts that are important when it comes to building a real-world application: binding, adopting the MVVM pattern, building a responsive UI, and more.

Now, it's time to put the knowledge we have acquired so far into practice. We can do this by using it to migrate real applications to the Windows App SDK and WinUI so that we achieve our modernization goal: building an enterprise application that can leverage all the capabilities of the Windows platform and that delivers a modern user experience, which can support new devices, new interaction modes, and accessibility.

In this chapter, we'll take a few sample applications that we've built with different technologies, and we'll migrate them to the Windows App SDK and WinUI, guiding you through all the required steps.

In this chapter, we'll cover the following topics:

- Getting an understanding of general migration guidance
- Updating your applications to .NET 6
- Migrating a Windows Forms application
- Migrating a WPF application
- Migrating a UWP application

Let's get started!

Technical requirements

The samples related to this chapter are available at the following URL:

```
https://github.com/PacktPublishing/Modernizing-Your-Windows-
Applications-with-the-Windows-Apps-SDK-and-WinUI/tree/main/
Chapter07
```

Getting an understanding of general migration guidance

Line-of-Business (LoB) applications are often developed in a three-tier architecture and, therefore, consist of three logical components:

- A user interface
- Business logic
- Data access

During the modernization of the application, you will need to touch upon all components, but not to the same extent.

If your application is still based on the .NET 4.x Framework, all the components will need to be lifted to the .NET 6 runtime. We'll cover this topic in the next section. It is not recommended that you move to .NET Standard 2.x anymore unless you plan to share your code with .NET Core 3.1 or Xamarin. This is because .NET Standard 2.1 has been superseded by .NET 6. You can find more information about how to build a class library in *Chapter 1, Getting Started with the Windows App SDK and WinUI*.

In general, the migration of the data access and business logic layers is relatively straightforward. You start by moving the entire project to the new framework and upgrading the dependencies to the latest version. There will likely be some API changes in the dependencies that need to be addressed. The tools we're going to cover in the next section will help you to identify those changes.

If your current project targets .NET Core, there is no need to worry right now; .NET Standard will still be around for a while – but active development is happening on the succeeding frameworks, for example, .NET 6. If you're already using the new .NET runtime to build your application, just remember to change the `TargetFramework` property in the `.csproj` file, as follows:

```
<TargetFramework>net6.0-windows10.0.19041.0</TargetFramework>
```

This target framework enables the usage of the Windows-specific APIs, which are required to support the Windows App SDK and WinUI.

The complexity of migration varies a lot based on the complexity of the original project. However, at a high level, the best approach is to create a new Windows App SDK project and to migrate the code with the least dependencies first. For example, if your solution is made by a project with the main application and multiple class libraries, you should start from the libraries and, only at the end, move to the application's project. Usually, these are the models, followed by the business logic (data transformation) classes and the data access layer classes. This helps you to focus on a single class at a time without needing to jump back and forth on each dependency. Assets can simply be copied over. Unless you're coming from a UWP application, UI elements such as pages, apps, and windows need to be recreated. This is because the source UI definition is either incompatible (Windows Forms) or subtly different (WPF). For UWP, it's a good start to simply copy over the entire file and adapt its namespaces.

We'll learn more about these best practices in the upcoming sections. However, first, let's see more details regarding the sample application we're going to migrate.

Exploring the sample application

Our sample application is a typical, data-driven application to manage employee records. Applications such as this can be found in many enterprises as part of their digital heritage. They have reached an almost feature-complete state and are only maintained on-demand.

The application covers the most widely used requirements, such as implementing **Create-Read-Update-Delete** (**CRUD**) operations, navigation, and validation. Additionally, users can interact with the application to display and select employee records in a table to display, edit, or delete them on a different page. The following screenshot shows the UWP version in action:

Figure 7.1 – Two views of the sample application: a list of all the records and a form to create a new one

We'll start from the same application that is built with different technologies (that is, Windows Forms, WPF, and UWP), and we'll migrate it to the Windows App SDK and WinUI, using technologies such as **Entity Framework** (**EF**) Core 6 for the data layer, .NET 6, and the Windows Community Toolkit. Also, we will apply the recommendations from the previous chapter, including the MVMM architecture and **Inversion of Control** (**IoC**).

All services and dependencies are registered in the code-behind file of the `App` object. We have added `NavigationService` (which is a stripped-down version of the Template Studio implementation we discussed in *Chapter 6, Building a Future-Proof Architecture*) to allow commands that are registered to view models to initiate navigation events. Also, we have added a `DialogService` class, as the delete command should issue a dialog box to confirm the deletion of an entry. The `ContentDialog` control that is used to display the dialog requires a reference to `XamlRoot`. We'll highlight that when migrating from UWP.

From a UI perspective, we're going to mostly use built-in WinUI controls, except for the DataGrid one, which is included in the Windows Community Toolkit. We're going to use many of the features we have learned about in previous chapters, such as converters, templates, and styles.

However, before we start migrating the application, we need to be sure that our code is fully supported by .NET 6. As such, let's see what the steps are to migrate a solution from .NET Framework to .NET 6.

Updating your applications to .NET 6

Upgrading your .NET Framework applications to .NET 6 is the best way to ensure you have a path forward to keep adding new features, integrating more and more with the Windows ecosystem, and leveraging the latest enhancements in the language and the tooling.

Migrating an application based on the full .NET Framework to .NET 6 can have a very different outcome based on many factors. There isn't a golden rule, but it depends on scenarios such as the following:

- The complexity of the solution
- The usage of libraries or third-party controls that haven't been migrated to the new ecosystem
- The usage of APIs or features that are considered deprecated and, as such, haven't been ported to the new .NET ecosystem, such as Application Domains, **Windows Communication Foundation (WCF)** servers, and **Windows Workflow Foundation (WF)**

Let's see what strategies we can adopt to migrate our .NET Framework projects.

The .NET Portability Analyzer

An effective way to understand the potential complexity of the migration process is to use a tool, created by Microsoft, called the **.NET Portability Analyzer**. It comes as a Visual Studio extension that can scan your code and prepare a detailed report of which APIs you are using that aren't supported in .NET 6.

> **Note**
>
> At the time of writing, the .NET Portability Analyzer hasn't been updated to support Visual Studio 2022 and .NET 6. The steps ahead will require Visual Studio 2019 and .NET 5 until the tool is updated. However, since .NET 6 doesn't include many breaking changes compared to .NET 5, running an analysis with the current version of the tool is still a good starting point to understand the complexity of the porting.

To use it, first, you have to download and install the extension from the Visual Studio Marketplace at `https://marketplace.visualstudio.com/items?itemName=ConnieYau.NETPortabilityAnalyzer`.

Once you have installed it, with Visual Studio, you can open the solution based on the full .NET Framework that you want to migrate. The first step is to right-click on it, and choose **Portability Analyzer Settings** to configure it:

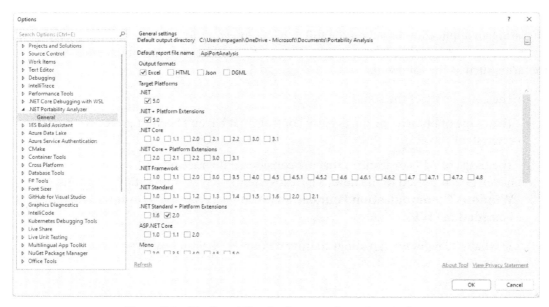

Figure 7.2 – The Portability Analyzer configuration window

From here, you can pick the platforms of the .NET ecosystem that you would like to target. The most important ones for our migration scenario are listed as follows:

- .NET 5.0.

- .NET 5.0 + Platform Extensions (which includes the Windows Compatibility Pack; this is a special package we're going to explore later that expands the surface of supported APIs).

- .NET Standard 2.0 (in case you need to continue sharing one of the libraries that is part of your solution with projects that are still based on the full .NET Framework).

The tool supports generating reports in multiple formats. The default one is Excel. However, if you prefer, you can also switch to HTML, JSON, or DGML.

Once you have completed the configuration, you can right-click on the solution in Visual Studio again and choose **Analyze Assembly Portability**. The tool will build your solution and then create the report using the format you have configured in the settings. Once the scan has been completed, the **Portability Analyzer Results** window will appear, allowing you to interact with the newly generated report along with the older ones.

This is what the report looks like (in this case, using the Excel format):

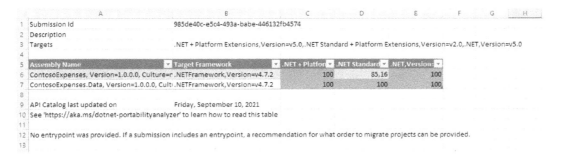

Figure 7.3 – The Portability Analyzer report

In the previous example, we achieved a successful result. All the APIs that we are using in our .NET Framework 4.7.2 project are supported by .NET 5. This means that we shouldn't expect any trouble during the migration. However, if we need to migrate the project to support .NET Standard 2.0, we can see that we might expect some challenges since approximately 15% of the APIs we're using aren't compatible.

If you hit a similar scenario, the Excel file includes a second sheet named **Details**, which will give you an overview of the APIs that are failing along with a potential workaround:

Target type	Target member	Assembly name	.NET Core + Platform Extensions	Recommended
T:System.AppDomain	M:System.AppDomain.get_SetupInformation	svcutil	Not supported	Remove usage.
T:System.AppDomainSetup	T:System.AppDomainSetup	svcutil	Not supported	Remove usage.
T:System.AppDomainSetup	M:System.AppDomainSetup.get_ConfigurationFile	svcutil	Not supported	Remove usage.
T:System.Data.DataSetSchemaImporterExtension	T:System.Data.DataSetSchemaImporterExtension	svcutil	Not supported	
T:System.Data.Design.TypedDataSetSchemaImporterExtension	T:System.Data.Design.TypedDataSetSchemaImporterExtension	svcutil	Not supported	
T:System.Data.Design.TypedDataSetSchemaImporterExtensionFx3!	T:System.Data.Design.TypedDataSetSchemaImporterExtensionFx:	svcutil	Not supported	

Portability Summary   Details   ⊕

Figure 7.4 – The Details section of the Portability Analyzer report

For example, in this screenshot, we can see that the application is using APIs related to the **Application Domain** feature, which has been removed from .NET 5.

Now that we have an idea of how easy (or complex) the migration will be, we can start the process.

Upgrading your solution

Typically, upgrading a solution to .NET 6 involves the following steps:

1. Upgrade the projects included in the solution to use the new SDK-style format.
2. Migrate the `packages.config` file. In .NET Framework applications, NuGet packages were declared inside a file called `packages.config`. With .NET 6, they have instead been moved directly inside the project's file using a new property, called `PackageReference`, as shown in the following snippet:

```
<Project Sdk="Microsoft.NET.Sdk">
  <PropertyGroup>
    <TargetFramework>net6.0-windows</TargetFramework>
    <OutputType>WinExe</OutputType>
    <UseWPF>true</UseWPF>
  </PropertyGroup>

  <ItemGroup>
    <PackageReference Include="CommunityToolkit.Mvvm"
      Version="7.0.3" />
  </ItemGroup>

</Project>
```

3. Upgrade the NuGet packages used in the solution to the latest version, since it's more likely they include support for .NET Standard or .NET 6.

To automate these tasks, and give you a better starting point for your migration, Microsoft has released a tool called **.NET Upgrade Assistant**. This is a global .NET tool. To install it, you just have to open Command Prompt and, assuming you have installed the latest .NET SDK, run the following command:

```
dotnet tool install -g upgrade-assistant
```

Once it's installed, you can launch it using the following command:

```
upgrade-assistant upgrade .\ContosoExpenses.sln
```

The preceding sample command assumes that we want to migrate a solution called ContosoExpenses.sln, which is stored in the current folder. Of course, you can replace the last parameter with the path of any project or solution on your machine that is based on .NET Framework.

The tool will guide you through the process by proposing different actions to take to migrate your solution, which you can either apply or skip based on your scenario and requirements. The tool is command-line based, as you can see in the following screenshot:

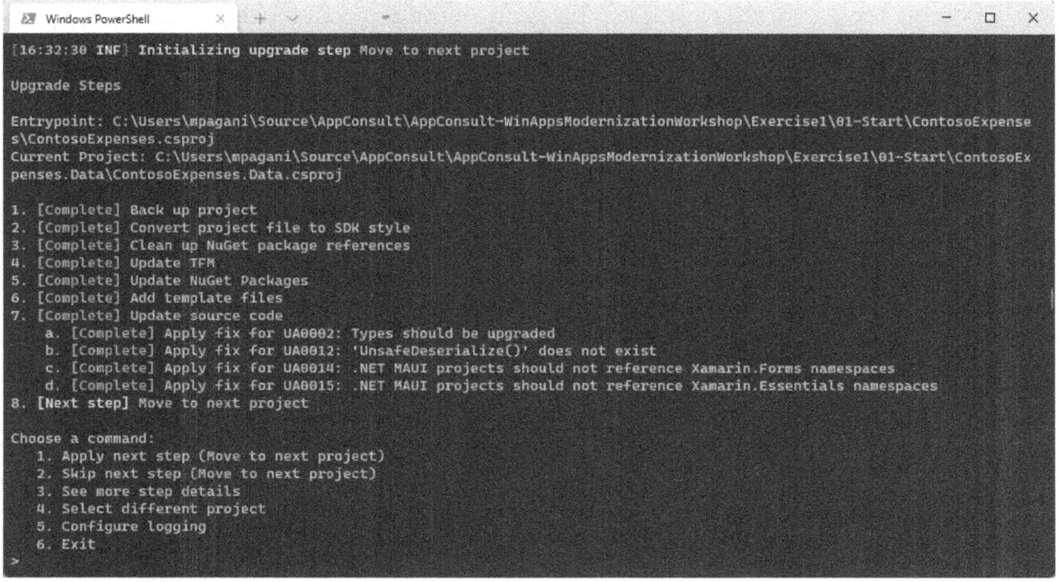

```
Windows PowerShell       ×    +  ∨                                            —   □   ×
[16:32:30 INF] Initializing upgrade step Move to next project

Upgrade Steps

Entrypoint: C:\Users\mpagani\Source\AppConsult\AppConsult-WinAppsModernizationWorkshop\Exercise1\01-Start\ContosoExpense
s\ContosoExpenses.csproj
Current Project: C:\Users\mpagani\Source\AppConsult\AppConsult-WinAppsModernizationWorkshop\Exercise1\01-Start\ContosoEx
penses.Data\ContosoExpenses.Data.csproj

1. [Complete] Back up project
2. [Complete] Convert project file to SDK style
3. [Complete] Clean up NuGet package references
4. [Complete] Update TFM
5. [Complete] Update NuGet Packages
6. [Complete] Add template files
7. [Complete] Update source code
    a. [Complete] Apply fix for UA0002: Types should be upgraded
    b. [Complete] Apply fix for UA0012: 'UnsafeDeserialize()' does not exist
    c. [Complete] Apply fix for UA0014: .NET MAUI projects should not reference Xamarin.Forms namespaces
    d. [Complete] Apply fix for UA0015: .NET MAUI projects should not reference Xamarin.Essentials namespaces
8. [Next step] Move to next project

Choose a command:
    1. Apply next step (Move to next project)
    2. Skip next step (Move to next project)
    3. See more step details
    4. Select different project
    5. Configure logging
    6. Exit
>
```

Figure 7.5 – The .NET Upgrade Assistant tool in action

Once you have launched the tool, before using it, first you will have to configure it:

1. In the first step, the tool will ask you for the entry point of your solution (typically, it's the project that contains the main application).

2. In the second step, based on the previous choice, the tool will suggest a migration order. For example, if your solution is made by two projects, the main app and a class library, the tool will suggest first migrating the class library and then the main app. As we mentioned at the beginning of the chapter, the fewer dependencies a project has, the higher the chances that the conversion will be successful. The suggested approach is to just press *Enter* so that the tool can follow the suggested order; otherwise, if for any reason you prefer to start from another project, you can just select which one you want to use.

After the first configuration, the tool will start a step-by-step process that will take care – for each project that is part of the solution – of the following:

- Performing a backup
- Converting it from the old `.csproj` format into the new SDK style one
- Moving all the NuGet packages references from the `packages.config` file to the project's file
- Updating the `TargetFramework` property in the project's file to use the most appropriate one
- Updating the NuGet packages to the latest version
- Applying a series of fixes in the source code for known blockers

Additionally, the tool will install a special library called **Windows Compatibility Pack** (which is represented by the `Microsoft.Windows.Compatibility` NuGet package). Its goal is to give you access to a broader set of Windows APIs in a .NET 6 or .NET Standard library. The package supports features such as the following:

- Interacting with the Windows Registry
- Using the Windows event log
- Setting Windows **Access Control Lists (ACLs)**
- Working with Windows services

It's important to highlight that these features will result in your libraries only being supported by Windows. As such, if you're planning to also use them on another platform (for example, you want to reuse a class library with a web or mobile application), you will need to adopt platform checking techniques to make sure that you don't use specific Windows APIs on another operating system. For example, consider the following code:

```
public bool IsFirstTimeLaunch()
{
    if (RuntimeInformation.IsOSPlatform(OSPlatform.Windows))
    {
        var regKey =
          Registry.CurrentUser.OpenSubKey(
          @"SOFTWARE\Contoso\ContosoExpenses", true);
        if (regKey != null)
        {
            string isFirstRun =
                regKey.GetValue("FirstRun").ToString();
            return isFirstRun == "true";
        }
    }
    else
    {
        //use an alternative approach
    }
}
```

The preceding snippet returns the information if the application has been launched for the first time. By using the RuntimeInformation.IsOSPlatform() method provided by .NET, we can tailor this code based on the platform where the application is running. If it's running on Windows, we can use the registry to store and retrieve this information; if we're running on another platform, we can use an alternative approach instead (for example, using a local file).

To make the developer's life easier in finding all the places in code where you need to pay attention, .NET 6 integrates an analyzer that will give you warnings every time it finds one or more APIs that are platform-specific. For example, consider the earlier snippet of code. If you try to compile it with Visual Studio, the build will be successful, but you will see the following warnings:

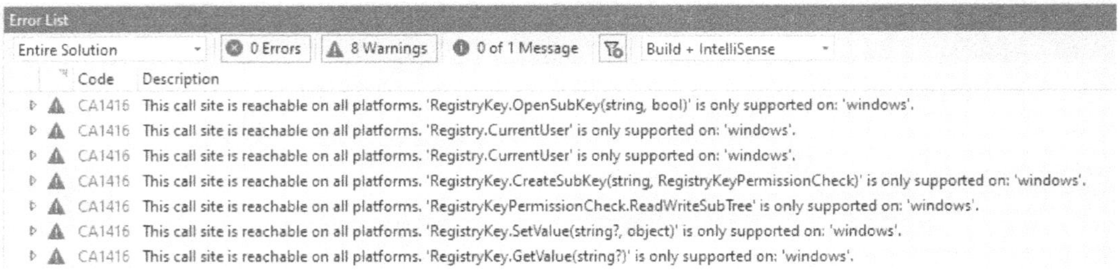

Figure 7.6 – The warnings generated by the .NET analyzer

Visual Studio is warning us that the APIs that work with the registry are only supported on Windows. So, if we're planning to reuse the same library on a web project running on Linux, we need to pay attention and make sure this code isn't hit.

Once the work of the .NET Upgrade Assistant tool has been completed, you're ready to start the next phase of the migration. As mentioned earlier, migrating an application from the full .NET Framework to .NET 6 can vary from being an easy task to a very challenging process. The .NET Upgrade Assistant tool is a great way to automate some of these tasks and give you a better starting point, but it cannot take care of the entire process. It will be up to you to confirm the quality of the migration, by running unit tests, integration tests, UI tests, and making sure that the application still behaves in the same way.

Now that you know how to migrate your applications to .NET 6, we're ready to explore how to migrate a Windows Forms application.

Migrating a Windows Forms application

Before we get started, let's take a brief inventory of the Windows Forms application. The data access layer is provided by the DBContext class of EF6.

> **Note**
>
> EF is a modern **object-database mapper** (**ORM**) for .NET, which simplifies working with databases. Thanks to EF, you can map all the database concepts (such as tables, rows, and columns) to the equivalent C# concepts (such as classes, properties, and collections). We won't cover EF in detail, since it's beyond the scope of this book. You can learn more at https://docs.microsoft.com/en-us/ef/.

To keep the sample focused and concise, we are not using additional patterns to decouple the context. The main window hosts two tabs, one containing the **Overview** view and the other with the detailed **Item** view:

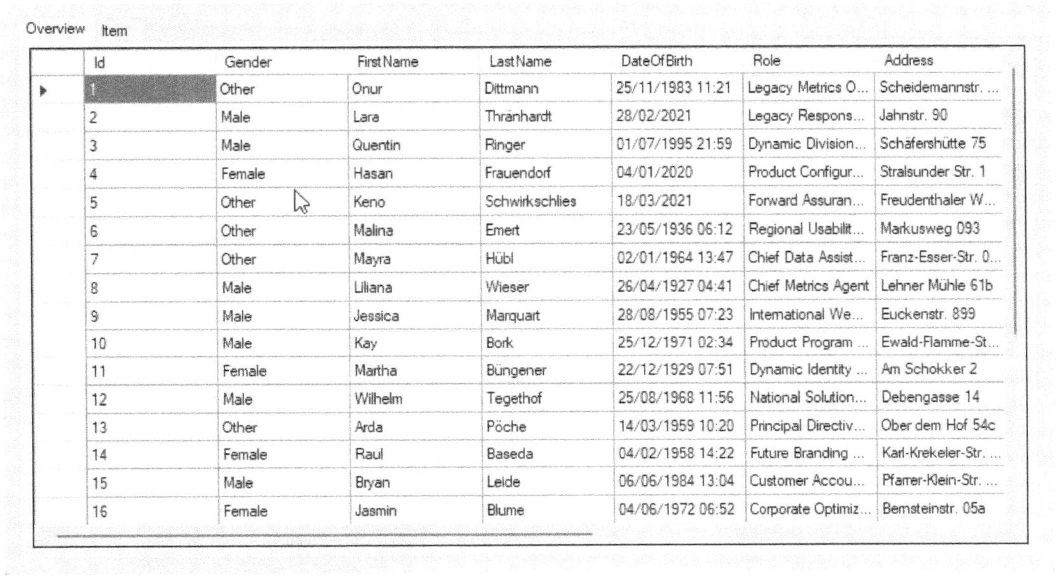

	Id	Gender	First Name	Last Name	DateOfBirth	Role	Address
▶	1	Other	Onur	Dittmann	25/11/1983 11:21	Legacy Metrics O...	Scheidemannstr. ...
	2	Male	Lara	Thränhardt	28/02/2021	Legacy Respons...	Jahnstr. 90
	3	Male	Quentin	Ringer	01/07/1995 21:59	Dynamic Division...	Schäfershütte 75
	4	Female	Hasan	Frauendorf	04/01/2020	Product Configur...	Stralsunder Str. 1
	5	Other	Keno	Schwirkschlies	18/03/2021	Forward Assuran...	Freudenthaler W...
	6	Other	Malina	Emert	23/05/1936 06:12	Regional Usabilit...	Markusweg 093
	7	Other	Mayra	Hübl	02/01/1964 13:47	Chief Data Assist...	Franz-Esser-Str. 0...
	8	Male	Liliana	Wieser	26/04/1927 04:41	Chief Metrics Agent	Lehner Mühle 61b
	9	Male	Jessica	Marquart	28/08/1955 07:23	International We...	Euckenstr. 899
	10	Male	Kay	Bork	25/12/1971 02:34	Product Program ...	Ewald-Flamme-St...
	11	Female	Martha	Büngener	22/12/1929 07:51	Dynamic Identity ...	Am Schokker 2
	12	Male	Wilhelm	Tegethof	25/08/1968 11:56	National Solution...	Debengasse 14
	13	Other	Arda	Pöche	14/03/1959 10:20	Principal Directiv...	Ober dem Hof 54c
	14	Female	Raul	Baseda	04/02/1958 14:22	Future Branding ...	Karl-Krekeler-Str. ...
	15	Male	Bryan	Leide	06/06/1984 13:04	Customer Accou...	Pfarrer-Klein-Str. ...
	16	Female	Jasmin	Blume	04/06/1972 06:52	Corporate Optimiz...	Bernsteinstr. 05a

Figure 7.7 – The data displayed in the Windows Forms application

The **Overview** tab displays a `DataGrid` control, which is used for displaying and sorting the data. When you select an entry, the application will load all the information about the employee into the second tab. Clicking on the column header will sort the entries by this column, which is a convenient built-in functionality of the `DataGrid` control. Other features such as in-line editing and saving have been deactivated on purpose.

In comparison, the **Item** details tab consists of a few `TextBox`, `Label`, `Button`, and `ComboBox` controls: these are all basic controls that are available in the Windows Forms framework. The `ComboBox` control is used for selecting the gender. The list of supported values is hardcoded and matches the `Gender` model class:

Figure 7.8 – The tab that displays the list of selected employees

We're now aware of the UI components of our existing application, and the first step is to find a suitable replacement in the Windows App SDK control library. Unless you already have some experience with the WinUI controls, the best and easiest way to become familiar with the available controls is via the XAML Controls Gallery application, which is available in the Microsoft Store or on GitHub. We have mentioned this application in *Chapter 5, Designing Your Application*. The application lists all the controls that are available in WinUI along with the ones included in the Windows Community Toolkit. This is especially helpful in our case, as the WinUI 3 DataGrid control is only available through the `CommunityToolkit.WinUI.UI.Controls` namespace. All the remaining controls are part of the official control library, but we are changing the masked `TextBox` control to the date into a `DatePicker` control.

Now that we have a list of all the required controls, we can take a look at the project structure:

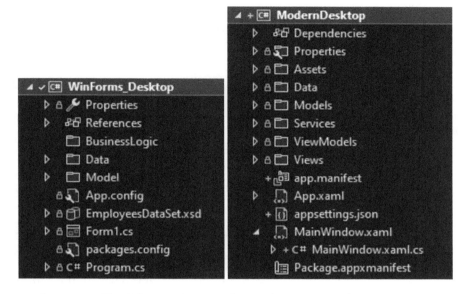

Figure 7.9 – The differences between the Windows Forms and WinUI projects

In the Windows Forms application, we have a static `Program` class that acts as an entry point for the application. The singleton application object is being used to enable visual styles, register handles, and set the form that was initially called upon. In the Windows App SDK project, we don't have a `Program` class but a singleton `Application` object class. By default, the App class is a partial class – where some parts are described in XAML and others are described in C#. Usually, global resources are described in `ResourceDictionaries` in XAML, while more complex configurations, such as the IoC container setup, are implemented in code.

A noteworthy difference is the App object life cycle. In Windows Forms, the commonly used life cycle events are provided through the forms – the `Application` object itself does not expose any life cycle events nor does the `Program` class. This approach has changed with the introduction of the App object (which was first added in WPF), which now exposes its own set of life cycle events such as `OnLaunched` or `UnhandledException`.

As discussed in *Chapter 5, Designing Your Application*, WinUI 3 uses a `Frame` object inside a `Window` container object to display content. If you are using a single-page application, you will use a single window that is initialized in the `OnLaunched` event of the `App` object. Any subpage content will be rendered on the `Frame` object within this window.

This approach is being used in our sample application. You can find the call to the window in the `App` class, as follows:

```
private Window;
protected override void OnLaunched(Microsoft.UI.Xaml.
  LaunchActivatedEventArgs args)
{
    this.window = new MainWindow();
    this.window.Activate();
}
```

In comparison, this is the frame definition in the `MainWindow.xaml` file:

```
<Frame x:Name="ContentFrame"/>
```

The Windows App SDK windowing APIs introduce new integrations with the Windows user experience, enabling interesting scenarios. The new set of APIs can be found in the `Windows.UI.WindowManagement.AppWindow` class, abstracting the container in which the content is hosted – making it comparable to an abstraction of the window's handle (HWND). This is particularly interesting for scenarios where you want to interact with other native windows, such as the `FileOpenPicker` class or the `MessageDialog` class. You just need to expose a handle in your app class and pass the reference to the initialization of the interoperability to parent the `dialog` handle, as shown in the following snippet:

```
public App()
{
    this.InitializeComponent();
    WindowHandle =
      WinRT.Interop.WindowNative.GetWindowHandle(this);
}
public static IntPtr WindowHandle { get; private set; }
private async Task ShowParentDialog()
{
    var dialog = new Windows.UI.Popups.MessageDialog(
```

```
        "Message here");
    WinRT.Interop.InitializeWithWindow.Initialize(dialog,
        App.WindowHandle);
    await dialog.ShowAsync();
}
```

We'll look at other examples of this approach in *Chapter 8*, *Integrating Your Application with the Windows Ecosystem*.

Now that we have defined the basic building blocks of the user interface, let's see how to properly implement the navigation.

Supporting navigation

We start the actual migration by selecting the right navigation pattern. The NavigationView control we learned to use in *Chapter 5*, *Designing Your Application*, is the best choice and can be configured in two modes through PaneDisplayMode: Top and Left. Left can be further customized with a much smaller footprint. Displaying the navigation on top of your content makes sense when your application only has a few fixed top-level navigation items that are of equal importance. Additionally, less important items can be hidden in overflow menus. The look will be very similar to the = TabControl menu from the starting Windows Forms application sample.

Another display option for NavigationView is the Left display mode. Here, the pane is expanded on the left-hand side of the screen, pushing the content further right. Using the left-aligned navigation is useful when you have more than a few (that is, more than five) navigation items or you want to divide the navigation items into distinct categories.

While the top alignment works well without the use of icons, it becomes very important to provide suitable icons to the navigation items when using the left-hand alignment. As this mode uses a considerable portion of the screen, it often provides an option to resize the navigation into a small bar. In this LeftCompact mode, the icons are displayed. If no icons are defined, the first character from the menu description is used, resulting in a less-than-great navigation experience. You can go one step further and hide the navigation entirely by setting the PanelDisplayMode control to LeftMinimal. This will only show the menu button (the one called hamburger) until the pane is opened. Once it opens, the navigation pane overlays the content, similar to the LeftCompact mode.

Before we go into the details of navigation using the `NavigationView` control, we have to understand how the content is being displayed. The content is hosted by pages, which inherit from the `Page` class. `Page` itself is `UserControl`, so it can be used as the root for all our content elements. Additionally, it brings its own life cycle events that allow us to bind to view models, load data from external sources, and more. The `Page` object is assigned to the `Content` property of the `Frame` object, which is part of our main window and defined next to the `NavigationView` control. The `Content` property of the `Frame` object is `null` upon initialization of the object in XAML, hence we are assigning the `Overview` page directly in the `OnLoaded` event of the menu. In all subsequent navigation actions, we can rely on the `Navigate()` method of the `Frame` object. The implementation details of all these patterns have been addressed in *Chapter 5, Designing Your Application*.

Following the open-closed principle of good software engineering, we don't want to hardcode the navigation targets into the `ItemInvoked` navigation event, as this would require us to modify the method every time a change in the navigation occurs, for example, a new item is added or removed. Instead, it's helpful to deliver the navigation target within the definition of the navigation item itself. This is where the `Tag` property comes to the rescue: it's a property that can get or set any arbitrary object, even complex ones. Additionally, it can be used to store additional information about the element. If you are familiar with Android development, a similar concept has been available for many years.

We are leveraging the `Tag` property to store our navigation in a fully qualified notion, as shown in the following snippet:

```
Tag="ModernDesktop.Views.OverviewPage"
```

Thanks to `Tag`, we can use reflection to find the type of the page we want to redirect the user to. This is provided to the `Navigate()` method, together with the content of `NavigationItem` to initiate the navigation to the content:

```
ContentFrame.Navigate(Type.GetType(item.Tag.ToString()),
  item.Content);
```

As you can see, there's a big leap between the Windows Forms navigation concept and the Windows App SDK navigation concept.

Based on what we have learned so far, this is the definition of the `NavigationView` control we have included in the XAML of our main page:

```xml
<NavigationView x:Name="MainMenu"
                PaneDisplayMode="Top"
                IsBackButtonVisible="Collapsed"
                Loaded="MainMenu_OnLoaded"
                ItemInvoked="MainMenu_OnItemInvoked"
                Background="Transparent"
                IsSettingsVisible="False"
                Margin="10" >
    <NavigationView.MenuItems>
        <NavigationViewItem  x:Name="Overview"
                            Icon="Home"
                            Content="Overview"
                            Tag="ModernDesktop.Views.O
                            verviewPage" />
        <NavigationViewItem x:Name="Create"
                            Icon="Add"
                            Content="Create"
                            Tag="ModernDesktop.Views
                            .UpsertPage" />
    </NavigationView.MenuItems>
    <NavigationView.FooterMenuItems>
        <NavigationViewItem x:Name="About"
                            Icon="More"
                            Content="About"
                            Tag="ModernDesktop.Views
                            .AboutPage" />
    </NavigationView.FooterMenuItems>
    <NavigationView.Content>
        <Frame x:Name="ContentFrame"/>
    </NavigationView.Content>
</NavigationView>
```

In comparison, the following code shows the code behind implementation, which includes the handling of the OnLoaded event (which sets the Content property of the Frame object with a new instance of the OverviewPage page) and the ItemInvoked event (which handles the selection of the item):

```csharp
private void MainMenu_OnLoaded(object sender,
    RoutedEventArgs e)
{
    ContentFrame.Content = new OverviewPage();
}


private NavigationViewItem lastItem;
private void MainMenu_OnItemInvoked(NavigationView
    sender, NavigationViewItemInvokedEventArgs args)
{
    var item = args.InvokedItemContainer as
        NavigationViewItem;
    if (item == null)
    {
        ContentFrame.Content = null;
        return;
    }
    else if (item == this.lastItem)
    {
        return;
    }

    ContentFrame.Navigate(Type.GetType(
        item.Tag.ToString()), item.Content);

    this.lastItem = item;

}
```

Now that the navigation has been set up, it's time to migrate the business logic. Since we want to build a future-proof application, we're going to move it to the ViewModel.

Migrating from the code-behind approach to the MVVM pattern

If we look at our sample Windows Forms application, we can see we have used the traditional code-behind approach. The Form1 class contains all the code to load the data, handle user interactions, and more. However, as we learned in *Chapter 6, Building a Future-Proof Architecture*, this isn't the right approach to build an application that must be easy to maintain and evolve. The MVVM pattern is the best approach to use since it helps to separate the business logic from the UI layer.

As such, we must follow the guidance we learned in *Chapter 6, Building a Future-Proof Architecture*, to create a ViewModel class that will take care of retrieving the data from the data layer, preparing it for being displayed, and connecting it to the UI layer using binding. The ViewModel classes are created using the helpers provided by the MVVM Toolkit library, which is part of the Windows Community Toolkit.

ViewModel is then connected to the UI layer by exposing it as a property on the code-behind class, which enables us to use the x:Bind markup expression:

```
public sealed partial class OverviewPage : Page
{
    public OverviewViewModel ViewModel { get; }

    public OverviewPage()
    {
        this.InitializeComponent();
        ViewModel =
            Ioc.Default.GetService<OverviewViewModel>();
    }
}
```

As part of the migration to a ViewModel class, we're going to do the following:

- Convert the data we must display in the UI (such as the collection of employees) into properties that implement the INotifyPropertyChanged interface to dispatch notifications to the binding channel.
- Convert event handlers into commands. Since they are exposed as a property, they can be connected to the UI layer using binding as well, which helps to keep the UI layer completely separate from the logic.

Let's see an example of both scenarios.

Migrating the data

In the `Form1` class of the Windows Forms application, we can see that we are using the `LoadEmployee()` method exposed by the `Form` class to retrieve a collection of all the employees using the APIs offered by EF:

```
protected override void OnLoad(EventArgs e)
{
    base.OnLoad(e);
    context = new EmployeeContext();
    employeeBindingSource.DataSource =
      context.Employees.Local.ToBindingList();
    employeeDataGridView.Refresh();
}
```

Since we're using a code-behind approach, we set the collection directly to the `DataSource` property of the `DataGrid` control using its name. When we move to MVVM, the collection must be connected to the `DataGrid` control using binding. As such, we need to expose it as a property of our `ViewModel` class, as shown in the following example:

```
private ObservableCollection<Employee> _employees;

public ObservableCollection<Employee> Employees
{
    get => _employees;
    set { SetProperty(ref _employees, value); }
}
```

This way, we'll be able to connect it to the View using the `x:Bind` markup expression.

In this section, we looked at one data migration example. In the final version of the application on GitHub, you will find that all the data has been migrated using the same approach.

Migrating the event handlers

Event handlers are used in the code-behind approach to define the logic you want to execute when a control raises an event. They're very powerful, but they also create a tight dependency between the logic and the UI layer: in fact, event handlers can only be defined in code-behind. The proper way to support handlers in the MVVM patterns is by using commands. As we learned in *Chapter 6, Building a Future-Proof Architecture*, commands are defined using `ViewModel` properties. This is so that they can be connected to the UI layer using the `Command` property available in the XAML and the binding.

For example, take a look at our Windows Forms application with the following handler, which is connected to the `CellClick` event of the `DataGrid` control. It's used to handle the selection of an employee from the grid:

```
private async void employeeDataGridView_CellClick(object
    sender, DataGridViewCellEventArgs e)
{
    if (e.RowIndex >= 0) // sorting returns -1
    {
        var value = employeeDataGridView.Rows[
            e.RowIndex].Cells[0].Value;
        selectedEmployee = await
            context.Employees.FindAsync(value);
        LoadEmployee();
    }
}
```

In comparison, in our `ViewModel` class, it will be defined as a command. This is thanks to the `RelayCommand` class offered by the Windows Community Toolkit:

```
public class OverviewViewModel : ObservableObject,
    INavigationAware
{
    public IRelayCommand<Employee> SelectCommand { get;
        private set; }

    public OverviewViewModel()
    {
        LoadEmployees();
    }
```

```
private void InitializeCommands()
{
    SelectCommand = new
      RelayCommand<Employee>(SelectEmployee);
}


private void SelectEmployee(Employee item)
{
    var selectedEmployee = new
      ObservableEmployee(item);
    this.navigationService.NavigateTo(typeof(
      DetailsPage), selectedEmployee);
}
}
```

Now we can connect the `SelectCommand` object to the View using the `x:Bind` markup expression, as we did with the data. Also, in this case, we repeat this process for every other event handler in our application.

As part of the migration from the code-behind approach, we also need to ensure that the business logic and the UI are clearly separated. In this section, you have seen an example that goes against this approach: the code to access the database using EF is directly called in code-behind, which creates a tight dependency between the data layer and the UI layer. As such, as part of the migration, you must move all the code that involves the data layer into separate classes, which will be used by the various `ViewModel` classes. The code to interact with the database is a good example: in the final version of the application that is available on GitHub, you can see that neither the code-behind nor the ViewModel contains any reference to the EF classes. Instead, all the operations with the database are centralized in a dedicated class, called `SqlDataService`, which is described by the `IDataService` interface.

Thanks to this approach, other than building a cleaner architecture, we can also set the fundaments to adopt the IoC pattern, which we're going to detail next.

Supporting the IoC pattern and dependency injection

Another thing we learned in *Chapter 6, Building a Future-Proof Architecture*, is the adoption of the IoC pattern to make the application easier to test and evolve. Instead of manually creating all of the dependencies that are required by a `ViewModel` (such as the `SqlDataService` class we learned about earlier) class at compile time, we register all of them inside a container, and we inject them into the `ViewModel` classes at runtime. This enables us to easily swap the implementation of a service (for example, if we want to move from SQL Server to a cloud service) or to mock them so that we can write unit tests for `ViewModel` (so that we don't need access to a real database to perform tests).

The migrated version of the application, which you can find on GitHub, follows the guidance we learned in *Chapter 6, Building a Future-Proof Architecture*, to support this feature. The following list details the required changes:

- In the `App` class, we use the `Ioc` helper provided by the toolkit to register all our `ViewModel` classes and services. The following code shows an excerpt of the full implementation, with the registration of the `SqlDataService` class:

```
public App()
{
    var serviceProvider = ConfigureServices();
    Ioc.Default.ConfigureServices(serviceProvider);
    this.InitializeComponent();
}
private static IServiceProvider ConfigureServices()
{
    var services = new ServiceCollection();
    . . .
    services.AddSingleton<INavigationService,
      NavigationService>();
    services.AddSingleton<IDataService,
      SqlDataService>();
    . . .
    return services.BuildServiceProvider();
}
```

- In the code-behind of every page, instead of manually creating a new instance of the corresponding `ViewModel`, we retrieve it from the container using the `Ioc` helper again, as shown in the following sample:

```
public sealed partial class OverviewPage : Page
{
    public OverviewViewModel ViewModel { get; }

    public OverviewPage()
    {
        this.InitializeComponent();
        ViewModel =
            Ioc.Default.GetService<OverviewViewModel>();
    }
}
```

- We add a dependency to all the required services in the constructor of each `ViewModel`. This way, the `Ioc` helper will give us an instance of `ViewModel` already populated with everything we need. For example, this is how we define that the `OverViewModel` class requires an `IDataService` object to interact with the database:

```
public OverviewViewModel(IDataService dataService,
    INavigationService navigationService)
{
    this.dataService = dataService;
    this.navigationService = navigationService;
}
```

Now we're all set with the architecture of the application. We can start working on the UI and migrate our `DataGrid` control.

Migrating the DataGrid control

Now that we have all the building blocks in place, let's migrate the `DataGrid` control. For Windows Forms, Visual Studio provides a tool called `DataSetGenerator` to define the `TableAdapter` types, entities, and properties connected to the `DataGrid` control via `DataSet`. While this is very convenient and straightforward, many settings are hidden from the wizards, making it quite a challenge to tweak or modify the binding. Using MVVM, we can now precisely control what data is bound to the `DataGrid` control. There is no wizard available, so migrating the MVVM architecture and exposing the data for the `DataGrid` control requires some plumbing work on our side.

The `DataGrid` control expects an `IEnumerable` collection as a data source, and in order to be able to track changes, an `ObservableCollection` type is exposed and bound to the `ItemSource` property. We learned about the `ObservableCollection` type in *Chapter 2, The Windows App SDK for a Windows Forms Developer*. It's a special type of collection that can dispatch notifications to the UI whenever something changes, such as a new item being added or removed. It's a perfect match for our binding purposes, which is achieved using the `x:Bind` markup expression we learned about in *Chapter 3, The Windows App SDK for a WPF Developer*. In the previous section, we learned how we can expose the list of employees from `ViewModel`.

However, unlike in Windows Forms, the WinUI `DataGrid` control isn't built into the framework, but comes with the Windows Community Toolkit. As such, before adding it to our page, we need to declare its namespace in the `Page` control, following the guidelines we learned about in *Chapter 4, The Windows App SDK for a UWP Developer*:

```
xmlns:controls="using:CommunityToolkit.WinUI.UI.Controls"
```

Connecting the `DataGrid` control with our collection of employees is a straightforward operation: thanks to what we learned in the previous section, we already have the collection declared as a property of the `OverviewViewModel` class, which is already exposed to the View through the `ViewModel` property that was declared in the code-behind:

```
<controls:DataGrid x:Name="dataGrid"
                   ItemsSource="{x:Bind
                       ViewModel.Employees, Mode=OneWay}"
                   IsReadOnly="True"
                   SelectionMode="Single"
                       Sorting="DataGrid_OnSorting" />
```

But what about handling the selection of an employee from the `DataGrid` control? We have already defined a command in `ViewModel` to support this scenario, called `SelectCommand`. However, there's a challenge: the `DataGrid` control doesn't expose a `Command` property, so how can we connect it using binding?

We must introduce the concept of **behaviors**, which is a XAML feature that enables you to attach complex logic to XAML controls directly in XAML. To use it, we must install a package from NuGet called `Microsoft.Xaml.Behaviors.WinUI.Managed`, which provides the `Interaction.Behaviors` collection and the `EventTriggerBehavior` and `InvokeCommand` classes. The former is used to define the event we want to manage, through the `EventName` property (in our case, this is `SelectionChanged`). The latter is used to bind a command to the event –we can bind it to our `ViewModel` command and even provide parameters. In our scenario, the selected entry can then be passed as a payload to the `NavigationService` to be used on the details page. Thanks to this feature, we can connect a command to virtually any event exposed by a control, enabling us to keep following the MVVM pattern even when we have to handle secondary events that are supported by the `Command` property.

This is what the full `DataGrid` implementation looks like:

```
<Page
    xmlns:controls="using:CommunityToolkit.WinUI.UI
      .Controls"
    xmlns:interactivity="using:Microsoft.Xaml
      .Interactivity"
    xmlns:core="using:Microsoft.Xaml.Interactions.Core">

    <Grid>
        <controls:DataGrid x:Name="dataGrid"
                           ItemsSource="{x:Bind
                           ViewModel.Employees,
                           Mode=OneWay}"
                           IsReadOnly="True"
                           SelectionMode="Single"
                           Sorting="DataGrid_OnSorting">
            <interactivity:Interaction.Behaviors>
```

```
<core:EventTriggerBehavior
    EventName="SelectionChanged">
    <core:InvokeCommandAction
        Command="{x:Bind
        ViewModel.SelectCommand}"
        CommandParameter="{x:Bind
        dataGrid.SelectedItem,
        Mode=OneWay}" />
</core:EventTriggerBehavior>
    </interactivity:Interaction.Behaviors>
    </controls:DataGrid>
</Grid>
</Page>
```

When the user selects an employee from the `GridView` control, we must trigger a navigation to the **Details** page, where we expect an entry as a parameter. In our Windows Forms app, we weren't switching between pages and could, therefore, rely on a backing field – now this information needs to be passed as a parameter instead.

We can achieve this goal by using the `NavigationService` approach we learned about in *Chapter 6, Building a Future-Proof Architecture*. This exposes methods that we can use in our `ViewModel` instances to navigate from one page to another with a parameter. The `NavigationService` class is injected inside the ViewModel using the IoC pattern, so we can simply perform the navigation using the following code:

```
private void SelectEmployee(Employee item)
{
    var selectedEmployee = new ObservableEmployee(item);
    this.navigationService.NavigateTo(typeof(DetailsPage),
        selectedEmployee);
}
```

We have completed the migration of the **Overview** page. Next, let's see how we can migrate the **Details** page.

Migrating the Details page

Migrating the **Details** page is an easy operation thanks to the knowledge we have acquired so far. This page displays all the properties of an employee, using input controls such as TextBox and DatePicker so that the user can also update them. As such, the operations to perform are the same as for the **Overview** page:

- We migrate the data using properties that implement the INotifyPropertyChanged interface. In this case, we have a property called Employee, where we store a reference for the selected employee.

- We migrate the event handlers to commands to support all the user interactions. For example, the UpsertCommand object is connected to the **Save** button, and it takes care of saving the changes made on the employee to the database.

Note that we still need to take extra steps to bind a TextBox control to a decimal property – however, the solution is more elegant with a converter, which we learned to use in *Chapter 2, The Windows App SDK for a Windows Forms Developer*. A DigitalToStringConverter object is used during binding, and it converts both, a string to decimal and a decimal to string whenever the source or target changes. The converter is registered in global ResourceDictionary of the application, in the App class.

This is an example of the syntax used to apply the converter to the binding channel:

```
<TextBox Text="{x:Bind ViewModel.SelectedEmployee.
Salary, Mode=TwoWay, Converter={StaticResource
DecimalToStringConverter}}" Header="Salary" />
```

Another difference to bear in mind, compared to the **Overview** page, is that most of the properties are connected to the UI using the TwoWay mode for binding. Since TextBox is an input control, TwoWay binding enables you to not only display the value of the property in the control but also set it.

And what about the Employee property? We can populate it using another helper that we saw in *Chapter 6, Building a Future-Proof Architecture*, where we talked about navigation: the INavigationAware interface. We can implement this in our ViewModel. Thanks to this, we can subscribe to the OnNavigatedTo event directly in ViewModel. This is so that, when the **Details** page is loaded, we can retrieve the selected employee from the NavigationService class, as shown in the following example:

```
public class DetailsViewModel : ObservableObject,
    INavigationAware
{
```

```
public void OnNavigatedTo(object parameter)
{
    if (parameter is ObservableEmployee employee)
    {
        SelectedEmployee = employee;
        IsExisting = true;
    }
    else
    {
        SelectedEmployee = new ObservableEmployee(new
            Employee());
        IsExisting = false;
        }
    }
}
```

That's it! We have completed an overview of the steps required to move our Windows Forms application to a modern WinUI application based on the MVVM pattern.

Now, let's explore the migration effort when the starting application is built, using WPF.

Migrating a WPF application

Migrating from a WPF application usually requires a lot less effort than for Windows Forms. This is due to at least two reasons:

- WPF applications already leverage XAML as the UI description language. While the actual XAML syntax might be different, it can easily be transformed into the WinUI XAML syntax. Concepts such as converters or resources are usually well understood.

- Given the effectiveness of XAML binding, many WPF applications are already built upon the MVVM architecture. While there were many MVVM frameworks for WPF available, such as MVVM Light, Prism, or lightweight DIY frameworks, they all follow the same principle. Not only that, MVVM is (almost) always accompanied by the use of IoC. This means that you might need to rework the XAML pages a bit, but you will be able to reuse almost all of your data layers and ViewModel classes without changes.

We are going to start by taking a quick inventory of the WPF application again. Our WPF application (which you can find in the GitHub repository at `https://github.com/PacktPublishing/Modernizing-Your-Windows-Applications-with-the-Windows-Apps-SDK-and-WinUI/tree/main/Chapter07/WPF`) is built upon the MVVM pattern just like the Windows App SDK app. It uses solutions and libraries that are extremely popular in the WPF ecosystem. For example, the IoC pattern is provided by the Unity library, which is a bit different from the modern IoC implementation. Back in the day, property injection was more common than constructor injection. Injected dependencies were exposed as properties and marked with an attribute, as shown in the following sample:

```
[Dependency]
 public IConcurrencyService ConcurrencyService { get; set; }
```

The implementation of the MVVM pattern is self-built and does not rely on any framework. All the advanced features, such as navigation and showing dialogs, are implemented manually. The same applies to base classes, which manually implement the `INotifyPropertyChanged` interface, as shown in the following sample:

```
public abstract class NotifyPropertyChangedBase :
   INotifyPropertyChanged
```

Self-rolled MVVM frameworks require more manual plumbing, as you will explore in our WPF sample. There is abstract `DelegateCommandBase` that implements `ICommand`, while the `NavigationService` class is more than just a simple service and relies on *manual* context and parameter implementation.

Data access is still provided through EF6 but wrapped in a repository pattern. Repository and **Unit of Work** (**UoW**) patterns are a common sight in legacy .NET 4.x WPF applications, hence we have included them.

The UI is similar to the Windows Forms sample app in concept, but it's more powerful since it uses many of the advanced features that WPF brings to the table: themes, styles, validators, and busy indicators.

As you can see, the basic concepts are already in place, but migration still requires touching on many pieces. First of all, let's migrate from Unity to a modern IoC container. As we have seen in the Windows Forms migration, the new .NET runtime provides a built-in implementation of the IoC pattern, thanks to the `IServiceProvider` interface. Therefore, the registration of dependencies in the bootstrapper class needs to be moved to a `ServiceProvider` registration.

Other framework dependencies, such as EF, are easier to migrate to and just require updating and adapting to the new API. You can learn more about migrating from EF to EF Core in the official documentation at https://docs.microsoft.com/en-us/ef/efcore-and-ef6/porting/. UoW and repository patterns should be disposed of, as the DBContext class already provides us with repositories and UoW under the hood. Unless you're using a generic, detached UoW pattern with transactions, using specific UoW and repository implementations will actually increase the load on your database, as SaveChanges will be called multiple times, making rolling back impossible to achieve. A detached, generic UoW pattern does bring its own set of challenges, so it's recommended that you stay clear from these patterns in modern applications. Dispose of them entirely, and move the custom queries to the service classes.

Also, WPF applications based on Unity to implement the IoC pattern often adopted **Managed Extensibility Framework (MEF)**, which is a library introduced with .NET Framework 4 to simplify the creation of lightweight and extensible applications. You can learn more about it at https://docs.microsoft.com/en-us/dotnet/framework/mef/.

In this scenario, unfortunately, migrating a Unity and MEF-based WPF application requires almost complete refactoring since there is no simple upgrade path. MEF features have moved to the System.Composition and System.ComponentModel.Composition namespaces. While the general idea remains the same, not all features have been ported, so consider moving carefully.

Migrating the ViewModel instances is far easier. In the WPF sample application, we implemented many of the MVVM features, such as INotifyPropertyChanged or other commands, manually. This is still supported, and it will continue to work with the Windows App SDK and WinUI. However, thanks to the lightweight and open source nature of the Windows Community Toolkit, we recommend you move to it rather than relying on a self-developed MVVM implementation. For example, in our migrated application, the DelegateCommand class has been replaced by the RelayCommand (or AsyncRelayCommand for an asynchronous operation) class. Or the INotifyPropertyChangedBase class has been replaced by the ObservableObject class.

The task of migrating XAML user controls and pages mainly involves finding and replacing potential namespace changes and bindings. When you move to WinUI, it's a good idea to switch to the new x:Bind markup expression, which is more powerful and delivers better performances. However, the binding syntax is slightly different, so you will have to make a few changes. For example, a key difference between traditional binding and x:Bind is the context lookup. In the WPF sample, we can see that, through the ViewBase class (which is used as a base class for all the pages), the binding context is provided by the DataContext class, which is exposed by every XAML control. This means that to connect the OverviewPage view with its corresponding ViewModel class, we must create a new instance of the OverviewViewModel class and assign it to the DataContext property. When we use x:Bind, the XAML framework will look for the context in code-behind instead, as we learned in *Chapter 3, The Windows App SDK for a WPF Developer*. As such, instead of setting the DataContext property, we must expose a property in code-behind to hold a reference to the corresponding ViewModel. For example, the following is what the code-behind approach of the OverviewPage page looks like in the final WinUI version:

```
public sealed partial class OverviewPage : Page
{
    public OverviewViewModel ViewModel { get; }

    public OverviewPage()
    {
        this.InitializeComponent();
        ViewModel =
            Ioc.Default.GetService<OverviewViewModel>();
    }
}
```

Now we can use the ViewModel property to access all the properties and commands exposed by the ViewModel class using the binding channel, as shown in the following example:

```
<TextBox Text="{x:Bind ViewModel.SelectedEmployee.Role,
Mode=TwoWay}" />
```

Finally, many of the controls have been improved over time, so it's helpful to check out the XAML **Controls Gallery** application we learned about in *Chapter 5, Designing Your Application*, for new features of the existing controls.

Now, let's take a look at how to migrate one of the most frequently used WPF features: themes.

Themes

WinUI theming is different from WPF theming. In WinUI, you can leverage three different base themes: Light, Dark, and HighContrast. A theme is now defined by a ThemeDictionary type and applied using its x:Key attribute, as follows:

```xml
<ResourceDictionary.ThemeDictionaries>
    <ResourceDictionary x:Key="Default">
        <SolidColorBrush
        x:Key="TargetBackground" Color="Red"/>
        . . .
    </ResourceDictionary>
    <ResourceDictionary x:Key="HighContrast">
        <SolidColorBrush
        x:Key="TargetBackground"
        Color="Black"/>
        . . .
    </ResourceDictionary>
</ResourceDictionary.ThemeDictionaries>
```

ThemeDictionaries aren't meant to provide an entire app theme anymore. Instead, the system themes should be preferred, and specific theme resources should only be used in a few cases to override system values.

Now that we have learned how to migrate our WPF sample, let's see the last technology covered in this chapter: the UWP app.

Migrating a UWP app

The Windows App SDK and WinUI can be considered the direct successors of UWP. In fact, its goal is to detach many of the UWP APIs and features from the operating system so that they can be consumed also by Win32 applications.

Consequently, among the technologies we have seen in this chapter, UWP is, for sure, the easiest one to start a migration from.

The high-level migration approach is similar to the one we have adopted to migrate our WPF sample application: as the main architecture pattern is already in place, it's mostly a matter of updating to the latest version of the Windows Runtime APIs and adapting to changes in the API. The documentation often provides a suggested migration path. In fact, moving from UWP could even be an easier task than moving from a .NET 4.x-based WPF application, as IoC and MVVM are likely provided by the same frameworks and packages as their Windows App SDK counterpart.

Most Windows Runtime APIs can be used in Windows App SDK applications, but there are some noteworthy exceptions. They are listed as follows:

- APIs that have a dependency on features and are only available in the UWP UI layer can't be used. There is the Windows UI Library (that is, WinUI 3) for that now.

- APIs that require a package identity (a globally unique identifier for a package) are only supported in MSIX packaged apps. If you're planning to integrate deeper with the operating system (such as managing firewall rules), MSIX is the preferred deployment technology. This is because it makes it easier to control the integration through the manifest file.

Another significant difference is that splash screens aren't supported out of the box for the Windows App SDK and need to be substituted with a window transition. This is in the case that your application requires some time to be initialized.

One of the key differences between UWP and WPF is how the dispatcher works. Let's see how we can migrate it.

Migrating the dispatcher

In our UWP application, we are using the `Microsoft.UI.Xaml.Window.Dispatcher` class to dispatch tasks to the UI thread. This is required when you are performing some operations in the background, but at some point, you have to update the user interface. However, as we learned in *Chapter 4, The Windows App SDK for a UWP Developer*, this class is no longer available due to the different implementation of the `Window` container between UWP and Win32.

As such, we must leverage the `DispatcherQueue` property exposed by the `Window`, `Page`, or `Control` objects. To run a task on the UI thread using the dispatcher, enqueuing it to the `DispatcherQueue` property will ensure it's executed on the thread associated with the dispatcher queue. Refer to the following code:

```
If (this.DispatcherQueue.HasThreadAccess)
{
    textBlock.Text = msg;
```

```
    }
    else
    {
        var result = this.DispatcherQueue.TryEnqueue(
            Microsoft.UI.Dispatching
            .DispatcherQueuePriority.Normal,() =>
            textBlock.Text = msg);
    }
```

You can refer to *Chapter 4, The Windows App SDK for a WPF Developer*, to see more details regarding how to use the new dispatcher and how you can optimize it with helpers from the Windows Community Toolkit.

Now, let's see another key difference between UWP and WinUI: the activation logic.

Activation

Let's take a look at the App class of our UWP sample application. In UWP, the activation logic can be implemented by overwriting specific activation methods of the `Application` object, such as `OnFileActivated` or `OnBackgroundActivated`, or by using the `ActivationKind` property of the `IActivatedEventArgs` parameter of the `OnActivated` override. Casting this parameter to a specific implementation (for example, `FileActivatedEventArgs`) gives access to the activation parameters.

These events are made available thanks to the peculiar life cycle of the UWP: applications can be suspended when they are idle; terminated when Windows needs to release memory to let other applications run; or even activated in the background, without showing any user interface.

In comparison, Windows App SDK applications are based on the Win32 model, and as such, they have a simpler life cycle. There are no overridable activation-related methods available in the `Application` model except for `OnLaunched`. The entire activation is now simplified and handled entirely in `OnLaunched`. For packaged applications, all 44 activation kinds known from UWP are available with the `UWPLaunchActivatedEventArgs`. `Kind` property of the `LaunchActivatedEventArgs` parameter. Unpackaged apps only support four activation kinds:

- `Launch`: This is for command-line activation, used when a user clicks the app icon.

- `File`: This is for registered file type activation, used when the user opens a file of the registered type.

- **Protocol**: This is for protocol-based activation, used when a user opens the app via a URI.

- **StartupTask**: This is for user login-based activation, used when the app is started on user login.

For example, our UWP application can handle file type association, so we have the following code in the `App` class:

```
protected override void OnFileActivated(FileActivatedEventArgs
   args)
{
    var file = args.Files[0];
    //open a dedicated page to display the selected file
}
```

Through the activation arguments, we retrieve a reference to the file the user is trying to open, which is stored in the `Files` collection.

In our Windows App SDK application, we must move this code directly inside the `OnLaunched` event and use the `AppInstance` class to retrieve a reference to the selected file, as shown in the following sample:

```
protected override void OnLaunched(Microsoft.UI.Xaml.
   LaunchActivatedEventArgs args)
{
    var eventArgs =
      AppInstance.GetCurrent().GetActivatedEventArgs();
    if (eventArgs.Kind == ExtendedActivationKind.File)
    {
        var fileActivationArguments = eventArgs.Data as
          FileActivatedEventArgs;
        var file = fileActivationArguments.Files[0];
        // open a dedicated page to display the selected
        // file
    }

    var shellFrame = new Frame
    {
        Content = new MainPage()
```

```
    };

    MainWindow.Content = shellFrame;
    MainWindow.Activate();
}
```

Since we have mentioned how activation is impacted by the life cycle of the application, let's look at a little bit more information regarding the differences between UWP and the Windows App SDK.

Managing the life cycle

As we mentioned earlier, the life cycle of a UWP application is very different from the Win32 one. UWP applications are born to run on multiple devices, with potentially limited resources, and, as such, they have a conservative life cycle. For example, when the application is minimized, it gets suspended, and it can't perform any background operations so that it doesn't consume the CPU. In comparison, a Win32 application doesn't have this limitation and can continue to run as long as the process is in execution.

In our UWP sample application on GitHub, you can see an example of this behavior in the code-behind of the `OverviewPage` class:

```
public OverviewPage(OverviewViewModel viewModel)
{

    InitializeComponent();
    ViewModel = viewModel;
    Application.Current.Suspending += Current_Suspending;

}

private void Current_Suspending(object sender, Windows.
  ApplicationModel.SuspendingEventArgs e)
{

    ApplicationData.Current.LocalSettings.Values[
      "OverviewViewModel"] = ViewModel;

}
```

Here, we subscribe to the `Application.Current.Suspending` event, which is triggered when the UWP application is about to be suspended. This is so that we can save important data that should survive if Windows terminates the application because it's running out of resources.

In terms of migrating the application to the Windows App SDK and WinUI, we don't have to do this anymore since Windows will no longer try to suspend or terminate our application (unless we do something wrong, such as not catching an exception that leads to a crash). However, there might still be scenarios where we want to optimize the execution based on the workload. We can do this thanks to the Power Management APIs that we learned about in *Chapter 4, The Windows App SDK for a UWP Developer*.

Instead of overriding methods of the `Application` object, we can now subscribe to events on the `PowerManager` object, which is included in the `Microsoft.Windows.System.Power` namespace, using an event handler. This allows you to register to these events in headless apps or background apps but is usually implemented in `MainWindow`:

```
private void SetupPowerManagerHandler()
{
    PowerManager.DisplayStatusChanged +=
      PowerManager_DisplayStatusChanged;
    PowerManager.SystemSuspendStatusChanged +=
      PowerManager_SystemSuspendStatusChanged;
}


private void PowerManager_DisplayStatusChanged(
  object sender, object e)
{
    if (PowerManager.DisplayStatus == DisplayStatus.On)
      { /*do things*/ }
    else if (PowerManager.DisplayStatus ==
      DisplayStatus.Off) { /*stop things*/ }
}


private void PowerManager_SystemSuspendStatusChanged(
  object sender, object e)
{
    if (PowerManager.SystemSuspendStatus ==
      SystemSuspendStatus.Entering) { /* stop things*/ }
}
```

In this sample, we are suspending the execution of operations in the case that the screen goes off, meaning that the user is no longer actively using their device.

With this, we have completed our journey of migrating our UWP app to the Windows App SDK and WinUI.

Summary

In this chapter, we put what we have learned so far in this book into practice. By using the knowledge acquired around the XAML framework, binding, the application's life cycle, navigation, and the MVVM pattern, we have migrated real applications built with different technologies to .NET 6, the Windows App SDK, and WinUI.

However, this is only the beginning. Now we have an application that is reliable, more pleasant to use, and based on modern runtimes. But we haven't added any new features yet. This is the goal of the next chapters: we're starting on a journey to integrate our applications with many exciting features that are available in Windows 10 and Windows 11. We will start with technologies such as localization and **Windows Hello**, which will be covered in the next chapter.

Questions

1. When we migrate from WPF to WinUI, we can reuse the same UI layer since both technologies are based on XAML. Is this true or false?

2. When we move to WinUI and the `x:Bind` markup expression, we can keep using the `DataContext` property to connect the Views with the ViewModels. Is this true or false?

3. If your UWP application supports multiple activation points, you have to change their implementation when you move to the Windows App SDK and WinUI. Is this true or false?

8
Integrating Your Application with the Windows Ecosystem

Windows isn't just a platform that enables you to run your applications, it offers a wide range of features to increase your productivity, such as with **Windows Hello**, which you can use to log in in a safe and seamless way thanks to biometric systems such as face recognition and fingerprint readers, **location services**, to identify the position of the user, and **Sharing**, which you can use to easily transfer information from one application to another.

Windows offers a developer ecosystem that enables you to light up all these features directly in your applications to make them even more powerful. For example, the way that you can log in to your PC just by using your camera thanks to Windows Hello, you can enable that same experience in your applications. At the same time, since we're building a desktop application, it means that we can integrate with the operating system in ways that aren't supported by other types of applications, such as working with the filesystem, integrating web experiences, and much more.

In this chapter, we're going to explore a few of these techniques to integrate your application with the various features that Windows makes available to developers, such as the following:

- Integrating APIs from the Universal Windows Platform
- Working with files and folders
- Supporting the sharing contract
- Integrating web experiences into your desktop application

Let's start!

Technical requirements

The code for the chapter can be found here:

```
https://github.com/PacktPublishing/Modernizing-Your-Windows-
Applications-with-the-Windows-Apps-SDK-and-WinUI/tree/main/
Chapter08
```

Integrating APIs from the Universal Windows Platform

Since the introduction of Windows 10, the Windows team has deeply invested in the **Universal Windows Platform** (**UWP**). Consequently, UWP became the entry point for all developers to enable any new feature added to Windows. However, in the past, this meant that only by building UWP apps were you able to integrate new features such as Windows Hello, geolocation, the sharing contract, and so on. As we discussed in *Chapter 1, Getting Started with the Windows App SDK and WinUI*, this approach led to slow adoption of the platform, especially by enterprise developers, since most of the time it meant rewriting existing .NET applications almost from scratch.

In recent years, as such, the Windows team has worked on ways to enable the usage of the UWP ecosystem in existing .NET applications, leading developers to enhance their existing applications with new features, without the need to restart from scratch with new technology.

The most recent outcome of this effort is C#/WinRT, which is a library that enables C# applications to consume APIs that belong to Windows Runtime, which is the framework the UWP is based on. This library has been built with the same guiding principles as the Windows App SDK: lifting the C# projection for Windows Runtime from the operating system to a library, so that it can be evolved and supported independently from the operating system.

However, when it comes to .NET applications, the integration is even deeper than what we can achieve just with the Windows App SDK. You won't need, in fact, to manually install any NuGet package, since this integration is included in the specific .NET target framework for Windows. You have already seen this configuration in action when we looked at the project file of a WinUI application. If you remember, this is how the `TargetFramework` property is set when you create a new WinUI project using the dedicated Visual Studio template:

```
<TargetFramework>net6.0-windows10.0.19041.0</TargetFramework>
```

This property can have different values, based on the target SDK you want to leverage in your application:

* `net6.0-windows10.0.17763.0`
* `net6.0-windows10.0.18362.0`
* `net6.0-windows10.0.19041.0`
* `net6.0-windows10.0.22000.0`

It's important to highlight how the target Windows SDK version doesn't mean that the application will run only on that specific version of Windows, but that the application will be compiled against that specific SDK, enabling you to use a different set of APIs. The UWP, in fact, evolved over time and each SDK released included new features and APIs you can use in your applications. This means that, for example, an application that targets the 10.0.22000.0 SDK (which matches the Windows 11 release) can run without issues also on Windows 10, as long as you don't use specific Windows 11 features. This approach is empowered by the capability detection feature of the UWP, which you can use to detect if a specific feature is available before using it and, eventually, fall back to a different approach.

Since this implementation is based on the .NET target framework, the usage of UWP features isn't directly connected to the Windows App SDK. To use them, in fact, you aren't required to install the Windows App SDK NuGet package, you just have to switch your current target framework to one of the specific Windows 10/11 ones.

If you're using WinUI as a UI framework to build your application, you're already set up. As we saw in *Chapter 1, Getting Started with the Windows App SDK and WinUI*, the `TargetFramework` property of any new WinUI project is already set in the right way. If you have an existing WPF or Windows Forms project based on .NET 5 or .NET 6, instead, it's very likely that your `TargetFramework` will look like this:

```
<TargetFramework>net6.0-windows</TargetFramework>
```

As you will notice, this target doesn't explicitly declare a specific version of Windows. This means that you will indeed be able to use specific Windows APIs, but that belong to the broader Win32 ecosystem, such as access to the Windows Registry, communication with Windows services, and so on. With this target, your application will also run on older versions of Windows, such as Windows 8 or Windows 7. However, you won't have access to any of the new features introduced in Windows 10. As such, the first step to integrate new APIs is to change the current `TargetFramework` to one of the specific Windows 10 ones, as outlined before.

Regardless of the UI platform of your choice, once you switch to a specific Windows 10 or 11 target framework, the C#/WinRT library will be automatically installed in your project. You can see that by expanding the **Dependencies** section of your application:

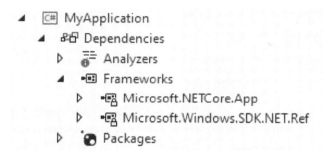

Figure 8.1 – The C#/WinRT library included in a .NET application

Inside the **Frameworks** node, other than the basic .NET 6 runtime (identified by the `Microsoft.NETCore.App` framework), we can see also a dependency called `Microsoft.Windows.SDK.NET.Ref`. This is the projection that enables us to use APIs that are part of the Windows Runtime inside our .NET application.

Now that we have learned how we can integrate Windows APIs into our .NET application (independently from the UI framework), let's see a couple of real examples.

Getting the location of the user

Windows includes a powerful set of APIs that you can use to get the location of the user, which makes it a precious companion for many consumer and enterprise scenarios: displaying the position of the user on a map in a delivery app; getting their location when they complete a task in a to-do application for first-line workers, and many more.

The heart of this feature is the `Geolocator` class, which belongs to the `Windows.Devices.Geolocation` namespace. Let's see a brief example of how to use it:

```
private async Task GetCurrentPositionAsync()
{
    Geolocator = new Geolocator();
    Geoposition position = await geolocator.
        GetGeopositionAsync();
    Console.WriteLine($"{position.Coordinate.Point.Position.
        Latitude} - {position.Coordinate.Point.Position.
        Longitude}");
}
```

The `Geolocator` class exposes an asynchronous method called `GetGeopositionAsync()`, which returns a `Geoposition` object that includes a lot of information about the current position of the user. The most important one is the `Coordinate` property: through the `Point.Position` object, you can access the latitude and longitude coordinates, as in the previous example.

This code, however, has a flaw. Windows gives full control to the user around privacy and security. As such, users have the option, through the Settings app, to disable the location services entirely or for a specific application. This is one of the areas where there's a significant difference compared to when you use the same APIs in the UWP application. The UWP, in fact, due to the sandbox in which applications run, has a granular capability model, which allows you to opt in when you want to use Windows features that have special implications around privacy and security. As such, UWP apps that use location services are not enabled by default and you must do the following:

1. Declare the location capability in the manifest.
2. Call the `RequestAccessAsync()` method exposed by the `Geolocator` class as the first step, which will trigger a popup asking for the user's permission to use the location services.

Win32 applications aren't able to leverage the same capabilities model but they can access all the features exposed by Windows. As such, .NET applications that use the location APIs have the opposite behavior: by default, they are granted access to the location services. The user has the following different options to disable them:

- Completely disable the location services in Windows, blocking every application from retrieving the position (option **1** in the following screenshot).

- If the application is deployed as packaged, it will show up in the list of apps available in the **Privacy & security | Location** section of the **Settings** app (option **2** in the following screenshot, which shows an example of a packaged .NET application called **MyApplication**).

- If the application is deployed as unpackaged, instead, it doesn't have an identity. As such, Windows isn't able to manage the permissions in a granular way. The application will show up in the generic list of desktop apps and, through a toggle, you can enable or disable access to all of them (option **3** in the following screenshot shows a list of unpackaged apps that have used the location services).

This is the screenshot showing an example of the three scenarios:

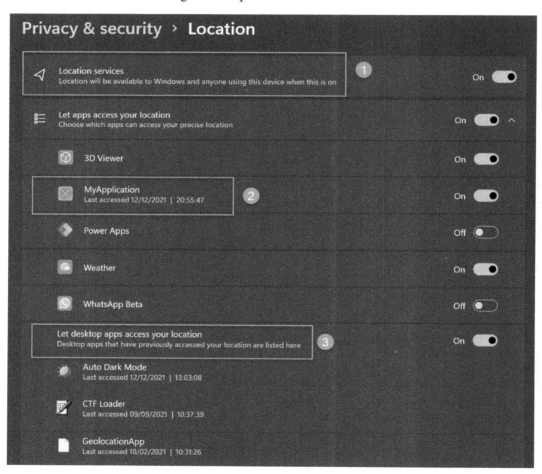

Figure 8.2 – The location services configuration included in Windows 11

Now that we know how Windows manages the security and privacy of location services, we can look at a better way to use the Geolocation APIs:

```
private async void OnGetPosition(object sender,
  RoutedEventArgs e)
{
    Geolocator geolocator = new Geolocator();
    if (geolocator.LocationStatus != PositionStatus.Disabled &&
      geolocator.LocationStatus != PositionStatus.NotAvailable)


    {
        Geoposition position = await geolocator.
          GetGeopositionAsync();
        Console.WriteLine($"{position.Coordinate.Point.
          Position.Latitude} - {position.Coordinate.
            Point.Position.Longitude}");
    }
}
```

The status of the location services is stored inside the `LocationStatus` of the `Geolocator` property, through the `PositionStatus` enumerator. We get the position of the user only if this status is different from `Disabled` or `NotAvailable`. It's very important to include this check, otherwise, you will get an exception if you try to invoke the `GetGeopositionAsync()` method when the location services are disabled.

The `Geolocator` class also supports the option to continuously detect changes in the position of the user, as GPS navigation apps do. The following code shows this feature in action:

```
private async void OnGetPositionChanges(object sender,
  RoutedEventArgs e)
{
    Geolocator geolocator = new Geolocator();
    if (geolocator.LocationStatus != PositionStatus.Disabled &&
      geolocator.LocationStatus != PositionStatus.NotAvailable)
    {
        Console.WriteLine(geolocator.LocationStatus);
        geolocator.PositionChanged += Geolocator_
          PositionChanged;
```

```
        }
    }

private void Geolocator_PositionChanged(Geolocator sender,
    PositionChangedEventArgs args)
{
    Console.WriteLine($"{args.Position.Coordinate.Latitude} -
        {args.Position.Coordinate.Longitude}");
}
```

All you have to do is to subscribe to the `PositionChanged` event, which, in the event arguments, gives you a `Position` object with all the information about the location change that has just been detected.

Now we know how to use the Geolocator API and how to retrieve the latitude and longitude of the current position. What can we do with this information? Let's see what we can achieve thanks to the Bing Maps service.

Taking advantage of the Bing Maps service

Bing Maps isn't just a mapping service for the consumer, it's also a set of APIs and controls that you can use in your applications to enhance them. Some of these APIs are built inside the UWP, so you can use them without having to manually perform network operations and parse JSON payloads. In this section, we're going to see how we can use these services to turn our coordinates into a human-readable address.

The first step is to register an application on the Bing Maps portal, which is available at `https://www.bingmapsportal.com/`.

After you have logged in with your Microsoft account, click on the label **Click here to create a new key** to access the registration form. You will be asked the following questions – make sure to set the fields in the same way as in the following screenshot:

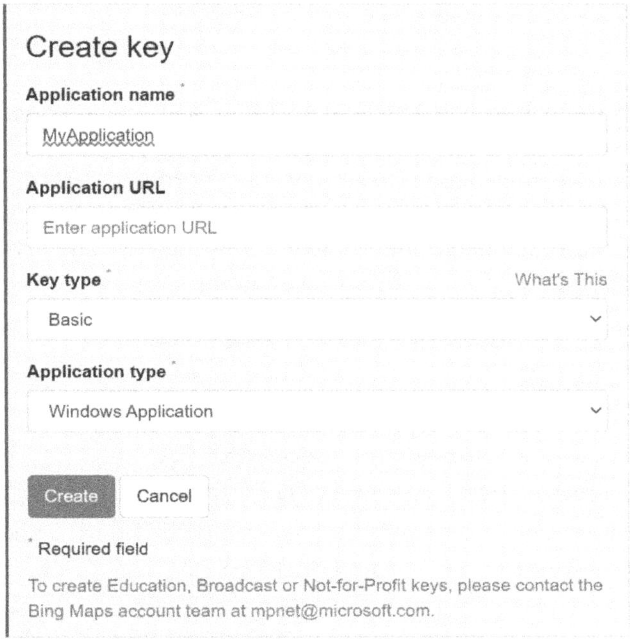

Figure 8.3 – The form to register a new Bing Maps key

After you have clicked the **Create** button, the key will show up in the following dashboard:

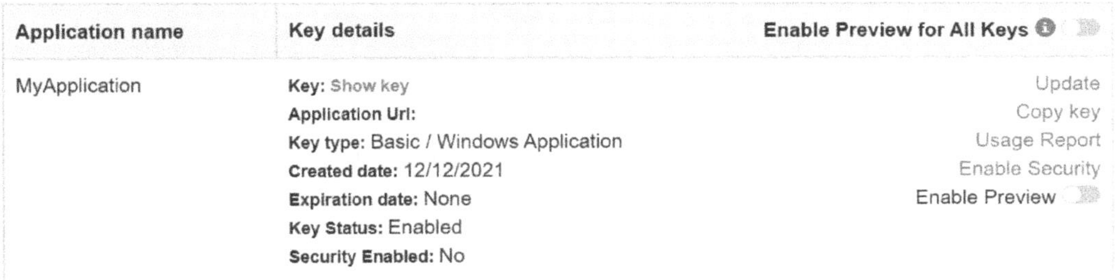

Figure 8.4 – A Bing Maps key in the dashboard

For security reasons, by default, the key won't be visible. You can either click on the **Show key** button and copy it, or directly click on the **Copy key** button. Either way, once the key is stored on your clipboard, you can go back to your application and set it inside the `ServiceToken` property of the `MapService` class (which belongs to the `Windows.Services.Maps` namespace), as in the following example:

```
public MainPage()
{
    this.InitializeComponent();
```

```
    MapService.ServiceToken = "<your key>";
}
```

Now, thanks to this key, you'll be able to use many other APIs included in the `Windows.Services.Maps` namespace. Let's see the usage of the `MapLocationFinder` class, which you can use to perform geocode (to retrieve the coordinates starting from an address) and reserve geocode (to retrieve the address starting from the coordinates) operations. In our scenario, thanks to the `Geolocator` class, we already have a set of coordinates, so we're going to perform reverse geocoding:

```
private async void myButton_Click(object sender,
   RoutedEventArgs e)
{
    Geolocator geolocator = new Geolocator();
    if (geolocator.LocationStatus !=
      PositionStatus.Disabled && geolocator.LocationStatus
        != PositionStatus.NotAvailable)
    {
        Geoposition position = await
          geolocator.GetGeopositionAsync();

        MapLocationFinderResult result =
        await MapLocationFinder.FindLocationsAtAsync(position.
          Coordinate.Point);

        Console.WriteLine(result.Locations[0].DisplayName);
    }
}
```

The `Geolocator` class is a great companion for the `MapLocationFinder` one since they use the same types to manage location data. As such, to perform reverse geocoding, it's enough to call the `MapLocationFinder.FindLocationsAtAsync()` method passing, as a parameter, the `Coordinate.Point` property of the `Geoposition` object returned by the `Geolocator` class.

When you do reverse geocoding, the result contains a property called `Locations`: it's a collection since the APIs can return multiple places for the same coordinates. In this example, we're showing the information about the first result. The `DisplayName` property contains a human-readable version of the full address, but you can also access more granular properties through the `Address` property, such as city, postcode, region, and so on.

> **Note**
>
> The UWP includes a control called `MapControl`, which is a great companion of the location APIs since you can use it to display the location of the user on a map, show the points of interest, calculate a path, and so on. Unfortunately, at the time of writing, `MapControl` hasn't been ported to WinUI and the Windows App SDK. It will be included in a future release.

Let's see another example of the integration of APIs from the UWP: Windows Hello.

Introducing biometric authentication

Windows includes a security platform called Microsoft Passport, which provides a safe way to store sensitive information such as your account credentials. One of the most interesting features of Microsoft Passport is Windows Hello, which is able to generate a token to sign in based on biometric parameters, such as face recognition or fingerprint readings. Windows Hello makes logging in to your computer safer in the following ways:

- It removes the possibility of credential theft. Since authentication happens with a two-factor authentication system based on a biometric component or on a PIN, you remove the requirement of using a password, which is usually the weak link in the security chain.

- The authentication system is based on a component that is only local and stored in a safe way, thanks to the usage of the **TPM** (short for **Trusted Platform Module**) chip, which enables hardware protection. Being only local means that the attacker would need physical access to your computer to steal your data, while with a regular password an attacker is normally able to access your data even remotely.

To work properly, Windows Hello requires your Windows account to be connected to a Microsoft account (in a consumer scenario) or to an Azure Active Directory account (in an enterprise scenario). Either way, when you set up Windows Hello, the operating system generates a public-private key pair on the device; the private key is generated and stored by the TPM, making it virtually impossible to access it. The keys are bound to a specific user so, if you have multiple users on the same machine, each of them will need to set up Windows Hello and the TPM still stores their own public-private key pair.

Windows Hello, from a developer's point of view, can be used for many advanced scenarios. For example, you can create a specific application key for the user, which gives you access to information such as the public key or the attestation. This way, you can link this data to a user profile, enabling you to use Windows Hello as an authentication system also for server-side scenarios (such as a backend API).

For the purpose of this book and to show you the integration with the Windows ecosystem, instead, we'll just focus on the simplest scenario: logging the user into an application. In this scenario, you typically authenticate the user first with a traditional system (such as a username and a password) and then you give them the option, after the first successful login, to enable Windows Hello to avoid inserting the credentials each time. In this scenario, you don't need to store any information on the backend, since you're using Windows Hello only as a local authentication system. If the same user needs to log in on another machine, they will need to repeat the process starting with their credentials.

This scenario is easy to implement, thanks to the APIs that are included in the `Windows.Security.Credentials` and `Windows.Security.Credentials.UI` namespaces:

```
private async Task AuthenticateUserAsync()
{
    bool keyCredentialAvailable = await KeyCredentialManager.
      IsSupportedAsync();
    if (keyCredentialAvailable)
    {
        var result = await UserConsentVerifier.
          RequestVerificationAsync("Checking if it's
            really you");
        if (result == UserConsentVerificationResult.Verified)
        {
            //continue the operation
        }
        else
        {
            //access is denied
        }
    }
}
```

As a first step, we must check if Windows Hello is enabled on the machine; otherwise, we need to fall back to a traditional authentication system. This information is returned by the `IsSupportedAsync()` method exposed by the `KeyCredentialManager` class, which returns a simple `true` / `false`.

If Windows Hello is available, we can proceed and trigger a user verification, by calling the `RequestVerificationAsync()` method of the `UserConsentVerifier` class. As a parameter, we pass a message that we want to display inside the authentication popup, as you can see in the following screenshot:

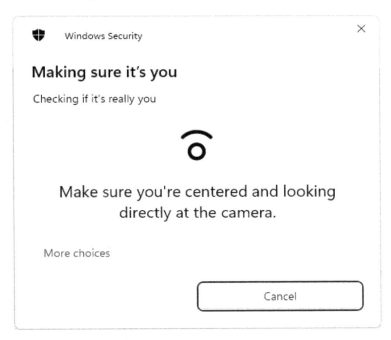

Figure 8.4 – The Windows Hello popup with a custom message

As an outcome, you will get a value of the `UserConsentVerificationResult` enumerator. If it's equal to **Verified**, it means that the authentication has been completed successfully, so we can move on with the operation we are protecting. Otherwise, we block the execution since we haven't been able to properly authenticate the user.

Windows Hello can be used for more advanced scenarios, including server-side ones. The key class to support them is `KeyCredentialManager`, which we have already seen, which offers you methods such as `RequestCreateAsync()`, which you can use to generate a private-public key pair for the user. The following example shows this scenario:

```
private async Task AuthenticateUserAsync()
{
    bool keyCredentialAvailable = await KeyCredentialManager.
        IsSupportedAsync();
    if (keyCredentialAvailable)
```

```
        {
            var keys = await KeyCredentialManager.
              RequestCreateAsync("username",
                KeyCredentialCreationOption.ReplaceExisting);
            if (keys.Status == KeyCredentialStatus.Success)
    {
            var publicKey = keys.Credential.RetrievePublicKey();
        var attestation = await keys.Credential.
          GetAttestationAsync();
    }
        }
    }
```

We pass to the `RequestCreateAsync()` method a unique identifier for the user (in the preceding code, it's a fixed string named `username`, but it could be the user email address or account identifier). If the operation completes successfully, we can use the `Credential` property of the returned object to perform tasks such as retrieving the public key or the attestation, so that we can send it to our backend.

As mentioned at the beginning of the section, we aren't going to look at this scenario in detail in this book. If you want to know more, you can read the official documentation at `https://docs.microsoft.com/en-us/windows/uwp/security/microsoft-passport`, which also contains a step-by-step tutorial on how to build a client-server architecture based on Windows Hello.

Thanks to the integration with the UWP, we also have access to a new set of APIs to work with files and folders. Let's take a deeper look!

Working with files and folders

One of the most common requirements of a desktop application is to work with files and folders to support a wide range of scenarios: from writing and reading a file to accessing the content of a folder. .NET includes a namespace dedicated to this scenario, called **System.IO**, which includes many APIs to work with files, directories, and streams. However, we won't cover them in this chapter, because these APIs have been mostly unchanged over the years and, in this book, we're assuming you have a basic knowledge of the .NET ecosystem and Windows desktop development.

The UWP, instead, has introduced a new set of APIs to work with files and folders that are especially important if you have adopted WinUI as a platform. The classic .NET APIs will continue to work but there are some scenarios (such as enabling users to pick a file from a folder) that will require you to use these new APIs. Let's explore them!

Working with folders

Folders are represented by the `StorageFolder` class, which belongs to the `Windows.Storage` namespace. Thanks to this class, you can perform common operations such as getting all the files inside a folder, creating a subfolder, creating a file, and so on.

The `StorageFolder` class exposes a static method that is particularly useful with the Windows App SDK called `GetFolderFromPathAsync()`. This API represents one of the key differences between UWP applications and Windows App SDK applications: the first run inside a sandbox, so we can't freely access any file or folder inside the system; the latter, instead, run in full thrust and, as such, we can use this API to convert any path into a `StorageFolder` object, as in the following sample:

```
private async Task GetFolderAsync()
{
    var path = Environment.GetFolderPath(Environment.
      SpecialFolder.LocalApplicationData);
    var folder = await StorageFolder.
      GetFolderFromPathAsync(path);
    Console.WriteLine(folder.Path);
}
```

In this example, we are combining the .NET APIs (the ones exposed by the `Environment` class) with the Universal Platform APIs to get a `StorageFolder` object that maps the local application data folder of the user.

We can use the `StorageFolder` object to work with every sub-folder of the current folder, either by opening a specific one (using the `GetFolderAsync()` method passing, as a parameter, the folder name) or by querying all the available sub-folders (using the `GetFoldersAsync()` method). The following example expands the previous snippet by retrieving a list of all the sub-folders of the local application data folder and printing their paths on the console:

```
private async Task GetFolderAsync()
{
    var path = Environment.GetFolderPath
```

```
            (Environment.SpecialFolder.LocalApplicationData);
        var folder = await StorageFolder
          .GetFolderFromPathAsync(path);
        var folders = await folder.GetFoldersAsync();
        foreach (var folderItem in folders)
        {
            Console.WriteLine(folderItem.Path);
        }
    }
```

However, once we have a `StorageFolder` object, the most interesting opportunity is to work with files that belong to the folder.

Working with files

The `StorageFolder` object, in a similar way we have seen for folders, exposes methods to get access to the stored files, either directly (by calling `GetFileAsync()` and passing the name) or by querying all the files (using the `GetFilesAsync()` method). Either way, files are represented by another class of the UWP called `StorageFile`. The following sample shows how to use the second option to list all the files that are stored inside the local application data folder and to print, on the console, their full paths:

```
private async Task GetFilesAsync()
{
    var path = Environment.GetFolderPath(Environment.
      SpecialFolder.LocalApplicationData);
    var folder = await StorageFolder.
      GetFolderFromPathAsync(path);
    var files = await folder.GetFilesAsync();
    foreach (StorageFile file in files)
    {
        Console.WriteLine(file.Path);
    }
}
```

Since we are working on a Win32 application, we can also directly convert a file path to a `StorageFile` object, as we have seen with the `StorageFolder` class:

```
private async Task GetFilesFromPathAsync()
{
```

```
    var path = Environment.GetFolderPath(Environment.
      SpecialFolder.LocalApplicationData);
    string filePath = @$"{path}\MyFile.txt";
    StorageFile file = await StorageFile.
      GetFileFromPathAsync(filePath);
}
```

Another way to get a reference to a file is to create one by using the `CreateFileAsync()` method passing, as a parameter, the name of the file and, optionally, the behavior you want to adopt if the file already exists, as in the following example:

```
private async Task CreateFileAsync()
{
    var path = Environment.GetFolderPath
      (Environment.SpecialFolder.LocalApplicationData);
    var folder = await StorageFolder
      .GetFolderFromPathAsync(path);

    StorageFile file = await folder.CreateFileAsync("file.txt",
      CreationCollisionOption.ReplaceExisting);

}
```

Once you have a `StorageFile` object, you have access to multiple asynchronous APIs to perform basic operations on files like the following:

- Using `CopyAsync()` and `CopyAndReplaceAsync()` to copy the file to another location. You must pass, as a parameter, a `StorageFolder` object that maps the folder you want to use as the destination.

- Using `MoveAsync()` and `MoveAndReplaceAsync()` to move the file to another location. Also, in this case, you must pass as a parameter a `StorageFolder` object that maps the folder to use as the destination. Optionally, you can also pass a new name for the file.

- Using `RenameAsync()` to rename the file, passing as a parameter the new name.

- Using `DeleteAsync()` to delete the file.

One of the most common operations when you work with files is treating them as streams so that you can read their content or write data inside them. The UWP uses an interface called `IRandomAccessStream()` to manipulate streams, which you can obtain by calling the `OpenAsync()` method exposed by the `StorageFile` class. The following sample shows how you can use this type of stream to write some content into a text file:

```
private async Task WriteFileAsync()
{
    var path = Environment.GetFolderPath(Environment.
        SpecialFolder.LocalApplicationData);
    var folder = await StorageFolder.
        GetFolderFromPathAsync(path);

    StorageFile file = await folder.CreateFileAsync("file.txt",
        CreationCollisionOption.ReplaceExisting);

    IRandomAccessStream randomAccessStream = await file.
        OpenAsync(FileAccessMode.ReadWrite);

    using (DataWriter writer = new
        DataWriter(randomAccessStream.GetOutputStreamAt(0)))
    {
        writer.WriteString("Hello world!");
        await writer.StoreAsync();
    }
}
```

We open the file in write mode (by passing, as a parameter of the `OpenAsync()` method, the value `ReadWrite` of the `FileAccess` enumerator) and then we use the `DataWriter` class as a wrapper for the stream (starting from the beginning of the file). Thanks to this class, we can write some text inside. The example shows the usage of the `WriteString()` method, but the `DataWriter` class exposes many other methods for every other data type, such as `WriteBytes()`, `WriteBoolean()`, `WriteTimeSpan()`, and many more.

If instead, you prefer to work with the traditional `Stream` class exposed by .NET (which belongs to the `System.IO` namespace), you can use one of the two available additional methods, based on the type of operation you want to perform: `OpenStreamForReadAsync()` and `OpenStreamForWriteASync()`. The following snippet shows the same writing example, but implemented using the .NET streams:

```
private async Task WriteFileAsync()
{
        var path = Environment.GetFolderPath(Environment.
          SpecialFolder.LocalApplicationData);
    var folder = await StorageFolder.
      GetFolderFromPathAsync(path);
    StorageFile file = await folder.CreateFileAsync("file.txt",
      CreationCollisionOption.ReplaceExisting);
    StorageFile file = await StorageFile.
      GetFileFromPathAsync(filePath);
    using (var stream = await file.OpenStreamForWriteAsync())
    {
        StreamWriter writer = new StreamWriter(stream);
        writer.WriteLine("Hello world");
        await writer.FlushAsync();
    }
}
```

In some cases, however, you won't need to work with streams to read and write data to a file, thanks to a series of helpers that are included directly in the UWP or in the Windows Community Toolkit. Let's start with the first scenario.

The `Windows.Storage` namespace includes a class called `FileIO`, which exposes a series of static methods to read and write the most common data types to a file, like the following:

- `ReadTextAsync()` and `WriteTextAsync()`, to read and write text content.

- `WriteBuffer()` and `ReadBuffer()` to read and write binary data.

- `AppendTextAsync()` to append text to an existing file.

All these methods work with a `StorageFile` object. The following sample shows how the previous writing sample can be simplified by using the `FileIO` class instead of streams:

```
private async Task WriteFileAsync()
{
    var path = Environment.GetFolderPath(Environment.
      SpecialFolder.LocalApplicationData);
    var folder = await StorageFolder.
      GetFolderFromPathAsync(path);
    StorageFile file = await folder.CreateFileAsync("file.txt",
      CreationCollisionOption.ReplaceExisting);
    //write the content to the file
    await FileIO.WriteTextAsync(file, "Hello world");
    //read the content from the file
    string content = await FileIO.ReadTextAsync(file);
}
```

Let's see now, instead, how the Windows Community Toolkit can help you to improve working with files and folders. As the first step, you must install the `CommunityToolkit.WinUI` package in your project from NuGet. Please note that if you have a WPF or Windows Forms application, this will introduce a dependency to the Windows App SDK, which wasn't required to perform all the other operations we have seen so far.

A scenario that makes the usage of the toolkit a great help is downloading a file from the network into your application. This scenario, typically, involves the usage of two chained operations: using the `HttpClient` class to download the stream of the file and then using the storage APIs to save it into a file. Thanks to the `StreamHelper` class, which belongs to the `CommunityToolkit.WinUI.Helpers` namespace, we can instead easily perform the task with a single line of code:

```
private async Task WriteFileAsync()
{
    var path = Environment.GetFolderPath(Environment.
      SpecialFolder.LocalApplicationData);
    var folder = await StorageFolder.
      GetFolderFromPathAsync(path);
    StorageFile file = await folder.CreateFileAsync("image.
      jpg", CreationCollisionOption.ReplaceExisting);
    await StreamHelper.GetHttpStreamToStorageFileAsync(new
```

```
                Uri("https://www.mywebsite.com/image.jpg"), file);
}
```

We use the `GetHttpStreamToStorageFileAsync()` method, passing as parameters the URL of the file and the `StorageFile` object where we want to save the downloaded content.

If you need to work with the content of the file directly in memory, instead, you can use the `GettHttpStreamAsync()` method, which will return an `IRandomAccessStream` object.

Another important scenario that isn't supported by the native APIs, but is provided by the toolkit, is checking whether the file already exists. Once you add the `CommunityToolkit.WinUI.Helpers` namespace to your file, the `StorageFile` class will expose a new extension method called `FileExistsAsync()`, which you can use as follows:

```
private async Task WriteFileAsync()
{
    var path = Environment.GetFolderPath
      (Environment.SpecialFolder.LocalApplicationData);
    var folder = await StorageFolder
      .GetFolderFromPathAsync(path);
    bool isExisting = await folder.FileExistsAsync("image.
      jpg");
    if (!isExisting)
    {
        StorageFile file = await folder.CreateFileAsync
          ("image.jpg", CreationCollisionOption
            .ReplaceExisting);
        await StreamHelper.GetHttpStreamToStorageFileAsync
          (new Uri("https://www.mywebsite.com/
            image.jpg"), file);
    }
}
```

In this example, we download an image from the web only if a file named `image.jpg` doesn't already exist in the local application data.

The Windows Community Toolkit also gives you some other interesting helpers, but before talking about them, we need to introduce a new concept: local storage.

Using the local storage in packaged apps

When you package your application with MSIX, you get an identity, which gives you access to a broader set of features and integration with the Windows ecosystem. One of them is access to special storage, which is local and belongs only to the application itself. This storage is created in the %LOCALAPPDATA%\Packages folder. Each packaged application will have its own sub-folder with a name something like the **Package Family Name** of the application, which is the unique identifier assigned by Windows to packaged apps. This means that, if you try to use this special storage in an unpackaged application, you will get an exception.

The advantage of this storage is that it follows the same life cycle as your application, making it a great fit for storing all the files and folders that it wouldn't make sense to keep once the application is removed: configuration files, logs, local databases, and so on. When the application is uninstalled, Windows will also take care of removing the local storage.

This storage can be accessed using the Windows.Storage.ApplicationData class, which exposes a static instance called Current. Windows supports several types of local storage, which are mapped with different sub-folders. Currently, the two relevant types are the following:

- Local, which is exposed through the ApplicationData.Current. LocalFolder class. This is a general-purpose folder to store any kind of file or folder that belongs to your application and that must stay available until the application is uninstalled.

- Temporary, which is exposed through the ApplicatonData.Current. TemporaryFolder class. This folder is a great fit for all the types of data that can be temporary, such as images that are cached to save bandwidth. When Windows is running low on disk space (or when the user performs a disk cleanup using the integrated Windows tool), it will clean up the temporary storage of all the applications. As you can imagine, this isn't the right place to put critical data.

Both these folders are mapped using the StorageFolder class, which means that you can use all the APIs we have seen so far to work with them. The following example shows the creation of a text file inside the local storage of the application:

```
private async Task WriteFileAsync()
{
    StorageFile file = await ApplicationData.Current.
      LocalFolder.CreateFileAsync("file.txt",
        CreationCollisionOption.ReplaceExisting);
    await FileIO.WriteTextAsync(file, "Hello world");
}
```

When you decide to use this special type of storage, the Windows Community Toolkit can help you with various methods exposed by the `StorageFileHelper` class. For example, the previous example can be simplified with a single line of code:

```
private async Task WriteFileAsync()
{
    //write a text file in the local storage
    await StorageFileHelper.WriteTextToLocalFileAsync
       ("Hello world", "file.txt", CreationCollisionOption
          ReplaceExisting);
    //write a text file in the temporary storage
    await StorageFileHelper.WriteTextToLocalCacheFileAsync
       ("Hello world", "file.txt", CreationCollisionOption
          .ReplaceExisting);
}
```

As you can see, the `StorageFileHelper` class exposes methods that you can use to directly write data to specific storage. In the previous example, we're using `WriteTextToLocalFileAsync()` to write text content directly into a file in the local storage, while `WriteTextToLocalCacheFileAsync()` does the same, but on a file in the temporary storage.

If instead, you need to create files that should survive the uninstallation of your application (for example, when you uninstall Office, Windows doesn't delete all your existing Word documents and Excel spreadsheets), you should use other system folders, such as the `Documents` library. Since your application is based on the Win32 ecosystem, the APIs we have seen so far don't have the same limitations as the sandbox used by UWP apps anymore. The Windows Community Toolkit can help you to simplify these scenarios as well, thanks to the other methods offered by the `StorageFileHelper` class. Consider the following sample:

```
private async Task WriteFileAsync()
{
    string path = Environment.GetFolderPath(Environment.
       SpecialFolder.MyDocuments);
    var folder = await StorageFolder.
       GetFolderFromPathAsync(path);
    await StorageFileHelper.WriteTextToFileAsync(folder, "Hello
       world", "File.txt");
}
```

We are creating a text file inside the `Documents` folder of the user, retrieved using the `Environment` class offered by .NET. Thanks to the `StorageFileHelper` class, we don't have to create the file first and then write the text, but we can do it with one line of code by using the `WriteTextToFileAsync()` method, passing as parameters, the `StorageFolder` object where we want to create the file, the text, and the filename. All these samples are based on text files but, of course, you can store binary data (such as an image) using the equivalent methods prefixed by `WriteBytes` (such as `WriteBytesToFileAsync()` or `WriteBytesToLocalFileAsync()`).

When your application is packaged, you also get the choice to use a special application URI to identify a file:

- If you want to reference a file inside the package, you can use the `ms-appx` protocol. For example, let's say that in your Visual Studio project, you have a configuration file called `config.json`, placed inside the root. You can access this file using the URI `ms-appx:///config.json`.

- If you want to reference a file inside the local storage, you can use the `ms-appdata` protocol. For example, if the same `config.json` file is stored inside the root of the local storage, you can reference it with the URI `ms-appdata:///local/config.json`.

The `StorageFile` class offers a static method to retrieve a reference to a file starting from the application URI called `GetFileFromApplicationUriAsync()`, which you can use as in the following sample:

```
StorageFile file = await StorageFile.GetFileFromApplicationUriA
sync("ms-appx:///config.json");
```

In the next section, we're going to see another useful feature provided by local storage: the ability to store settings.

Using local storage to store settings

Another common requirement for applications when it comes to working with storage is to manage settings. It's very likely, in fact, that your application enables your users to customize one or more options. If your application is packaged, local storage gives you an excellent way to store settings, since it removes all the complexity of having to support a complex data structure, such as a database or a JSON file.

Local settings, in fact, can be easily managed through a dictionary, which stores key-value pairs. The following example shows how we can use this feature to store the theme selected by the user:

```
private void OnWriteSettings()
{
    ApplicationData.Current.LocalSettings.Values["Theme"] =
      "Dark";
    var value = ApplicationData.Current.LocalSettings.
      Values["Theme"].ToString();
}
```

The collection of settings is exposed by the `ApplicationData.Current. LocalSettings.Values` property. In the first line of code, we create a new item with `Theme` as a key and `Dark` as a value. In the next line, we do the opposite operation: we retrieve the value of the item identified by the `Theme` key and we store it in a variable.

As you will notice, when you work with the local settings, you must manually manage the casting, since the `Values` collection can store any generic object. To simplify the code (and make it more readable), in this case, you can also use a helper provided by the Windows Community Toolkit called `ApplicationDataStorageHelper`. Let's see how we can simplify the previous code:

```
private void OnWriteSettings()
{
    var helper = ApplicationDataStorageHelper.GetCurrent();
    //write the setting
    helper.Save<string>("Theme", "Dark");
    //read the setting
    var value = helper.Read<string>("Theme");
}
```

We get a reference to the `ApplicationDataStorageHelper` object by calling the `GetCurrent()` method. Then, we can save and read settings by using the `Save()` and `Read()` methods. Both methods support generics, so you don't have to manually cast the data type anymore. In the previous example, since the information about the theme is a string, we just use `Read<string>()` to directly get back a string object.

So far, we have seen the options we have to work with files in a programmatic way, without needing any user intervention. However, there are scenarios where we must ask for input from the user, such as the location where they want to save a file or to choose a file to import into the application. In the next section, we'll learn how to manage this scenario with pickers.

Working with file pickers

File pickers are important in applications since they provide a way for the user to load or save files in a specific location. When you use a picker, the application will open a dialog that enables users to explore their hard disk and look for a file to import into the app or a location to save a file created by the app. WPF and Windows Forms have their own classes to support file pickers. For example, in WPF you use the OpenFileDialog class. But what about WinUI? In this case, you can use the picker APIs that belong to the Windows.Storage.Pickers namespace. Let's see how to implement both open and save scenarios.

Using a picker to open a file

To import a file into your application, you can use the FileOpenPicker class. The usage compared to the UWP is a bit different since, in the case of a Win32 app, we must manually link the picker to our current window, through its native handle (HWND):

Here is an example of using the FileOpenPicker class in a WinUI application:

```
private async void OnPickFile(object sender, RoutedEventArgs e)
{
    var filePicker = new FileOpenPicker();

    // Get the current window's HWND by passing in the Window
       object
    var hwnd = WinRT.Interop.WindowNative.
       GetWindowHandle(this);

    // Associate e HWND with the file picker
    WinRT.Interop.InitializeWithWindow.Initialize(filePicker,
       hwnd);

    filePicker.FileTypeFilter.Add("*");
    StorageFile file = await filePicker.PickSingleFileAsync();
```

```
    if (file != null)
{
        Console.WriteLine(file.Path);
}
```

First, you retrieve the HWND using the `WinRT.Interop.WindowNative.GetWindowHandle()` method. In this example, we're assuming the code is running directly inside a `Window`, so we can pass this as a parameter. If you are executing this code in another class (for example, a page or a helper), you must pass a reference to the `Window` object you have declared in the `App` class.

Other than that, the usage of the class is straightforward:

1. First, you use the `FileTypeFilter` collection to add all the extensions you want to support. The dialog will display only files that respect these criteria. In the example, we're passing a wildcard (*), which means that we're going to display all the files.

2. Then, you call the `PickSingleFileAsync()` method, which will return a `StorageFile` object. If you need to open multiple files at once, you can use the `PickMultipleFilesAsync()` method, which, in this case, will return a collection of `StorageFile` objects, one for each selected file.

Once you have a reference to the `StorageFile` object (or the objects, if you picked multiple files), you can use the APIs we have seen in the other sections of this chapter to perform further operations, such as reading its content. In the example, we're just printing on the console output the full file path.

The previous example is based on a WinUI application, but the same API can also be used in other platforms. The only difference is the way you retrieve the HWND, since `WinRT.Interop.WindowNative.GetWindowHandle()` is a specific WinUI API. In WPF, for example, you would use the following code:

```
private async void OnPickFile(object sender, RoutedEventArgs e)
{
    var filePicker = new FileOpenPicker();
    var hwnd = new WindowInteropHelper(this).EnsureHandle();
    WinRT.Interop.InitializeWithWindow.Initialize(filePicker,
        hwnd);
    filePicker.FileTypeFilter.Add("*");
    var file = await filePicker.PickSingleFileAsync();
    if (file != null)
```

```
{
        Console.WriteLine(file.Path);
}
```

As you will notice, everything is the same except for the highlighted line: in WPF, you use the `EnsureHandle()` method of the `WindowInteropHelper` class to retrieve the HWND of the window passed as a parameter.

The UWP also has an equivalent API for folders called `FolderPicker`, which exposes the `PickSingleFolderAsync()` method. The API works in the same way, with only the following differences:

- You don't need to specify the `FileTypeFilter` collection since folders don't have a type.

- The object type you get in return is `StorageFolder` instead of `StorageFile`.

Here is an example:

```
private async void OnPickFolder(object sender, RoutedEventArgs
    e)
{
    var folderPicker = new FolderPicker();

    var hwnd = WinRT.Interop.WindowNative.
      GetWindowHandle(this);
    WinRT.Interop.InitializeWithWindow.Initialize(folderPicker,
      hwnd);

    StorageFolder folder = await folderPicker.
      PickSingleFolderAsync();

    if (file != null)
    {
        Console.WriteLine(file.Path);
}
```

Once you have a `StorageFolder` object, you can perform all the operations we have seen in this chapter, such as listing the sub-folders or the included files or creating a new file.

Let's see now how we can use a picker to save a file, instead.

Using a picker to save a file

With a picker, you can enable users to choose where to save a file with content generated by your application. The class to use is called `FileSavePicker` and its usage is similar to the `FileOpenPicker` one we have just learned how to use. The only difference is that this time, the picker will take care of creating a new file and it will give us back a reference to it with a `StorageFile` object, which we can use to save the content.

Let's take a look at the following example:

```csharp
private async void OnSaveFle(object sender, RoutedEventArgs e)
{
    var filePicker = new FileSavePicker();

    var hwnd = WinRT.Interop.WindowNative.
      GetWindowHandle(this);

    WinRT.Interop.InitializeWithWindow.Initialize(filePicker,
      hwnd);

    filePicker.SuggestedFileName = "file.txt";
    filePicker.FileTypeChoices.Add("Text", new List<string>() {
      ".txt" });
    StorageFile file = await filePicker.PickSaveFileAsync();

    await FileIO.WriteTextAsync(file, "Hello world");
}
```

As for `FileOpenPicker`, in this case, we also have to first retrieve the native handle of the window and use it to initialize a new `FileSavePicker` object. The `FileSavePicker` class offers a few properties that we can use to customize the experience, such as `SuggestedFileName` to set the default name of the file. However, the critical one (without setting it, the picker will return an exception) is `FileTypeChoices`, which is a collection of the file types that your application can generate. In this example, we're assuming that the application can generate text files. In the end, we call `PickSaveFileAsync()`, which will trigger the Windows dialog. Once we have chosen the location to save the file and a name, we will get back a `StorageFile` object that we can fill with the content. In this example, we use the `FileIO` APIs to store sample text.

With this, we have completed our journey into exploring the storage APIs offered by the UWP, which can be a great companion for the .NET APIs, especially in WinUI applications. Now we can explore another option offered by Windows integration: sharing data across multiple applications.

Supporting the sharing contract

Sharing is one of the most common actions you perform on your computer. You don't always realize it because you do it multiple times per day, but when you copy some content and you paste it into another application, you're effectively sharing data across two different applications. However, there might be scenarios where a copy and paste might be fully effective: for example, you're copying a URL or some other complex data and you would like the other application to understand it and treat it in the proper way.

To make this scenario more effective, Windows 10 has introduced a sharing contract, which acts as a standard way to share specific types of data (such as images, texts, and links) across different applications. Being a contract, it means that the two applications don't need to know each other:

- The source application will use the sharing APIs to share data, using one of the specific formats supported by the contract.
- The target application will register itself in the system as eligible to receive one or more types of data from the sharing contract.

Let's build a suite of applications to implement this scenario: a source one and a target one. We're going to use a WPF application based on .NET 6.0, to show you that you don't need to move to WinUI as a UI layer to leverage this feature. However, the same exact code will work also in a Windows Forms or WinUI application.

Let's start!

Building the source app

The source app is the simplest one to build since it doesn't need any special dependency. The only requirement is that, like for the other samples we have seen in this chapter, we must set `TargetFramework` of our WPF application a specific Windows 10 or 11 one, since the sharing contract is part of the UWP. As such, make sure that the `TargetFramework` property of your WPF (or Windows Forms project) looks like this:

```
<TargetFramework>net6.0-windows10.0.19041</TargetFramework>
```

The next step is to create a helper class, which will enable us to use the sharing APIs from a Win32 application. If you already have experience building UWP applications, you will realize this is a different experience. This is what the helper class looks like:

```
public static class DataTransferManagerHelper
{
    static readonly Guid _dtm_iid = new Guid(0xa5caee9b,
        0x8708, 0x49d1, 0x8d, 0x36, 0x67, 0xd2, 0x5a, 0x8d,
        0xa0, 0x0c);

    static IDataTransferManagerInterop
      DataTransferManagerInterop => DataTransferManager
        .As<IDataTransferManagerInterop>();

    public static DataTransferManager GetForWindow(IntPtr hwnd)
    {
        IntPtr result;
        result = DataTransferManagerInterop.GetForWindow(hwnd,
            _dtm_iid);
        DataTransferManager dataTransferManager =
          MarshalInterface<DataTransferManager>
            .FromAbi(result);
        return (dataTransferManager);
    }

    public static void ShowShareUIForWindow(IntPtr hwnd)
    {
        DataTransferManagerInterop.ShowShareUIForWindow(hwnd);
    }

    [ComImport]
    [Guid("3A3DCD6C-3EAB-43DC-BCDE-45671CE800C8")]
    [InterfaceType(ComInterfaceType.InterfaceIsIUnknown)]
    public interface IDataTransferManagerInterop
    {
        IntPtr GetForWindow([In] IntPtr appWindow, [In] ref
            Guid riid);
```

```
        void ShowShareUIForWindow(IntPtr appWindow);
    }
}
```

In a UWP application, this helper isn't required because the platform already provides the required hooks to access these APIs via Windows Runtime. We won't look at the class implementation in detail, since it's all boilerplate code required to import the proper COM interfaces and objects required by the sharing APIs.

What's interesting to see is how to use this API in our source application:

```
private void OnShareData(object sender, RoutedEventArgs e)
{
    var myHwnd = new WindowInteropHelper(this).EnsureHandle();
    var dataTransferManager = DataTransferManagerHelper.
      GetForWindow(myHwnd);
    dataTransferManager.DataRequested += (obj, args) =>
    {
        args.Request.Data.SetText("This is a shared text");
        args.Request.Data.Properties.Title = "Share Example";
        args.Request.Data.Properties.Description = "A
          demonstration on how to share";
    };

    DataTransferManagerHelper.ShowShareUIForWindow(myHwnd);
}
```

The way we trigger the preceding code is based on how and which kind of data we want to share from our application. In the previous example, we're assuming our application has a share button, which invokes the preceding event handler.

The first step is to retrieve the HWND, which is the native handler of the window. As we have learned in the section dedicated to file pickers, this is the only code that must be different based on the UI platform you're using: the preceding example, based on the WindowInteropHelper class, works with WPF and Windows Forms. If you're using WinUI, instead, you will need to use the following code to retrieve the HWND:

```
var myHwnd = WinRT.Interop.WindowNative.GetWindowHandle(this);
```

The rest of the code is the same regardless of the UI platform. First, we obtain a `DataTransferManager` object for the current window, using the `DataTransferManagerHelper` class we previously created. Then we subscribe to the `DataRequested` event, which is triggered when the sharing operation is invoked. Inside the event handler, we must supply to the request the data we want to share, through the `Request.Data` property exposed by the event arguments.

In the previous example, we're sharing some text using the `SetText()` method. Other supported options are `SetUri()` to share a link or `SetStorageItems()` to share a file (passing, as a parameter, a collection of `StorageFile` objects, which we have learned how to use in this chapter). We also customize the properties of the sharing operation, by setting its `Title` and `Description`.

In the end, we trigger the sharing UI by calling the `ShowShareUIForWindow()` method exposed by the `DataTransferManagerHelper` class passing, once again, the HWND of the current window, which we have previously retrieved.

Even if we haven't developed the target application, we can already see the sharing contract in action, since in Windows there are many pre-installed apps that can act as a target, such as **Mail**:

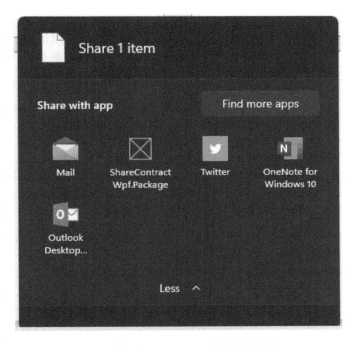

Figure 8.5 – The Share UI provided by Windows 11

Thanks to the Windows integration, you can also see a glimpse of how powerful this feature is. If you choose **Mail** as the target, for example, you can see how the information you have shared is automatically used to compose a new message: the title is set as a subject of the mail, while the text becomes the body of the mail.

Let's see now how we can build a similar experience.

Building the target app

Building the target app requires a bit more work since it's the one that will effectively share the content. If we take as an example the previous test we did, we realize that it's the Mail app that is in charge of the most complex work, since it needs to take the data shared by the source app, turn it into an email, and then provide a way for the user to send it.

> **Note**
>
> At the time of writing, acting as a share target is supported only by packaged apps. As such, if you're using WPF or Windows Forms, make sure to add the Windows Application Packaging Project to your solution; if you're using WinUI, instead, make sure to use the packaged model. If you're integrating the sharing contract into a WPF or Windows Forms application, you must also install the Windows App SDK NuGet package, as described in *Chapter 1, Getting Started with the Windows App SDK and WinUI*. We'll need to use the Activation APIs, which are included in the runtime, to manage the application activation from a sharing operation.

To recreate a similar experience in our application, as such, we need to have a dedicated window or page for the sharing operation. When the application is opened normally (by clicking on the icon in the **Start** menu, for example), the main page of the application will be displayed. When the application is opened through a sharing contract, instead, we'll redirect the user to a specific page of the application, which gives the user the option to share the content. The page could display a preview of the shared content; provide extra fields to add additional information; or have a button to complete the share operation.

For example, in our WPF application, we have created a window called `ShareWindow` dedicated to the sharing operation, which contains two `TextBlock` controls to display a preview of the sharing data and a button to perform the sharing:

```
<Window>
    <StackPanel>
        <TextBlock x:Name="txtTitle" />
        <TextBlock x:Name="txtSharedText" />
```

```
        <Button Content="Complete" Click="OnComplete"/>
    </StackPanel>
</Window>
```

The next step is to tweak our application's initialization code so that we can redirect the user to one page or another based on the scenario. In our example, since it's a WPF application, we can override the OnStartup() event of the App class. If it was a WinUI app, we would have leveraged the OnLaunched() event of the App class. Regardless of which is the entry point, however, the code we're going to execute is the same:

```
public partial class App : Application
{
    protected override void OnStartup(StartupEventArgs e)
    {
        Window Window;
        var instance = AppInstance.GetCurrent().
          GetActivatedEventArgs();
        if (instance.Kind == ExtendedActivationKind.
          ShareTarget)
        {
            window = new ShareWindow();
        }
        else
        {
            window = new MainWindow();
        }

        window.Show();
    }
}
```

Thanks to the AppInstance class, which comes from the Windows App SDK and belongs to the Microsoft.Windows.AppLifecycle namespace (we learned about it in *Chapter 4, The Windows App SDK for a UWP Developer*), we can identify the activation event that led our application to be opened by calling the GetActivatedEventArgs() method. In the response, we can use the Kind property to distinguish between different activation events through the ExtendedActivationKind enumerator. If the value is ShareTarget, we are in a sharing scenario: we redirect the user to new ShareWindow we have just created for this purpose; otherwise, we redirect the user to the standard MainWindow instance, which is the default one.

In a WinUI application, rather than opening different `Window` (which could lead to a few challenges, since WinUI doesn't support multiple windows at the time of writing), we could have leveraged same `MainWindow` for both scenarios and then, using the navigation techniques we learned in *Chapter 5*, *Designing Your Application*, we could have triggered navigation to a different page.

Now that the user is redirected to `ShareWindow`, let's see its implementation in code-behind:

```csharp
public partial class ShareWindow : Window
{
    private ShareOperation shareOperation;

    public ShareWindow()
    {
        InitializeComponent();
    }

    private async void Window_Loaded(object sender,
      RoutedEventArgs e)
    {
        var instance = AppInstance.GetCurrent()
          .GetActivatedEventArgs();
        if (instance.Kind ==
          ExtendedActivationKind.ShareTarget)
        {
            var args = instance.Data as
              IShareTargetActivatedEventArgs;
            shareOperation = args.ShareOperation;
            string text = await args.ShareOperation
              .Data.GetTextAsync();
            string title = args.ShareOperation
              .Data.Properties.Title;
            txtSharedText.Text = text;
            txtTitle.Text = title;
        }
    }
}
```

```
    private void OnComplete(object sender, RoutedEventArgs e)
    {
        // share the data
        shareOperation.ReportCompleted();
        Application.Current.Shutdown();
    }
}
```

We use the `Loaded` event of the Window class to retrieve all the information about the sharing operation. We again use the `AppInstance` class to retrieve the activation arguments and check whether it's indeed a `ShareTarget` scenario. This time, however, we move on to the next step, which is casting the `Data` property of the arguments to an `IShareTargetActivatedEventArgs` object, which is specific for sharing scenarios. Through this object, we can access the `ShareOperation` property, which contains all the information coming from the source application. In the previous example, we used the following:

- The `GetTextAsync()` method is exposed by the `Data` property to retrieve the shared text. Of course, the `Data` property exposes multiple `Get()` methods for the different data types you can share, such as `GetUrl()` for links.

- The `Properties` object is exposed by the `Data` property to retrieve the title.

In the previous sample, we simply display this data in `ShareWindow` as a preview.

For the moment, however, we are just showing the user a glimpse of the data they have received from the source application. Now we need to build the logic to complete the sharing operation, connected to the `Button` control we have added in XAML. You won't find this logic in the previous sample, because it totally depends on the kind of application you have built: if it's a Twitter client, the button will post the tweet on your timeline; if it's a photo application, the image will be added to your gallery; and so on. However, regardless of the logic, it's important that you call the `ReportCompleted()` method exposed by the `ShareOperation` object at the end of the task so that Windows can complete the sharing operation.

If you have experience with the UWP, you will notice a different behavior in your Win32 application: it won't be automatically terminated once the sharing operation is completed. As such, it's up to you to implement the proper behavior based on your expectations. In the previous sample, we are terminating the WPF application by calling `Application.Current.Shutdown()`. In another scenario, you might want to keep the application alive and redirect the user to the main page.

We aren't done yet. We have implemented all the code we need, but we haven't told Windows that our application can act as a share target. We do this through the manifest of the application, by registering a specific extension. This is why to become a share target, your application must be packaged with MSIX: the extension is supported only by the manifest, unlike other activation paths (such as file and protocol), which are also supported by unpackaged apps, as we learned in *Chapter 4, The Windows App SDK for a UWP Developer*.

Double-click on the `Package.appxmanifest` file in your Windows Application Packaging Project and move to the **Declarations** section. From the dropdown, add the **Share Target** declaration and, in the **Data formats** section, add a new entry for each data type you support as part of the sharing. In our example, where we support text, we add **Text** as a data format as shown in the following screenshot:

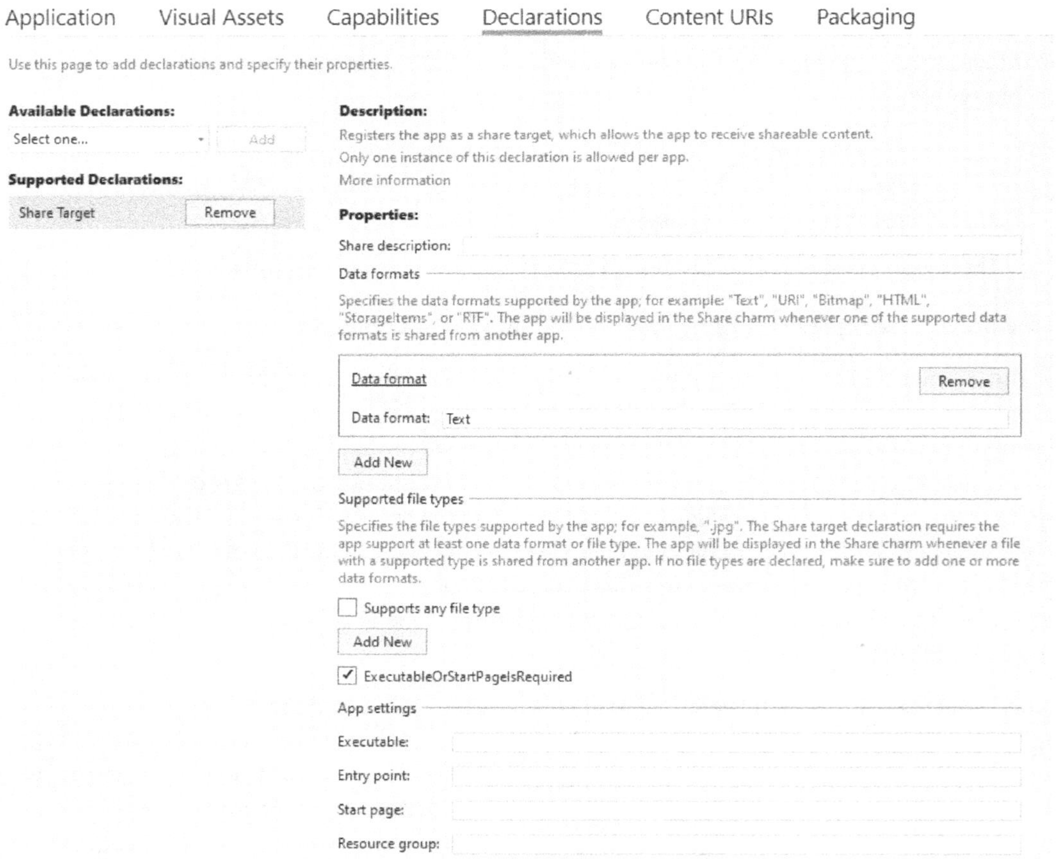

Application   Visual Assets   Capabilities   Declarations   Content URIs   Packaging

Use this page to add declarations and specify their properties.

Available Declarations:

Select one...   ▾   Add

Supported Declarations:

Share Target   Remove

Description:

Registers the app as a share target, which allows the app to receive shareable content.
Only one instance of this declaration is allowed per app.
More information

Properties:

Share description:

Data formats

Specifies the data formats supported by the app; for example: "Text", "URI", "Bitmap", "HTML", "StorageItems", or "RTF". The app will be displayed in the Share charm whenever one of the supported data formats is shared from another app.

Data format   Remove

Data format: Text

Add New

Supported file types

Specifies the file types supported by the app; for example, ".jpg". The Share target declaration requires the app support at least one data format or file type. The app will be displayed in the Share charm whenever a file with a supported type is shared from another app. If no file types are declared, make sure to add one or more data formats.

☐ Supports any file type

Add New

☑ ExecutableOrStartPageIsRequired

App settings

Executable:

Entry point:

Start page:

Resource group:

Figure 8.6 – Registering an application in the manifest to act as a share target

Now, after having deployed the share target application, you can test the outcome of your work by again launching the shared source one. If you have properly implemented all the components, you will see your application is listed as a potential target in the share UI.

We have completed our journey into the sharing contract. Thanks to this section, we have learned how we can share data across multiple applications roughly: thanks to the contract, we just need to define the type of data we want to share or that we can manage, and Windows will take care of connecting the dots by opening a communication channel.

Let's now move on to the final topic of this chapter: creating hybrid experiences between the web and the desktop ecosystem.

Integrating web experiences in your desktop application

The importance of the web has significantly increased over the last decade, and it keeps growing day by day. Daily, we perform multiple tasks using web applications: from making a money transfer from our bank account to sharing a document with a co-worker using a collaborative platform.

As such, developers often need to create hybrid applications that can deliver the best of both worlds: reusing the investments they made in web applications but enhancing them with the powerful capabilities offered by a native platform. Today, in your everyday job, you use many of these apps. A good example of this approach is Microsoft Teams: being a native app, it can take advantage of the native capabilities of Windows, such as push notifications and audio and video integration. At the same time, it can integrate a wide range of web experiences, from editing Office documents to getting real-time information from GitHub; from hosting Power Apps to displaying complex Power BI dashboards.

.NET platforms such as WPF and Windows Forms have provided the option to integrate web experiences for a long time thanks to the WebBrowser control, which you can use to host web applications inside your desktop application. However, there's a catch that still holds true even with the most recent revisions of the .NET ecosystem, such as .NET 6: the WebBrowser control uses the Internet Explorer engine to render web content. This means that the control isn't a good fit to host modern web experiences, since Internet Explorer uses an outdated engine that doesn't support all the latest innovations in web platforms, including the latest HTML, JavaScript, and CSS features.

For this reason, Microsoft has created a new control called WebView2, based on the same Chromium engine used by the new Microsoft Edge, which supports all the latest features of the web ecosystem. This control is available for all the Microsoft development platforms: in the case of WinUI, it's already integrated, while for the others, you must install a dedicated package.

Let's learn more in the next section!

Adding the WebView2 control to your page

The starting point of building a hybrid application with the `WebView2` control is different based on the development platform you have chosen:

- If you are building a WinUI application with the Windows App SDK, `WebView2` is already integrated. You won't have to do anything special to use it; you can just add it to your page as you do with any other standard control like `Button` or `TextBlock`, as in the following example:

```
<Window>
    <Grid>
        <WebView2 Source="http://www.packtpub.com"
            x:Name="MyWebView" />
    </Grid>
</Window>
```

- If you are building a WPF or Windows Forms application, you must first install a dedicated NuGet package. Please note that `WebView2` is supported both by .NET 5/6 applications and full .NET Framework ones. Right-click on your project, choose **Manage NuGet packages**, and install the package with **Microsoft.Web.WebView2** as the identifier. The main difference with WinUI is that, in this case, since the control comes from an external library, you will have to add an explicit reference. In the case of Windows Forms, you will find it in the designer toolbox; in the case of WPF, you must declare the `Microsoft.Web.WebView2.Wpf` namespace in your `Window` and then use it as a prefix, as in the following example:

```
<Window
        xmlns:wv2="clr-namespace:Microsoft.Web.WebView2.
        Wpf;assembly=Microsoft.Web.WebView2.Wpf"
    <Grid>
        <wv2:WebView2 Source="http://www.packtpub.com"
            x:Name="MyWebView" />
    </Grid>
</Window>
```

Once you have added the control to your page, you can start working with it. The APIs exposed by the control are the same regardless of the UI platform of your choice.

In the previous snippet of code, you have already seen a glimpse of the basic usage of the control. By setting the `Source` property to a URL, the corresponding web application will be rendered inside the control. The following shows an example of such usage:

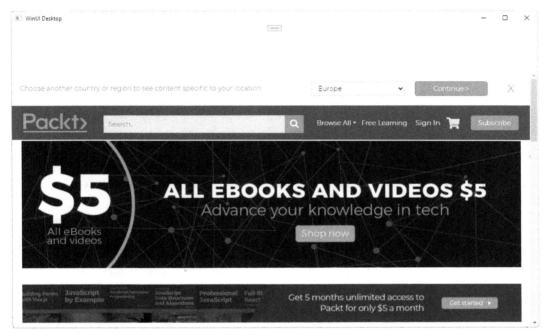

Figure 8.7 – A desktop WinUI application that is hosting the Packt website using the WebView2 control

The same goal can be achieved in code if you want to recreate a browser-like experience in your application:

```
private async void OnNavigateUrl(object sender,
    RoutedEventArgs e)
{
    await MyWebView.EnsureCoreWebView2Async();
    MyWebView.CoreWebView2.Navigate("http://www.packtpub.com");
}
```

The `WebView2` control (`MyWebView` is the name assigned to the control in XAML) exposes a property called `CoreWebView2`, which gives you access to all the advanced APIs exposed by the control. Before using it, it's always important to call the asynchronous `EnsureCoreWebView2Async()` method, which makes sure that the property has been correctly initialized. The `Navigate()` method is the one you can use to trigger navigation to a specific URL. You can also use the `WebView2` control to load local HTML content, by passing the HTML string to the `NavigateToString()` method:

```
private async void OnNavigateString(object sender,
  RoutedEventArgs e)
{
    await MyWebView.EnsureCoreWebView2Async();
    WebView.CoreWebView2.NavigateToString("<p>Hello world!</
      p>");
}
```

Lastly, you can manage the life cycle of your web applications directly from the control by subscribing to one of the many available events, such as the following:

- `NavigationStarting`, when you trigger navigation to another URL (or when the page itself triggers a redirect).
- `SourceChanged`, when the `Source` property is set to another value.
- `ContentLoading`, when the content of the web page begins to load.
- `NavigationCompleted`, when the navigation process has been completed.

Many of these events give you access to advanced information about the HTTP request. For example, you can use the `NavigationStarting` event to inspect the destination URL, the request headers and, eventually, cancel the operation. Look at the following example:

```
<WebView2 x:Name="MyWebView" NavigationStarting="WebView_
  NavigationStarting" />
```

In XAML, we have subscribed to the `NavigationStarting` event, which is managed by the following event handler:

```
private void MyWebView_NavigationStarting(WebView2
    sender, Microsoft.Web.WebView2.Core.
      CoreWebView2NavigationStartingEventArgs args)
{
    if (args.Uri == "https://www.microsoft.com")
```

```
    {
        args.Cancel = true;
        Console.WriteLine("This website is blocked!");
    }
}
```

Through the event arguments, we can inspect the `Uri` property to see where the user is being redirected. In this example, we're simulating a policy that blocks non-allowed websites. If the URI matches with the Microsoft website, we set the `Cancel` property to `true`, which will abort the navigation.

These are some of the basic concepts that you can implement with the `WebView2` control. However, to really support hybrid scenarios, we need to move to the next level and learn how we can deeply integrate our web applications with the native experience.

Enabling interactions between the native and web layers

To fully support the creation of true hybrid apps, the `WebView2` control supports two powerful scenarios:

- Enabling web applications to send information to the native layer
- Enabling native applications to send information to the web layer

Let's see how we can implement both scenarios.

Communicating from the web app to the native app

The `WebView2` control can receive messages from the web application through a JavaScript API exposed by the browser called `window.chrome.webview.postMessage()`. Let's use this API to build a simple application that, given a specific city, returns its coordinates. The city's name will be sent by the web application, while the geocoding operation will be performed using the location APIs we have learned how to use in this chapter.

Let's start with the web example:

```
<a onclick="window.chrome.webview.postMessage('Milan')">Get
    coordinates</a>
```

This is a snippet of HTML code that renders a hyperlink that, when it's clicked, invokes the `postMessage()` function passing, as a parameter, the city name (in this example, it's hardcoded; in a real app, it would be coming from a text field in a web page). When your web application is running in a traditional browser, this API won't be effective: you will get an error because the `postMessage()` function is `undefined`. As such, if your application is expected to also run in a browser, you should check if this function really exists and supply an alternative path.

When the web application is running inside a `WebView2` control, instead, it will trigger an event that you can intercept in the native layer, as the following example shows:

```
public sealed partial class MainWindow : Window
{
    public MainWindow()
    {
        this.InitializeComponent();
        MyWebView.EnsureCoreWebView2Async();
        MyWebView.CoreWebView2.WebMessageReceived +=
            CoreWebView2_WebMessageReceived;
    }

    private async void CoreWebView2_WebMessageReceived
        (CoreWebView2 sender,
            CoreWebView2
            WebMessageReceivedEventArgs args)
    {
    string city = args.TryGetWebMessageAsString();
    var result = await MapLocationFinder.FindLocationsAsync(city,
        null);
    Console.WriteLine($"Latitude: {result.Locations[0]
        .Point.Position.Latitude} - Longitude: {result
        .Locations[0].Point.Position.Longitude}");
    }
}
```

The event is called `WebMessageReceived` and it's exposed by the `CoreWebView2` property of the `WebView2` control. Inside the event handler, we can use the `TryGetWebMessageAsString()` method exposed by the event arguments to retrieve the message coming from the JavaScript function. In our example, the application will receive the name of a city (Milan), which we have passed to the `postMessage()` function.

Since the `WebMessageReceived` event is handled in the native layer, we have access to the entire Windows API ecosystem. In the example, we're using the `MapLocationFinder` class (included in the `Windows.Services.Maps` namespace) to perform a geocoding: we call the `FindLocationAsync()` method passing the name of the city and we get back a collection of results (inside the `Locations` property). We grab the first item and we output to the console its latitude and longitude.

This was just an example, but the possibilities are unlimited since we are running this code in the client application. An action executed on the web page can execute any native Windows operation, from displaying a notification to connecting to a Bluetooth device; from saving a file on the disk to triggering an operation in the background.

Let's now move on to the opposite scenario.

Communicating from the native app to the web app

There are two approaches you can use to let your .NET application send information to the web application. The first one is by using messages, in a comparable way to what we did in the previous section (but in the opposite direction).

In your C# code, you can send a message using the `PostWebMessageAsString()` method (if you need to send a string) or the `PostWebMessageAsJson()` one (if you need to send a more complex structure described with JSON). The following example sends a string to the web application:

```
private async void OnButtonClicked(object sender,
  RoutedEventArgs e)
{
    Geolocator locator = new Geolocator();
    var result = await locator.GetGeopositionAsync();
    string message = $"Latitude: {result.Coordinate.Point
      .Position.Latitude} - Longitude: {result.Coordinate
```

```
        .Point.Position.Longitude}";
    await WebView.EnsureCoreWebView2Async();
    webView.CoreWebView2.PostWebMessageAsString(message);
}
```

The first part of the code is using the `Geolocator` class we learned how to use in this chapter to retrieve the position of the user and to compose a message with the latitude and the longitude returned by location APIs. Then, by calling the `PostWebMessageAsString()` method exposed by the `CoreWebView2` object, we send this information to the web application.

On the web side, we can listen to incoming messages from the native layer with an event listener, as in the following example:

```
<script>
    window.chrome.webview.addEventListener('message', arg => {
        window.alert(arg.data);
    });
</script>
```

We subscribe to the event called `message` exposed by the `window.chrome.webview` object. Inside the event handler, we can retrieve from the `data` property of the arguments the message sent by the native layer. In this example, we're just displaying it to the user using a web popup, by passing it to the `window.alert()` function supported by every browser.

Through the CoreWebView2 object, you can also directly invoke JavaScript APIs that are exposed by the page or the browser. The following sample will produce the same outcome as the previous one, but implemented differently:

```
private async void OnButtonClicked(object sender,
    RoutedEventArgs e)
{
    Geolocator locator = new Geolocator();
    var result = await locator.GetGeopositionAsync();
```

```
    string message = $"Latitude: {result.Coordinate.Point
        .Position.Latitude} - Longitude:
        {result.Coordinate.Point.Position.Longitude}";
    await WebView.EnsureCoreWebView2Async();
    await WebView.CoreWebView2.ExecuteScriptAsync($"window.
        alert('{message}')");
}
```

In this case, we don't need to add any code to our web application. We use the `ExecuteScriptAsync()` method exposed by the `CoreWebView2` object to directly execute the `window.alert()` function exposed by the browser:

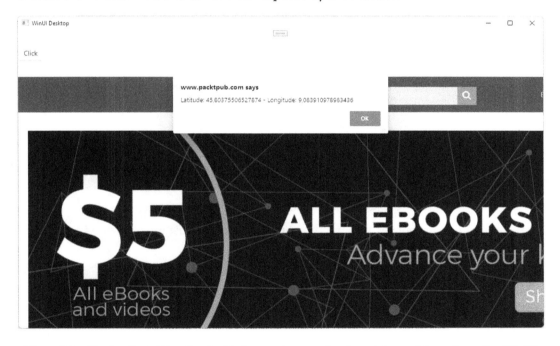

Figure 8.8 – The JavaScript function to display a web popup has been triggered from the native WinUI application

When you compile your application in debug mode, the `WebView2` control will give you access to the same developer tools that are available in Microsoft Edge. When your application is running, make sure the focus is on the web application and press *F12*: the developer tools will open up in another window, giving you the option to debug the JavaScript code, inspect the HTML, capture network traces, and much more:

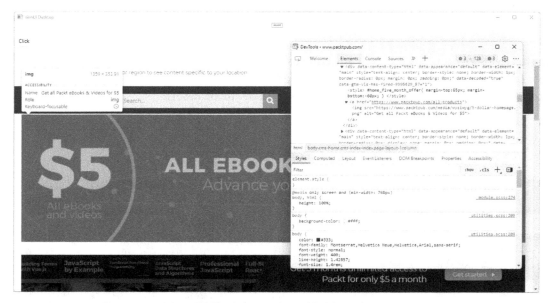

Figure 8.9 – The web developer tools enabled for a WebView2 control

Before completing our journey into building hybrid apps, we must introduce an important topic: the `WebView2` runtime.

Distributing the WebView2 runtime with your applications

Despite the `WebView2` control using the same engine as Edge, it isn't the browser doing the rendering, but a dedicated runtime that is optimized for hybrid scenarios called `WebView2` Runtime.

If your customers are running Windows 11, you're good to go: the `WebView2` runtime comes preinstalled, removing from you any responsibility for installing it and keeping it up to date. If your customers are running Windows 10, instead, users might not have the `WebView2` runtime already installed on their machine. As such, it's important to make sure that the runtime is installed before your application starts.

The WebView2 runtime supports two deployment techniques:

- **Evergreen version**: This is the suggested approach for most developers. The runtime is installed system-wide, and it's automatically kept updated by Windows.

- **Fixed version**: In this scenario, a specific version of the runtime is hard-linked to your application, and it's distributed as part of your binaries. This approach is a good fit for a critical application that must stay unchanged most of the time since you are in control of updates. The WebView2 runtime will never be updated unless you choose to do so by releasing an updated version of the application that includes a newer version. As such, there are no chances that a WebView2 update might potentially break or change the behaviors you have implemented.

Let's see some more details about the two distribution techniques.

Distributing the evergreen version runtime

The evergreen runtime comes as a standalone installer, which you can integrate into your deployment workflow. If you remember what we learned in *Chapter 1, Getting Started with the Windows App SDK and WinUI*, it's the same approach that we must adopt when we want to distribute unpackaged apps that use Windows App SDK. Since it's a traditional setup, which can run unattended by using the `/silent /install` switches, you can easily integrate it into your MSI installer or your PowerShell deployment script.

There are three options to perform the integration:

- You can use a bootstrapper, which is a very small setup that will download the most recent version directly from the internet as part of the installation. The bootstrapper itself can be downloaded as part of the installation process. All you have to do is to go to the official WebView2 runtime download page at `https://developer.microsoft.com/en-us/microsoft-edge/webview2/#download-section` and, under the **Evergreen Bootstrapper** section, click on **Get The Link** button. You will get a URL, which you can integrate into your installer to download the bootstrapper and execute it, before installing the main application.

- You can use a bootstrapper but include it in your installer instead of downloading it in real time. In this case, from the same WebView2 runtime download page, you can click the **Download** button to download the `MicrosoftEdgeWebView2Setup.exe` installer, which you can integrate into your existing deployment process. The runtime will continue to be downloaded directly from the internet (as in the first scenario), but you won't need to download the bootstrapper as well during the setup.

- Next is using an offline installer. In this case, from the same `WebView2` runtime download page, you must choose one of the options under the **Evergreen Standalone Installer** section, based on the CPU architecture you're targeting. This installer is tailored for offline scenarios since the runtime won't be downloaded in real time from the internet, but it's fully included. As a downside, your installer will be bigger since, other than your application, you're going to include the full runtime (potentially in multiple versions, if you need to target different CPU architectures). The bootstrapper, instead, automatically takes care of downloading the correct version for the CPU architecture of the target machine. It's important to highlight, however, that, even if the first installation is done in offline mode, you will continue to benefit from the Evergreen distribution model: as long as your machine can connect to the internet, you will automatically get updates when they're available.

This distribution model is a good fit for applications that are deployed on machines that are connected to the internet.

Distributing the fixed version runtime

The fixed version runtime can be downloaded, as well, from the official `WebView2` Runtime download page at `https://developer.microsoft.com/en-us/microsoft-edge/webview2/#download-section`. You must choose a specific version and architecture: what you will get is a CAB file (with an approximate size of 160 MB) with the entire runtime.

In this scenario, there isn't a real installation, since the runtime won't be installed system-wide: it will be specific only for your application and it will never be automatically updated. You will need to embed the files that compose the runtime directly inside your package, along with the other binaries of your application. The first step, as such, is to unpack the CAB file (using a utility such as 7-Zip or WinRAR) inside your application's package. It can be the package itself or a subfolder. There isn't any specific rule; you can use the approach that works best for your needs.

If the evergreen version runtime doesn't require any specific configuration in the application (it will be picked up automatically), in the case of the fixed version you will instead need to make a few code changes, since we must instruct the `WebView2` control where to find the binaries. In a .NET application, you can do this by setting the `BrowserExecutableFolder` property of the `CreationProperties` object exposed by the `WebView2` control, as in the following example:

```
MyWebView.CreationProperties.BrowserExecutableFolder =
    $"{Environment.CurrentDirectory}\\WebView2Runtime";
```

In this code, we're assuming that the runtime will be stored in a folder called `WebView2Runtime`, which is included in the root folder where the application has been deployed. You must use this initialization code before setting the `Source` property in XAML or calling the `EnsureCoreWebView2Async()` method.

Thanks to this section, now we have all the tools we need to start building hybrid applications and to reuse our investments in the web ecosystem in our Windows applications.

Summary

In this chapter, we have explored many ways in which Windows isn't just a host for our applications, but can provide many features that can really make the difference between a native application and a web one. We have learned how .NET makes it easy to integrate APIs from the UWP, something that in the past was available only if you wanted to start building a new application from scratch. The UWP unlocks endless opportunities, and, in this chapter, we have only touched on a few of them, such as location services, Windows Hello, the sharing contract, and the new filesystem capabilities.

Another thing we have learned in this chapter is that yes, it's true, this book is dedicated to building desktop applications but, as developers, we can't deny the impact of the web ecosystem. Therefore, Microsoft started from scratch with a new engine to build the new Microsoft Edge, which enables developers to get the best out of the web, thanks to the latest HTML, CSS, and JavaScript capabilities, not to mention the support for new features such as WebAssembly, WebVR, and so on. Through the `WebView2` control, included in the Windows App SDK, we can bring these features directly into our desktop applications and create hybrid experiences that can deliver the best of both worlds. In this chapter, we have learned about the features that make it possible, such as enabling communication between a web application and a desktop one.

We have just scratched the surface since the integration opportunities with the Windows ecosystem are endless. In the next chapter, we're going to focus on a powerful feature offered by Windows: the ability to integrate artificial intelligence directly into our applications, without using an online service.

Questions

1. To use the `WebView2` control in desktop applications, you just need to have the latest version of the Microsoft Edge browser installed on your machine. True or false?

2. To use UWP APIs in a .NET application you must use a library called C#/WinRT, which you must install from NuGet as the first step. True or false?

3. The new `StorageFile` and `StorageFolder` APIs can also be used in unpackaged applications. True or false?

9
Implementing Notifications

In our professional and personal lives, we perform multiple tasks every day, often relying on a variety of applications to help get them done: messaging apps, mail clients, productivity tools are just a few examples. Continuously checking these tools to see if there's an important update would be very tedious. Therefore, especially with the advent of smartphones, notifications have become critical to manage our digital life. Instead of forcing us to continuously check what's going on, it's the application itself that will notify us when something important is happening and we need to pay attention.

Notifications play a critical role in all kinds of applications, including desktop ones. As such, Windows 10 has introduced (and evolved release after release) a rich notification ecosystem, that we can use to interact with our users even when they aren't actively working with our application. As we're going to learn in this chapter, Windows doesn't enable only a passive notification ecosystem: you can enrich them with buttons, text boxes, dropdowns, and many other ways to make them interactive.

Windows 10 has also introduced a dedicated area for notification, called **Action Center**, which was redesigned in Windows 11. It's a panel that opens up on the right of the screen when you click on the time and date, which acts as an inbox for all the notifications generated by the installed applications. Even if you miss a notification because you weren't in front of your computer when it was displayed, you'll be able to find it in this section.

In this chapter, we'll address the following topics:

- Sending notification from a Windows application

- Working with toast notifications

- Adding interaction to a toast notification

- Displaying a progress bar

- Displaying a badge in the taskbar

- Implementing push notifications

By the end of this chapter, you will have an understanding of the available notification types, how to integrate them, and how to send them. Let's start!

Technical requirements

The code for the chapter can be found here:

```
https://github.com/PacktPublishing/Modernizing-Your-Windows-
Applications-with-the-Windows-Apps-SDK-and-WinUI/tree/main/
Chapter09
```

Sending a notification from a Windows application

Notifications are another one of the features which are part of the **Universal Windows Platform** (**UWP**) ecosystem and that, originally, wasn't available to the Win32 ecosystem. However, expanding this feature to the whole developer ecosystem very soon became a priority for Microsoft, since more and more apps wanted to participate in the notification ecosystem to take advantage of all the new features and to provide a more streamlined experience. When you participate in the native ecosystem, rather than building your own custom notification solution, you get advantages like the following:

- A familiar look and feel for the user, which follows the Windows UX guidelines

- Integration with features such as **Focus Assist**, which automatically hides all the notifications when your screen is mirrored or when you're sharing your screen with other users through applications such as Microsoft Teams

- Integration with the Action Center, so that users can review notifications even if they were sent while they were away from the computer

Therefore, even before Microsoft started to work on the Windows App SDK, the team decided to provide helpers in the Windows Community Toolkit, which makes it easier to add notifications to any kind of application, either built with UWP, WPF, Windows Forms, or WinUI. The Windows Community Toolkit also made it easier to work with the notification payloads. The content of a notification, in fact, is described with an XML payload to better support push notification scenarios, in which notifications aren't generated locally by the Windows application but can be generated, for example, from a backend written in Java, Node.js, or Python. However, when it comes to local notifications, manipulating XML introduces some challenges, since it isn't a strongly typed language. The Windows Community Toolkit, instead, offers APIs that will generate the XML for you, enabling you to use C# objects to define the notification's content.

Windows supports multiple types of notifications:

- **Toast notifications**: These are the most common ones. They are rendered as a popup displayed at the bottom right corner of the screen. They are the richest ones since they can include text, images, actions, and so on. Toast notifications are moved to the Action Center if the user does not dismiss them.

- **Badge notifications**: These are the ones you can use to display and update a counter on the application's icon. They are used, for example, by messaging applications to display the number of unread messages directly in the Taskbar.

- **Raw notifications**: These contain any kind of payload, and, as such, it's up to your application to manually parse the content and handle them. Raw notifications are usually coupled with push notifications, which are generated by an online service (such as a web application or a job running in the cloud) and then handled by your local application. We will cover this topic in this chapter only very briefly.

- **Tile notifications**: Windows 10 offers the concept of tiles, which are special icons for your applications that you could pin to the Start menu. With tiles notifications, an application can update the icon to display relevant information without asking the user to open it. An example is the built-in Weather application, which can update the tile with the forecast for the current location. Tiles, however, are now considered deprecated since they have been removed in Windows 11. As such, we won't cover them in this chapter.

In this chapter, we're going to use a WinUI application as an example, but the same exact set of APIs works in a **WPF** or **Windows Forms application**). The only difference in the way you use the features comes from the deployment model: if you choose to distribute your application as packaged with MSIX, in fact, there will be an extra step to make.

Let's start!

Working with toast notifications

Regardless of the platform and the deployment model of your choice, the first step is to install the Windows Community Toolkit package dedicated to notifications. Right-click on your project, choose **Manage NuGet packages**, and look for the package with **CommunityToolkit.WinUI.Notifications** as an identifier.

> **Note**
>
> Despite the package name containing the **WinUI** keyword, it isn't specific only for this platform, and can work also with WPF, Windows Forms, or UWP.

The other requirement is to use a dedicated `TargetFramework` for Windows 10 or 11 since the toolkit needs to consume some APIs from the Universal Windows Platform. In our scenario, we're building a WinUI application, so we're good to go. If you're using Windows Forms or WPF, instead, make sure to set the `TargetFramework` at least as follows:

```
<TargetFramework>net6.0-windows10.0.19041.0
   </TargetFramework>
```

In the case of a WinUI project, there's also another change to make. By default, when you create a new WinUI application, the minimum supported Windows version is Windows `10 10.0.17763.0`, which is the October 2018 update. This is controlled by the `TargetPlatformVersion` property of the `.csproj` file:

```
<TargetPlatformMinVersion>10.0.17763.0</TargetPlatform
   MinVersion>
```

The notification APIs from the Windows Community Toolkit, however, use features that have been introduced in Windows `10 10.0.18362.0`, which is the May 2019 update. If you want to remove the warnings that Visual Studio will give you otherwise, you should change the `TargetPlatformMinVersion` property as follows:

```
<TargetPlatformMinVersion>10.0.18362.0</TargetPlatform
   MinVersion>
```

Now you're all set. Let's start with a very basic scenario, which requires the usage of APIs included in the `CommunityToolkit.WinUI.Notifications` namespace:

```
private void SendNotification()
{
    new ToastContentBuilder()
        .AddText("New message")
```

```
        .AddText("There's a new message for you!")
        .Show();
}
```

Simple, isn't it? You just need to create a new `ToastContentBuilder()` object, which exposes different methods to customize the notification's template. In this case, we're using the `AddText()` one to add two text elements. Once we have created the notification payload, we can display it by calling the `Show()` method. Under the hood, the library will generate the following XML payload:

```
<binding template="ToastGeneric">
    <text hint-maxLines="1">New message</text>
    <text>There's a message for you!</text>
</binding>
```

As you can see, the C# APIs are much more convenient to use than having to manually generate the XML content.

If you're going to test this code in an unpackaged scenario (a WPF, Windows Forms, or an unpackaged WinUI app), everything will work. If, instead, you are using a WinUI packaged app or you have added a Windows Application Packaging Project to your solution so you could package your WPF or Windows Forms app, you will get an exception. In a packaged scenario, in fact, we have to register the notification activation in the manifest first. Let's see how to do it.

Supporting toast notifications in packaged apps

To enable notifications in a packaged app, we need to register the `ToastNotificationActivation` extension in the manifest, otherwise, we would get an exception every time we try to display a notification. Additionally, this extension is required to support the activation scenarios we're going to see later in this chapter: the user interacts with the notification, and we want to trigger an action.

Right-click on the `Package.appxmanifest` file of your application (it's either in the main project if it's a WinUI app, or in the Windows Application Packaging Project if it's a WPF or Windows Forms app) and choose **View code**. The extensions we need to add, in fact, aren't supported by the visual editor.

The first step is to declare two custom namespaces in the XML, as in the following sample:

```
<Package
xmlns:com="http://schemas.microsoft.com/appx/manifest/com/w
  indows10"
```

```
xmlns:desktop="http://schemas.microsoft.com/appx/manifest/
  desktop/windows10"
  IgnorableNamespaces="com desktop">
  . . .
</Package>
```

Then we must declare two extensions inside the Application instance node, as in the following snippet:

```
<Application Id="App"
  Executable="$targetnametoken$.exe"
  EntryPoint="$targetentrypoint$">
  . . .
  <Extensions>
    <desktop:Extension Category="windows
      .toastNotificationActivation">
      <desktop:ToastNotificationActivation ToastActivator
        CLSID="535A532A-C79C-4B9D-BDE4-FCD0B3E3C171" />
    </desktop:Extension>
    <com:Extension Category="windows.comServer">
      <com:ComServer>
        <com:ExeServer Executable="WinUINotifications.exe"
          Arguments="-ToastActivated" DisplayName="Toast
            activator">
          <com:Class Id="535A532A-C79C-4B9D-BDE4-
            FCD0B3E3C171" DisplayName="Toast activator"/>
        </com:ExeServer>
      </com:ComServer>
    </com:Extension>

  </Extensions>
</Application>
```

From the preceding code, we see that the first extension is called windows. toastNotificationActivation. It's a unique identifier for our application, so it's important to replace the ToastActivatorCLSID property with a unique GUID. You can generate one from Visual Studio (you'll find the option under the menu **Tools | Create GUID**) or using one of the many available online tools, such as https://www. guidgenerator.com/.

The second extension is `windows.comServer` and it's required to register the toast activator as a COM extension so that Windows can invoke it when needed. Compared to the preceding sample, there are two properties to change:

- The `Executable` property of the `ExeServer` entry must match the main executable of your application (which, in most of the cases, matches with the name of the project).

- The `Id` property of the `Class` entry must match the same GUID that you have previously generated and assigned to the `ToastActivatorCLSID` property.

Thanks to these changes, the exception will go away, and toast notifications will start to work. Let's explore now other ways to enrich a notification.

Adding images

Toast notifications support multiple types of images, which are translated into a different placement inside the template. The simplest scenario is having an image placed on the left of the text content. This approach is also called **app logo** since, on older versions of the operating system, if you didn't specify any image, Windows used the default logo of the application. On the most recent versions, including Windows 11, instead, if you don't specify this option, the image will simply be left out, leaving all the space for the text content. The logo will be displayed, instead, at the top, right before the name of the application that generated the notification.

You can use the `AddAppLogoOverride()` method, passing as a parameter the `Uri` of the image. If your application is packaged, the `Uri` can be an HTTP URI (so an image available on the internet) or an application URI (so an image included in the package). If your application is unpackaged, instead, you can't use an HTTP URI, but you will have to download the image first on your local machine and then create a URI with the full path of the file. The following sample shows all the supported scenarios:

```
private void ShowNotification()
{
    //packaged app, from the network
    new ToastContentBuilder()
        .AddText("New message")
        .AddText("There's a new message for you!")
        .AddAppLogoOverride(new Uri
            ("https://www.mywebsite.com/myimage.png"))
        .Show();
```

```
    //packaged app, from the package
    new ToastContentBuilder()
        .AddText("New message")
        .AddText("There's a new message for you!")
        .AddAppLogoOverride(new Uri("ms-appx:
          ///Assets/myimage.png"))
        .Show();

    //unpackaged app, from the disk
    //packaged app, from the network
    new ToastContentBuilder()
        .AddText("New message")
        .AddText("There's a new message for you!")
        .AddAppLogoOverride(new Uri(@"C:\myimage.png"))
        .Show();
}
```

This is how this type of image is rendered in a notification:

Figure 9.1 – A toast notification with an app logo image

Another choice to display images in a notification is using the concept of a **hero image**. In this scenario, the image is displayed with a bigger size at the top of everything else (including the logo and the name of the application). This feature is enabled by the AddHeroImage() method and it has the same requirements we have seen for the logo: if the application is packaged, the image can be either on the internet or local; if it's unpackaged, instead, it must be local. Let's take a look at the following example:

```
private void SendNotification()
{
    new ToastContentBuilder()
    .AddText("New message")
    .AddText("There's a new message for you!")
    .AddHeroImage(new Uri
```

```
    ("https://www.mywebsite.com/myimage.png"))
    .Show();
}
```

The following screenshot shows the usage of a hero image in a notification:

Figure 9.2 – A toast notification with a hero image

Finally, you can also include the image inline, which in this case will be displayed below the main content. This feature works like the others related to images (with the same restrictions), except that the method to use is called AddInlineImage(), as in the following sample:

```
private void SendNotification()
{
    new ToastContentBuilder()
    .AddText("New message")
    .AddText("There's a new message for you!")
    .AddInlineImage(new Uri
        ("https://www.mywebsite.com/myimage.png"))
    .Show();
}
```

The preceding code will generate the following result:

Figure 9.3 – A toast notification with an inline image

Let's see now how to customize the basic information in a notification.

Customizing the application's name and the timestamp

As we have seen in the previous examples, by default, a notification will use, as a header, the logo, and the name of the application that generated it. Another piece of information that is generated automatically is the timestamp: when the notification gets stored in the Action Center, it will display the date and time it was generated.

We can override these two values if needed. For example, our application might have a complex and long name and we want to use a more friendly one; or the notification is about an event that happened in the past or that it's going to happen in the future.

The name can be changed using the `AddAttributionText()` method, while the timestamp can be changed using the `AddCustomTimeStamp()` one. The following sample shows both in action:

```
private void SendNotification()
{
```

```
new ToastContentBuilder()
.AddText("New message")
.AddText("There's a new message for you!")
.AddAttributionText("via my amazing WinUI app")
.AddCustomTimeStamp(new DateTime(2022, 12, 25,
    15, 0, 0))
.Show();
}
```

All the examples we have seen so far took care of creating a new notification on top of all the existing ones. What if, instead, you want to manipulate the existing notifications created by your application? Let's introduce the concept of tagging.

Tagging a toast notification

Every time your application sends a toast notification, by default it will be automatically added on top of the others. If your application sends multiple notifications, this is what you're going to see in the Action Center:

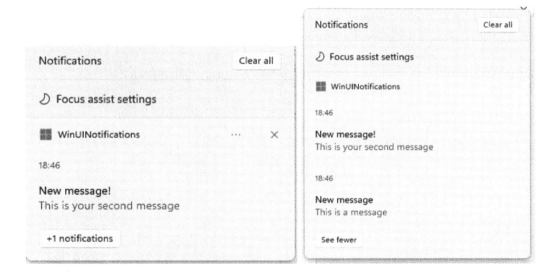

Figure 9.4 – Multiple toast notifications generated by the same applications

Windows will display only the most recent one by default (on the left), but the button below will tell you that there are more notifications available. If you click it, the Action Center will display all of them.

This approach is a great fit, for example, for a messaging app. But what if you're building a football application, which delivers a notification every time the score of the match changes? In this case, retaining all the old notifications might confuse the user. To better manage this scenario, Windows supports the concept of tagging a notification with a keyword. If your application sends a new notification with the same tag, Windows will replace the existing one, rather than adding it on top of the others. This goal is achieved by setting the `Tag` property on the notification:

```
private void SendNotification2()
{
    new ToastContentBuilder()
    .AddText("Goal!")
    .AddText("The new score is 2-0")
    .Show(toast =>
    {
        toast.Tag = "match1";
    });
}
```

When we call the `Show()` method, we also get a reference to the generated `ToastNotification` object, so that we can further customize other options, including the `Tag` one.

Tagging is important also to better support another scenario: scheduling notifications.

Scheduling a notification

In all the examples we have seen so far we have used the `Show()` method, which immediately triggers the notification. Another option is to schedule them. If you're building a calendar application, for example, you might want to schedule a reminder when the user creates an appointment.

The way you build a scheduled notification is the same as we have seen for the traditional ones. The only difference is that, instead of calling the `Show()` method, you're going to call the `Schedule()` one passing, as a parameter, a `DateTimeOffset` object that represents the date and time when you want to display the notification. The following sample shows scheduling a notification 5 minutes from now:

```
private void SendNotification()
{
    new ToastContentBuilder()
```

```
.AddText("New message")
.AddText("This is a message")
.Schedule(DateTimeOffset.Now.AddMinutes(5));
}
```

When you're using the scheduling approach, tagging the notification becomes even more important, since it enables you to retrieve it at a later stage so that we can delete it or change it. In our calendar application, if the user cancels the appointment from the calendar, we don't want the notification about the event to be still displayed. To tag a scheduled notification, we must use the same approach we have used for a regular one, as in the following sample:

```
private void SendNotification()
{
    new ToastContentBuilder()
    .AddText("Christmas party")
    .AddText("Party with family and friends")
    .Schedule(new DateTimeOffset(new DateTime(2021, 12, 25,
        10, 0, 0)), toast =>
        {
            toast.Tag = "event1";
        });
}
```

Thanks to the `ToastNotifierCompat` class, we can then retrieve the list of all the notifications scheduled by our application and, by using the tag, find the one we need to delete:

```
private void DeleteNotification()
{
    ToastNotifierCompat notifier = ToastNotification
      ManagerCompat.CreateToastNotifier();

    IReadOnlyList<ScheduledToastNotification>
      scheduledToasts = notifier.GetScheduledToast
        Notifications();

    var notification = scheduledToasts.FirstOrDefault(i =>
      i.Tag == "match1");
```

```
if (notification != null)
{
    notifier.RemoveFromSchedule(notification);
}
}
```

Once we have created a notifier using the `ToastNotificationManagerCompat.CreateToastNotifier()` method, we can use the `GetScheduledToastNotifications()` method to retrieve a list of all the scheduled notifications that belong to our application. Then, by using the `FirstOrDefault()` expression, we can retrieve the one with our tag (`match1`) and pass it as a parameter to the `RemoveFromSchedule()` method to delete it.

Now that we have learned the basic ways to customize and manage notifications, let's see one of the most interesting features: supporting interactions.

Adding interaction to a notification

What makes notifications so powerful is that they don't represent just a passive piece of information. The user can interact with them to perform actions without having to open the application. Let's explore the various ways a user can interact with a notification.

The most basic interaction is clicking on the notification, which is also the most common scenario. For example, users get a notification that they have received a new mail and when they click on it, they expect the mail client to open up so that they can read it.

This scenario can be accomplished thanks to the `AddArgument()` method, which you can use to pass a key /value pair that you can retrieve when the application is activated from a notification. Let's see the following example:

```
private void SendNotification()
{
    new ToastContentBuilder()
    .AddText("New message")
    .AddText("There's a new message for you!")
    .AddArgument("messageId", "1983")
    .Show();
}
```

We have added an argument called messageId, with as value the hypothetical ID of the message that was displayed in the notification.

Another scenario is using a notification to enable users to perform multiple actions without having to open the application. If we reuse the example of a mail client, we could allow the user to directly mark a mail as read or flag it as important directly from the notification. In this case, we need to add another layer of interactivity: buttons.

Buttons are represented by the ToastButton class and, to each of them, you can assign a different argument, which you can use when the application is activated to understand which button was pressed. Let's see the following example:

```
private void OnSendNotificationWithReply ()
{
    new ToastContentBuilder ()
    .AddText ("New message")
    .AddText ("There's a new message for you!")
    .AddButton (new ToastButton ()
        .SetContent ("Mark as read")
        .AddArgument ("action", "read")
        .AddArgument ("messageId", "1983"))

    .AddButton (new ToastButton ()
        .SetContent ("Flag as important")
        .AddArgument ("action", "flag")
        .AddArgument ("messageId", "1983"))
    .Show () ;
}
```

This notification includes two buttons, which are added using the AddButton () method. We set the label of the button (using the SetContent () method) and, with the same AddArgument () method we have learned in the previous example, we add two arguments that will help us to understand which button was pressed (through the action parameter) and which is the message where we must apply the action (through the messageId parameter).

The following screenshot shows the result of the preceding code:

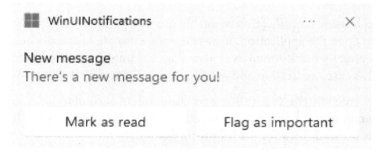

Figure 9.5 – A toast notification with two buttons

Now that our notification can pass activation arguments, how can we retrieve them? Let's learn about that in the next section.

Handling activation from a notification

In *Chapter 4, The Windows App SDK for a UWP Developer*, we have learned how we can manage application's activation using the `AppInstance` class. For toast notifications, instead, we must use a different approach, since as we have learned so far this feature isn't part of the Windows App SDK yet, but it's included in the Windows Community Toolkit.

The action is handled through an event called `OnActivated`, which is exposed by the `ToastNotificationManagerCompat` class included `CommunityToolkit.WinUI.Notifications` namespace. This event must be subscribed to in the startup event of the application, so the exact location changes based on the platform you're using to build your application. If it's a WinUI application, you're going to use the `OnLaunched()` event declared in the `App` class, as in the following example:

```
protected override void OnLaunched(Microsoft.UI.Xaml.
  LaunchActivatedEventArgs args)
{
    ToastNotificationManagerCompat.OnActivated
      += eventArgs =>
    {
        //manage the activation
    };

    var m_window = new MainWindow();
    m_window.Activate();
```

```
}
```

If it's a WPF application, instead, you use the `OnStartup()` event exposed by the `App` class, as in the following example:

```
public partial class App : Application
{
    protected override void OnStartup(StartupEventArgs e)
    {
        ToastNotificationManagerCompat.OnActivated
            += eventArgs =>
        {
            //manage the activation
        };
    }
}
```

Regardless of the platform of your choice, whenever the user interacts with a notification your application will be launched and the `OnActivated` event will be triggered. The heart of the activation is the event arguments, which you can use to retrieve the arguments you have set when you have created the notification. The way you retrieve the arguments is the same regardless of whether they are associated with the notification click or to the interaction with specific control, like a button. Let's start with the following notification, which supports both scenarios:

```
private void SendNotification()
{
    new ToastContentBuilder()
    .AddText("New message")
    .AddText("There's a new message for you!")
    .AddArgument("action", "click")
    .AddArgument("messageId", "1983")

    .AddButton(new ToastButton()
        .SetContent("Mark as read")
        .AddArgument("action", "read")
        .AddArgument("messageId", "1983"))
```

```
        .AddButton(new ToastButton()
            .SetContent("Flag as important")
            .AddArgument("action", "flag")
            .AddArgument("messageId", "1983"))
        .Show();
}
```

In this example, we have three types of interaction: through the click on the notification, through the `Mark as read` button, and through the `Flag as important` button. Each of them returns a different action as an argument, in addition to the `messageId` with the identifier of the message. The following code shows how we can handle this notification in the `OnActivated()` event:

```
protected override void OnLaunched(Microsoft.UI.Xaml
    .LaunchActivatedEventArgs args)
{
    ToastNotificationManagerCompat.OnActivated
        += eventArgs =>
    {
        ToastArguments args = ToastArguments.Parse
            (eventArgs.Argument);
        string action = args.Get("action");
        int messageId = args.GetInt("messageId");

        switch (action)
        {
            case "click":
                //display the message
                break;
            case "read":
                //mark the message as read
                break;
            case "flag":
                //mark the message as flagged
                break;
        }
    };
```

```
    var m_window = new MainWindow();
    m_window.Activate();
}
```

Thanks to the `ToastArguments` class, we can use the `Parse()` method to parse the activation arguments (stored in the `Argument` property) and work with them in an easier way. Specifically, the `ToastArguments` class exposes multiple methods to retrieve the value of an argument already converted into the expected type. You can see two examples in the code:

- The value of the `action` argument is a `string`, so we use the `Get()` method to retrieve it.
- The value of the `messageId` argument is a number, instead, so we use the `GetInt()` method to retrieve it.

There are many other methods available, such as `GetBoolean()`, `GetFloat()`, or `GetEnumerator()`.

Thanks to the notification arguments, we can perform the right task based on the action selected by the argument. For example, if the action is `click`, we could use the `messageId` to retrieve the specific message and display it to the user; or if the action is `read`, we could use the `messageId` to find the message in the database and change the field that contains the information if the message has been read.

We have just scratched the surface in this section. Buttons and clicks aren't the only ways we can use to add interactivity. Let's see some more advanced ways.

Supporting user input

Thanks to notifications, we can ask for input from our users without needing to open the application, thanks to the built-in support for text boxes and dropdowns. In the activation arguments, we can retrieve this input, even if we'll need to use some different APIs than the one, we have seen to retrieve the activation arguments.

Let's start with the text box scenario with the following example:

```
private void SendNotification()
{
    new ToastContentBuilder()
    .AddText("New message")
    .AddText("There's a new message for you!")
```

```
    .AddInputTextBox("reply", "Type your message",
      "Message")
    .AddButton(new ToastButton()
       .SetContent("Send")
       .AddArgument("action", "send")
  .SetTextBoxId("reply"))
    .Show();
}
```

To add a text box you can use the `AddInputTextBox()` method. The only required parameter is the identifier of the field (`reply`, in our example), which we'll need during the activation to retrieve the text inserted by the user. The other two are optional: the second is a placeholder text, which is displayed inside the field until the user types something; the third is a label that is displayed above the field.

When we add user input to a notification, we always need a button to trigger the action, otherwise, the notification will permanently stay on screen, waiting for interaction from the user. For this reason, we also add a button with the `Send` label. We also use the `SetTextBoxId()` method, passing as a parameter the identifier of the field to link the two controls. It's not required, but it's highly recommended: thanks to this setting, the button will be displayed at the right of the `TextBox`; additionally, the button will be disabled until the user starts to type something in the field.

This is the notification generated by the previous code:

Figure 9.6 – A toast notification with a TextBox as input field

To handle the activation, we still need to use the `OnActivated()` event exposed by the `ToastNotificationManagerCompat` class. However, the user input won't be passed as an argument, but to a special collection called `UserInput`. Let's take a look at the following example:

```
protected override void OnLaunched(Microsoft.UI.Xaml
   .LaunchActivatedEventArgs args)
{
    ToastNotificationManagerCompat.OnActivated
       += eventArgs =>
    {
        ToastArguments args = ToastArguments.Parse
           (eventArgs.Argument);
        string action = args.Get("action");

        var input = eventArgs.UserInput;
        string message = input["reply"].ToString();

        Console.WriteLine($"Action {action} with message
           {message}");
    };

    var m_window = new MainWindow();
    m_window.Activate();
}
```

`UserInput` is a collection of all the inputs provided by the user, categorized by the identifier we have assigned to the input control. In our scenario, we use the keyword `reply` as the identifier, which we have set when we have created the notification. As a value, we'll get in return the message typed by the user.

Another type of user input is the dropdown, which we can add to a notification using the `AddToastInput()` method and the `ToastSelectionBox` class, as in the following example:

```
private void OnSendNotificationWithDropdown ()
{
    new ToastContentBuilder()
    .AddText("New message")
```

```
    .AddText("There's a new message for you!")
    .AddToastInput(new ToastSelectionBox("response")
    {
        DefaultSelectionBoxItemId = "ok",
        Items =
        {
            new ToastSelectionBoxItem("ok", "Ok"),
            new ToastSelectionBoxItem("late", "I'll be
                late"),
            new ToastSelectionBoxItem("thanks", "Thanks!")
        }
    })
    .AddButton(new ToastButton()
        .SetContent("Ok")
        .AddArgument("action", "send"))
    .Show();
}
```

When we create a ToastSelectionBox object, we must initialize it with the field identifier, which we'll need later to retrieve the selected value in the activation. Then we can define the available choices through the Items collection, which contains multiple ToastSelectionBoxItem objects. Each of them is created with a key (which we'll receive in the activation to understand which option was selected) and a text, which will be displayed in the notification. We can also use the DefaultSelectionBoxItemId property to specify the identifier of the item that we want to use as default. Also, in this case, we must have at least one button to trigger the interaction.

This is how the notification looks like:

Figure 9.7 – A toast notification with a dropdown as user input

The way we retrieve the selected item in the activation is the same we have seen with the text field: through the `UserInput` collection exposed by the activation arguments.

```
protected override void OnLaunched(Microsoft.UI.Xaml
   .LaunchActivatedEventArgs args)
{
    ToastNotificationManagerCompat.OnActivated
       += eventArgs =>
    {
        ToastArguments args = ToastArguments.Parse
           (eventArgs.Argument);
        string action = args.Get("action");

        var input = eventArgs.UserInput;
        string message = input["response"].ToString();

        switch (message)
        {
            case "ok":
                //send ok message
                break;
            case "late":
                //send late message
                break;
            case "thanks":
                //send thanks message
                break;
        }
    };

    var m_window = new MainWindow();
    m_window.Activate();
}
```

In this case, the key we use to get the value from the `UserInput` collection is the identifier we have used to create the `ToastSelectionBox` object (`response`, in our case). As a value, we'll get the identifier of the option selected by the user, which we can use the way we prefer; in the sample, we'll compose and send a different predefined response, based on the selection of the user.

The last user input we're going to explore is snooze/dismiss, which is a great fit for scenarios where the notification is used as a reminder. Thanks to this option, we can easily enable the user to snooze the notification, so that the reminder will be postponed for later.

To create such a notification, it's enough to set some extra properties:

```
private void OnSendNotificationWithReminder ()
{
    new ToastContentBuilder()
    .SetToastScenario(ToastScenario.Reminder)
    .AddText("It's time!")
    .AddText("This is your reminder")
    .AddButton(new ToastButtonSnooze())
    .AddButton(new ToastButtonDismiss())
    .Show();
}
```

Compared to a traditional notification, we have made the following changes:

- We call the `SetToastScenario()` method, passing as a parameter the `Reminder` value of the `ToastScenario` enumerator. With this setting, the notification will stay on the screen until the user has snoozed or dismissed it. Another possibility is to use the `Alarm` value, which enables alarm scenarios. In this case, the notification will also have a looping sound, which will play until the user acts.

- We add the snooze and dismiss buttons but, instead of using the generic `ToastButton` class, we use the specific `ToastButtonSnooze` and `ToastButtonDismiss` ones. This way, Windows will automatically apply the correct behavior: the snooze button will postpone the reminder, while the dismiss one will make the notification disappear permanently.

In this scenario, you don't need to manually handle the activation, since Windows will take care of everything for you.

With this standard configuration, the snooze will use a default timer set by Windows. However, you can use a `ToastSelectionBox` to give the option to the user to choose after how much time the notification should appear again, as we have done in the following example:

```
private void OnSendNotificationWithReminder ()
{
    new ToastContentBuilder()
    .SetToastScenario(ToastScenario.Reminder)
    .AddText("It's time!")
    .AddText("This is your reminder")
    .AddButton(new ToastButtonSnooze() { SelectionBoxId =
      "snoozeTime" })
    .AddButton(new ToastButtonDismiss())
    .AddToastInput(new ToastSelectionBox("snoozeTime")
    {
        DefaultSelectionBoxItemId = "5",
        Items =
        {
            new ToastSelectionBoxItem("5", "5 minutes"),
            new ToastSelectionBoxItem("15", "15 minutes"),
            new ToastSelectionBoxItem("30", "30 minutes"),
            new ToastSelectionBoxItem("60", "1 hour"),
        }
    })
    .Show();
}
```

The way you set up the `ToastSelectionBox` object is the same we have previously seen for a regular dropdown. The only difference is that, through the `SelectionBoxId` property, we connect the `ToastButtonSnooze` control with the `ToastSelectionBox`, by passing the same identifier. This way, Windows will automatically handle the snoozing and the notification will be delayed for the time selected by the user. The time must be passed as the first parameter of each `ToastSelectionBoxItem`, expressed in minutes.

In this section, we have explored all the features that we can include in a notification, from the basic ones (such as text and image) to the most advanced ones (such as buttons and dropdowns). Let's move on now and see another advanced scenario: displaying the status of an operation.

Displaying a progress bar

A common requirement for applications is to notify the user about the progress of an operation. What if you want to keep the user up to date even when the application is minimized? Notifications can help you by integrating a progress bar that the main application can update when needed.

This scenario is supported by the concept of bindable data, which enables us to add one or more placeholders in the notification, which can be replaced at a later stage with actual data.

Let's see first how we can create such a notification:

```
string tag = "report-progress";


new ToastContentBuilder()
    .AddText("Report generation in progress...")
    .AddVisualChild(new AdaptiveProgressBar()
    {
        Title = "Report generation",
        Value = new BindableProgressBarValue
            ("progressValue"),
        ValueStringOverride = new
            BindableString("progressValueString"),
        Status = "Generating report..."
    })
    .Show(toast =>
    {
        toast.Data = new NotificationData();
        toast.Data.Values["progressValue"] = "0.0";
        toast.Data.Values["progressValueString"] = "0 %";


        toast.Data.SequenceNumber = 1;


        toast.Tag = tag;
    });
```

This time we're using a new method called `AddVisualChild()`, which enables us to add arbitrary elements to the notification. The object we create is called `AdaptiveProgressBar` that, other than simple information such as `Title` and `Status`, exposes two special properties:

- `Value`: This is a number between 0 and 1 that gets translated with the progress bar filling. This property is set with a `BindableProgressBarValue` object identified by the `progressValue` key. Thanks to this object, we'll be able to update the value even after the notification has been generated and displayed.

- `ValueStringOverride`: This is a text that is displayed right below the progress bar. It's typically used to provide a more descriptive status of the operation, such as a percentage or the number of processed items against the total. Since it's a text, we set the property with a `BindableString` object, identified by the `progressValueString` key. Also, in this case, we'll be able to update the text after the notification has been generated.

When we call the `Show()` method, we used the same approach we have learned before to further customize the toast notification. One of the fields should be familiar: it's the `Tag` property, which we'll need later to identify the notification that we must update.

The other field is new, and it's called `Data`. Its type is `NotificationData` and we can use it to replace the placeholders that we have included in the notification's template with actual data. The placeholders are stored inside the `Values` collection, and they are identified by the keys we have set in the template. When the notification is generated, it's likely that the operation we want to track with a progress bar hasn't started yet, so we set both values to 0. Then, we also set the `SequenceNumber` property to 1, to notify Windows that this is the first update.

If we execute this code, the notification will be displayed and it will stay visible, waiting for further updates. Let's see now how we can track progress by analyzing the following example:

```
for (uint cont = 0; cont <= 10; cont++)
{
    await Task.Delay(1000);

    var data = new NotificationData
    {
        SequenceNumber = cont
    };

    IFormatProvider provider =
```

```
        CultureInfo.CreateSpecificCulture("en-US");
    double progressValue = (double)cont / 10;
    string progressValueConverted = $"{(progressValue *
      100).ToString()} %";

    data.Values["progressValue"] =
      progressValue.ToString(provider);
    data.Values["progressValueString"] =
      progressValueConverted;

    ToastNotificationManagerCompat.CreateToastNotifier()
      .Update(data, tag);
}
```

We are simulating a long-running operation using a for statement, which will increment the counter from 1 to 10, with a 1-second delay at each cycle. At every execution, we create a new NotificationData object, and we do the same thing we did when we have created the notification: we update the SequenceNumber property and the two placeholders, progressValue, and progressValueString, through the Values collection. The difference is that this time, we're incrementing the progress bar, using the counter as a value. The only special scenario to handle is that the BindableProgressBarValue object requires a value from 0 (the progress bar is empty) to 1 (the progress bar is full), as such we must convert the counter into a percentage.

Once we have populated the NotificationData object, we can update the existing notification through the Update() method exposed by the ToastNotifier object created with the ToastNotificationManagerCompat class. Other than the data, we must also pass the tag of the notification, so that Windows can identify which is the notification to update. Refer to the following screenshot:

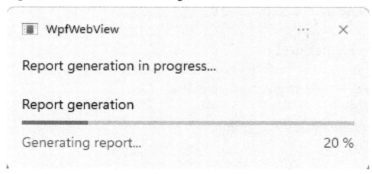

Figure 9.8 – A notification with a progress bar

The previous code is a simplification to help you learn the basic concepts of this feature. In a real application, the progress would be calculated out of a real operation, like downloading a file from the internet, loading data into a database, and so on.

With progress bars, we have explored all the opportunities that toast notifications give you to enhance the experience of your users. Now it's time to see another type of notifications supported by Windows: badge notifications.

Displaying a badge in the taskbar

If your application is packaged with MSIX, you can support it with another type of notification, called **badge notifications**. These notifications are helpful when the user decides to pin your application on the Windows taskbar: the badge, in fact, is displayed in an overlay of your application's icon. Badge notifications support two kinds of content: a number (for example, you want to quickly display the number of unread messages) or a symbol (for example, you want to notify users that the application needs their attention).

To use these type of notifications, other than the CommunityToolkit.WinUI. Notifications namespace, you must also add the Windows.UI.Notifications class.

Let's start with an example of creating badge notifications with a number:

```
private void OnSendBadgeNumberNotification ()
{
    BadgeNumericContent numeric = new
        BadgeNumericContent (15);
    var xml = numeric.GetXml ();

    BadgeNotification notification = new
        BadgeNotification (xml);

    var badgeManager = BadgeUpdateManager.CreateBadge
        UpdaterForApplication ();
    badgeManager.Update (notification);
}
```

Also, in this case, the Windows Community Toolkit offers a class that we can use to create badge numeric notifications without having to manually write the XML. We just need to create a new `BadgeNumericContent` object passing, as a parameter, the number we want to display. However, we don't have an equivalent of the `ToastContentBuilder` class, so to display the notification we must use the standard UWP classes `BadgeNotification` and `BadgeUpdateManager`, which can manipulate only the XML content.

To handle this scenario, the toolkit provides a helper method called `GetXml()` that converts the C# object into XML. Once we have it, we can wrap it inside a `BadgeNotification` object and pass it as a parameter to the `Update()` method exposed by the `BadgeUpdateManager` class.

Badge notifications with a symbol (called **glyph**) work in the same way, except that we use the `BadgeGlyphContent` class instead of the `BadgeNumericContent` one. In this case, when we initialize the object, we must pass one of the values of the `BadgeGlyphValue` enumerator, which represents different types of symbols. The following example shows an alert badge on top of the application's icon:

```
private void OnSendBadgeGlyphNotification ()
{
    BadgeGlyphContent glyph = new
      BadgeGlyphContent(BadgeGlyphValue.Alert);
    var xml = glyph.GetXml();

    BadgeNotification notification = new
      BadgeNotification(xml);

    var badgeManager = BadgeUpdateManager.CreateBadge
      UpdaterForApplication();
    badgeManager.Update(notification);
}
```

The following screenshot shows the two types of badge notifications in action:

Figure 9.9 – A numeric badge notification (on the left) and a glyph badge notification (on the right)

With this we covered one set of notifications, now let's move to the final section of this chapter which will be quite brief: as mentioned at the beginning of the chapter, we're going to talk about push notifications, but only as an overview.

Implementing push notifications

All the scenarios we have seen so far are based on local notifications, which are generated directly by your application when a specific event happens. However, there are scenarios where the event isn't generated locally, but remotely. A new message, a new mail, a new goal scored by your favorite football team are just a few examples. Sure, your application could continuously send messages to your backend to check if there are updates, but it wouldn't be very efficient. This approach would have a huge impact on the battery and the resources of your computer.

Therefore, the industry has introduced the concept of push notifications. Instead of the device continuously checking with the backend if there are updates, it's the backend itself that sends a message to the device whenever there's something new. This scenario is made possible thanks to the introduction of a cloud service, which is provided by the owner of the platform and acts as a middleman between the client application and the backend and enables the following flow:

1. The user registers their application against the cloud service, and they receive back a client ID and a client secret that they can use to perform authentication.

2. The first time the application starts, it contacts the cloud service to register the device and it receives back a channel, represented by an HTTP URL.

3. The application sends the HTTP URL to the backend so that it can store it in a database or another similar solution. The channel is stored with some additional information that the backend can use to identify the user. We don't want to send a notification about a message to the wrong user, or an update about the football match to people who aren't interested in that particular team.

4. Whenever there's a new notification to send, the backend authenticates against the cloud service and, if it's successful, it receives an authentication token.

5. The backend generates a payload with the content of the notification and sends it to the HTTP URL of the device, together with the authentication token.

6. If the token is valid, the cloud service processes the notification, and it dispatches it to the device which corresponds to the HTTP URL.

7. The device receives the message, and it uses the payload to generate a notification.

This architecture is adopted by all the players on the market who build mobile and desktop ecosystems: from Microsoft to Google to Apple. However, each platform has its own specific implementation of the following three key areas:

- The authentication system against the cloud service
- The payload's format (some services use XML, some other JSON)
- The type of notifications that are supported by the operating system

The Microsoft implementation of this cloud service is called **Windows Notification Service (WNS)**, which supports three types of notifications:

- Toast notifications
- Badge notifications
- Raw notifications

Technically, WNS supports a fourth type, which is **tile** notifications, but since they aren't supported anymore by Windows 11, we aren't considering them in this section.

The first two types don't require any special configuration from the application's side, other than eventually subscribing to the `OnActivated` event exposed by the `ToastNotificationManagerCompat` class to handle the activation. Your backend will send an XML payload, which describes the content of the notification, and Windows will handle it for you, by displaying a toast notification or by adding a badge to your application's icon.

Raw notifications, instead, give you the maximum flexibility since they can contain any arbitrary payload. However, Windows does not natively manage them, but it's up to your application to handle them most appropriately through the activation events provided by the push notification APIs. Raw notifications are used when the message doesn't necessarily need to be translated with visual content. For example, a raw notification can trigger a background sync, or a long-processing task performed by the client application.

But how can we send a notification from our backend?

Implementing the backend

Push notifications are based on a standard technology: it's an XML payload sent to an HTTP URI using a `POST` command. This means that it doesn't matter if your backend is implemented in .NET, Java, or Node.js, you will be able to easily integrate the sending of push notifications.

However, before sending a notification, you must authenticate against the WNS. Without this requirement, any malicious developer would be able to send notifications on your behalf just by spoofing the URI associated with a device. The authentication part is the most relevant new feature introduced by the Windows App SDK. Push notifications, in fact, are already supported by UWP applications and by MSIX packaged Win32 apps. However, there's a strict requirement to satisfy: the application must be published on the Microsoft Store. By registering your application with the Microsoft Store, you are granted the client identifier and the client secret that are needed by your backend to authenticate against the WNS. These identifiers and secrets, however, are tight to the identity assigned to your application from the Store, which means that the only way for your application to receive notifications is to publish it on the Store. Any other distribution technique would lead to the creation of a different identity, preventing the application to receive notifications.

This approach is a blocker for many scenarios (such as enterprise applications) where the application can't be submitted on the Microsoft Store. As such, the Windows App SDK is introducing (still experimental) an updated version of the WNS which, instead of being tied to the Microsoft Store, is connected to **Azure Active Directory** (**AAD**). The Azure portal provides a way to register an application that needs to interact with AAD, which is already used today in many scenarios: applications that need to authenticate the user using a Microsoft or Microsoft 365 account; the usage of the Microsoft Graph APIs to integrate your application with the Microsoft 365 ecosystem; and many more. When you register a new application on Azure, you will get the client ID and the client secret that your backend needs to authenticate against WNS, so that it can start to send push notifications. Since AAD isn't tight to the Microsoft Store anymore, this scenario enables the usage of push notifications in any kind of Win32 application, regardless of the way it's distributed.

At the time of writing, however, you must be aware of its limitations.

Limitations of the current implementation

As mentioned at the beginning of the chapter, we aren't going through the topic of push notifications in a detailed way because it's still in preview and, as such, subject to multiple changes.

If you want to test this feature, you can follow the guidance available at `https://docs.microsoft.com/en-us/windows/apps/windows-app-sdk/notifications/push/`, but keep in mind the following limitations:

- Push notifications are supported only by MSIX packaged apps.
- The minimum requirement is Windows 10 version 2004.
- There is no guarantee of latency or reliability.

- There's a limit of 1 million notifications per month.

- You can register an application on Azure, but you can't link it to the **Package Family Name** (**PFN**) (which defines the identity) of your application directly on the portal. You must provide it through a form (the link for which is available in the documentation) and wait for the team to manually associate the PFN to your application ID in the WNS database.

Push notifications are planned to be released with the Windows App SDK 1.1 (scheduled for Q2 2022), which will remove the current limitations and will streamline the authentication process.

Push notifications can help to level up your application, by adding even more engagement to your applications. In this section, we have learned just the basic concepts, which can be useful to understand the architecture to implement when the feature will be moved out of preview.

Summary

Notifications is one of the areas where a native application has a significant advantage compared to a web application, especially on Windows. Web applications and browsers provide some limited support for notifications, but it's only thanks to the rich Windows ecosystem that you can enable all the features we have seen in this chapter—images, interactions, progress bars, reminders. Thanks to these features, we can build more engaging applications, which can interact with the user even when they aren't actively using them.

However, be careful when you decide to implement them. Notifications can quickly turn from the most useful feature provided by an application to the most annoying one if they are abused. Don't use notifications for advertising or for delivering information that isn't critical or time sensitive.

In the next chapter, we're going to learn another exciting opportunity that the Windows ecosystem provides for our application: the ability to use **Artificial Intelligence** (**AI**) and **Machine Learning** (**ML**) through the computational power of our devices.

Questions

1. Toast notifications are supported only by WinUI applications. True or false?

2. Adding a tag to a toast notification is an essential requirement for Windows to display it. True or false?

3. There are no differences between implementing toast notifications in packaged and unpackaged applications. True or false?

10
Infusing Your Apps with Machine Learning Using WinML

Artificial intelligence is, without a doubt, one of the most relevant topics in the technology ecosystem, which unlocks new and exciting opportunities to automate tasks that, in the past, simply weren't possible, from recognizing objects in a picture or a video feed to understanding a user's behavior.

These scenarios are powered by machine learning, a technology that enables you to train a model (which is stored in a file) with a set of data. For example, you can train a model to recognize a specific object in a picture. Once you have a model, by identifying patterns, the algorithm will try to apply the same logic used during the training to any picture you provide as input.

The cloud is the technology that has enabled such a big growth of machine learning; thanks to its computing power capabilities, you can train a machine learning model with your data in hours instead of days or weeks. Or you can use many already trained models with a platform-as-a-service approach. Think, for example, of Cognitive Services (`https://azure.microsoft.com/en-us/services/cognitive-services/`), a service provided by Azure that you can use to integrate features such as speech recognition, face recognition, or anomaly detection. All these features are exposed through simple REST APIs or a dedicated SDK that you can consume in any type of application. You just need to pass to the API the input data (such as audio speech or an image) to get back a result already processed through the built-in models.

There are scenarios, however, where the cloud can introduce challenges. Think, for example, of an application that powers an industrial machine and uses a camera to detect anomalies in the pieces that are produced. In this context, you need to analyze a huge volume of pieces very quickly, so the cloud can become a bottleneck; as fast as your connection can be, there will always be a latency. Additionally, the cost can become prohibitive, since you must transfer a huge amount of data.

In this chapter, we're going to learn how to use WinML to solve these challenges through the following topics:

- A brief introduction to ONNX and WinML
- Evaluating a machine learning model in your Windows application
- Training and using your own machine learning model

Let's start!

Technical requirements

The code for the chapter can be found in the GitHub repository here:

`https://github.com/PacktPublishing/Modernizing-Your-Windows-Applications-with-the-Windows-Apps-SDK-and-WinUI/tree/main/Chapter10`

A brief introduction to ONNX and WinML

To solve the challenges we have outlined in the introduction of this chapter, the ecosystem has introduced the concept of Edge computing. Instead of delegating every operation to the cloud, machine learning models are evaluated directly on a local device, without needing an internet connection. The cloud still plays a critical role, but as a companion; for example, if the recognition rate of an anomaly is under a certain threshold, the application might choose to send the video feed to a cloud service for deeper analysis.

Windows 10 has introduced a platform called WinML to enable any Windows device to become part of the Edge ecosystem. Thanks to this feature, you can evaluate machine learning models directly on your device, using the computing power offered by the CPU and GPU of your machine. WinML exposes a set of APIs that you can integrate into your Windows applications so that you can evaluate a model in your application and process the results.

When you start working on Edge scenarios, the main difference compared to a cloud service is that the machine learning model must be hosted directly by your application. **ONNX** (which is the acronym for **Open Neural Network Exchange**) is one of the most popular standards for open machine learning interoperability. When you have a model based on ONNX, in fact, you can consume it with multiple languages and application types.

There are multiple ways to get access to an ONNX model:

- You can download an existing one published on the internet. In this chapter, we're going to build a Windows application based on WinML starting from an existing model.

- Many of the services and tools dedicated to training machine learning models give you the option to export them to ONNX. In this chapter, we're going to see this scenario first-hand with Azure Custom Vision.

- You can use tools such as ONNXMLTools (available at `https://docs.microsoft.com/en-us/windows/ai/windows-ml/onnxmltools`) to convert existing models created with other machine learning platforms, such as TensorFlow, Keras, or Apple Core ML.

An essential tool to work with ONNX is **Netron**, an open source application available at `https://netron.app/` that you can use to explore different types of machine learning models, including ONNX. The application requires a good knowledge of how machine learning algorithms work, but it's also useful for the scenarios we're going to explore in this chapter to determine the input and output we need to manipulate in the application. Netron can be installed on your machine as a desktop application or used directly from your browser through the web application.

ONNX has evolved over time and models can target a specific version of the standard. It's important to always verify the version before using a model, since it might not be supported by the Windows version you're targeting. For example, the latest version of ONNX, 1.7, is only supported by Windows 11. You can use the table available at `https://docs.microsoft.com/en-us/windows/ai/windows-ml/onnx-versions` as a reference to understand the compatibility level.

Once you have an ONNX model, you can integrate it into your application and, through the WinML APIs, you can pass the input data to perform an evaluation. WinML can be consumed in the following two ways:

- **Natively from the operating system**: Windows ships with a built-in version of WinML that doesn't require any dependency other than enabling your application to support Windows 10 APIs, as we learned in *Chapter 8, Integrating Your Application with the Windows Ecosystem*. This approach is the most flexible one since it doesn't require you to add any extra components to your application; additionally, updates and new features are automatically delivered when a new Windows version ships. You will need at least Windows 10, version 1809, to support this approach.

- **From a NuGet package**: In this case, the WinML runtime is directly packaged in your application, which expands the support for this platform back to Windows 8.1. With this approach, however, the size of your application grows (since you must include the runtime as part of the package), and you must take care of updating it when you want to use new features. The package to install is called `Microsoft.AI.MachineLearning`.

Regardless of the approach you choose, what makes WinML powerful is the deep connection with the operating system and the device features. WinML, in fact, abstracts the hardware by enabling developers to manually target the CPU or GPU or to automatically choose the best approach. WinML supports hardware acceleration to provide better performance, by using DirectML, which is a low-level set of APIs that is part of the DirectX family.

Let's start working with WinML now, by building a sample application that can locally evaluate an ONNX model. The features we're going to implement work in the same way regardless of whether you choose to use the native integration or the NuGet package.

Evaluating an ONNX model with WinML

In this section, we're going to start with an existing ONNX model. On the internet, in fact, you can find many pretrained models for some common scenarios, such as object detection or face analysis. A good and reliable source for these models is supported directly by the ONNX community, which curates an official repository on GitHub at `https://github.com/onnx/models`.

The model we're going to use in our application is for object recognition and it's called **SqueezeNet**. Given an image, it's able to recognize objects from 1,000 different categories. It's a great fit for edge scenarios since it provides reliable results with compact size; the model, in fact, weighs only 4 MB, leading to a very small increase in the size of the application. We can download the model directly from the URL `https://github.com/onnx/models/blob/master/vision/classification/squeezenet/model/squeezenet1.1-7.onnx`.

Thanks to this model, we're going to build an application that will enable users to load an image and get back a category that describes it. We can use the Netron application we previously downloaded to explore it. As mentioned in the previous section, unless we have a very good knowledge of machine learning algorithms, it won't be easy to decode the model. However, thanks to the application, we can discover two critical pieces of information we're going to need to use the model in WinML: the input and the output. If we choose **View | Properties** from the top-level menu, Netron will open a panel where we can see that the model expects a property called `data` as input, while the output is stored in a property called `squeezenet0_flatten0_reshape0`.

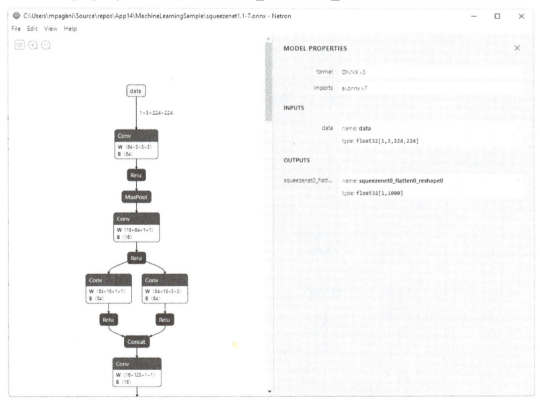

Figure 10.1 – The properties of an ONNX model

We can also see that the output type is an array of numbers, from **1** to **1000**. The model will return, in fact, a probability number for each of the 1,000 categories supported by the model. The category with the highest probability is the one that best describes the content of the image. The category labels, however, aren't included as part of the model. Thanks to Netron, we can, in fact, see that the output is numbers that identify the categories. As such, we also need the label that describes them that we can use to link the identifiers with the category name. In the case of the model we have chosen, this file is published at `https://github.com/onnx/models/blob/master/vision/classification/synset.txt`. Download this file on your computer, as we're going to need it later.

Now that we have everything we need, we can start building our Windows application. In this chapter, we're going to create a WinUI application, but the same APIs can also be used from a WPF or Windows Forms application. Just remember, as we learned in *Chapter 8*, *Integrating Your Application with the Windows Ecosystem*, that if you opt in for WPF or Windows Forms, you must change the `TargetFramework` property of your project to use one of the specific Windows 10 workloads (such as `net6.0-windows10.0.19041.0`).

The starting point of our journey will be the **Blank App, Packaged (WinUI in Desktop)** template in Visual Studio.

Once you have created a new project, the first step is to import the two files we previously downloaded from GitHub: the ONNX model and the text file with the labels. Make sure that, in the `.csproj` definition, these two files are included with **Content** as **Build Action**, as in the following sample:

```
<ItemGroup>
  <Content Include="sysnet.txt" />
  <Content Include="squeezenet1.1-7.onnx" />
</ItemGroup>
```

Now we're ready to start writing some code. Let's see, section by section, all the tasks we need to perform to integrate WinML.

Loading the ONNX model

The first action to take is to load the ONNX model using the WinML APIs so that we can start working with it. A model is represented by the `LearningModel` class, which belongs to the `Windows.AI.MachineLearning` namespace. This class exposes a few static methods that you can use to load a model, which support both unpackaged and packaged scenarios. Let's consider the following sample:

```
private async Task<LearningModelSession>
```

```
    CreateLearningSession()
{

        var modelFile = await
    StorageFile.GetFileFromApplicationUriAsync(new
    Uri("ms-appx:///squeezenet1.1-7.onnx"));

        LearningModel model = await
    LearningModel.LoadFromStorageFileAsync(modelFile);
        LearningModelSession session = new
    LearningModelSession(model);
        return session;
}
```

In our scenario, we're building a packaged application with WinUI, so we can use
the application URI approach we learned how to use in *Chapter 8, Integrating
Your Application with the Windows Ecosystem*. Since the model is part of our package,
we can load it using the `GetFileFromApplicationUriAsync()` method
exposed by the `StorageFile` class, passing as a parameter the URL `ms-appx:///`
`squeezenet1.1.-7.onnx`. We get back a `StorageFile` object, which we can pass to
the `LoadFromStorageFileAsync()` static method exposed by the `LearningModel`
class.

In the case of an unpackaged app, the `LearningModel` class also exposes methods that
support traditional file paths, such as `LoadFromFilePath()`:

```
LearningModel model = LearningModel.LoadFromFilePath
  (@"C:\model.onnx");
```

Once we have a model, we can create an evaluation session, by creating a new
`LearningModelSession` object, passing as a parameter the model we have just
loaded. By default, Windows will automatically create a session on the best available
device. However, you can force the session to run on a specific device, by passing
a `LearningModelDevice` object as a second parameter. The following sample shows
how to force the execution on a GPU powered by DirectX:

```
var session = new LearningModelSession(model, new
  LearningModelDevice(LearningModelDeviceKind.DirectX));
```

Now that we have the model, we also need the labels for the various categories.

Loading the labels

As we mentioned in the introduction, the model we have chosen doesn't include the labels of the categories it supports, but we must load them from an external file we have downloaded from the same GitHub repository. This file (which we have already included in the project) is just a plain text file: each line contains the category index, followed by its description, separated by a comma. The following snippet of code parses the text file to import it into a collection of the `Dictionary<long, string>` type:

```
private async Task<Dictionary<long, string>> LoadLabels()
{
        var file = await StorageFile
    .GetFileFromApplicationUriAsync(new Uri(
        "ms-appx:///sysnet.txt"));
        var text = await FileIO.ReadTextAsync(file);
        var labels = new Dictionary<long, string>();
        var records = text.Split(Environment.NewLine);
        foreach (var record in records)
        {
                var fields = record.Split(",", 2);
                if (fields.Length == 2)
                {
                        var index = long.Parse(fields[0]);
                        labels[index] = fields[1];
                }
        }
        return labels;
}
```

Also, in this scenario, since we're working with a WinUI-packaged application, we can load the file using an application URI, which is `ms-appx:///sysnet.txt` (since we have placed the text file at the root of the project). The code uses the storage APIs we learned about in *Chapter 8, Integrating Your Application with the Windows Ecosystem*, to read the entire text in memory, and then, for each line, it adds a new item to the dictionary called `labels`; the item key will be the category identifier, while the item value will be the category description.

Now that we have all the requirements ready, we can feed the model with the image we want to recognize.

Preparing the model input

Our model requires, as input, an image, which will then be evaluated against the categories supported by the model. As such, as the first step before performing the evaluation, we must enable the user to load an image. In our sample, we're going to allow users to load images from their computer, by using the `FileOpenPicker` class we learned how to use in *Chapter 8, Integrating Your Application with the Windows Ecosystem*. Let's see the full code snippet first:

```csharp
private async Task<ImageFeatureValue> PickImageAsync()
{
        var filePicker = new FileOpenPicker();

        // Get the current window's HWND by passing in the
    Window object
        var hwnd = WinRT.Interop.WindowNative
        .GetWindowHandle(this);

        // Associate the HWND with the file picker
    WinRT.Interop.InitializeWithWindow
        .Initialize(filePicker, hwnd);

        SoftwareBitmap softwareBitmap;

        filePicker.FileTypeFilter.Add("*");
        var file = await filePicker.PickSingleFileAsync();

        using (IRandomAccessStream accessStream = await
        file.OpenAsync(FileAccessMode.Read))
        {
                Windows.Graphics.Imaging.BitmapDecoder
            bitmapDecoder = await Windows.Graphics
            .Imaging.BitmapDecoder.CreateAsync(accessStream);
                softwareBitmap = await bitmapDecoder
            .GetSoftwareBitmapAsync();
                softwareBitmap = SoftwareBitmap.Convert
            (softwareBitmap, BitmapPixelFormat.Bgra8,
                BitmapAlphaMode.Premultiplied);
```

```
    }
      VideoFrame input = VideoFrame.CreateWithSoftwareBitmap
    (softwareBitmap);
      ImageFeatureValue image =
    ImageFeatureValue.CreateFromVideoFrame(input);
      return image;
}
```

The first part of the code should be familiar. It's the solution we learned how to use in *Chapter 8, Integrating Your Application with the Windows Ecosystem*, which enables the usage of the `FileOpenPicker` class in a Win32 application; the solution requires us to get the native handle (HWND) of the application's window and use it to initialize the picker object. Just remember that this sample is specific for a WinUI application. If you're working with WPF or Windows Forms, you'll need to retrieve the HWND using the platform-specific APIs we have seen in previous chapters.

Once we have a reference to the file selected by the user, we open its content through a stream of the `IRandomAccessStream` type. The next steps are needed to prepare the image in the right format expected by the ONNX model, which is BGRA8. It's a 32-bit-per-pixel format in which the blue, green, and red color components are placed before the alpha value. To obtain the image in this format, we use the `BitmapDecoder` class to convert the image into a `SoftwareBitmap` representation, passing `BitmapPixelFormat.Bgra8` as the format type. Once we have a `SoftwareBitmap` object, we can use it to create a `VideoFrame` object (which is a class offered by the `Windows.Media` namespace) thanks to the `CreateWithSoftwareBitmap()` method. Why do we need to perform another conversion? Because the `Windows.AI.MachineLearning` namespace provides a special type to handle images called `ImageFeatureValue`, which can be generated starting from a `VideoFrame` object.

We had to go through a few steps, but finally, now we have the proper input for our model: an image in BGRA8 format with the proper encoding.

Performing the evaluation

Now we have everything we need to perform the evaluation:

- A model, loaded in memory
- A list of labels for the various categories supported by the model
- An image to use as input for the classification

To perform this step, we need some help from the Netron application we installed at the beginning of this chapter. If you remember, by opening our model with this application, we were able to determine the input and output format. We're going to use this information in the next code snippet:

```
private async Task<IReadOnlyList<float>>
    EvaluateModelAsync(LearningModelSession session,
        ImageFeatureValue image)
{
        LearningModelBinding bind = new
        LearningModelBinding(session);

        bind.Bind("data", image);

        var results = await session.EvaluateAsync(bind, "0");

        // Retrieve the results of evaluation
        var resultTensor = results.Outputs
    ["squeezenet0_flatten0_reshape0"] as TensorFloat;
        var resultVector = resultTensor.GetAsVectorView();

        return resultVector;
}
```

Before performing the evaluation, we must first pass the input image to the model. We do this thanks to the LearningModelBinding class, which we initialize using the LearningModelSession object we previously created. Now, we can pass the input by calling the Bind() method, using the following:

- As the first parameter, the key identifies the input. This is the information we have retrieved from Netron; the expected name of the input is data.

- As a second parameter, the ImageFeatureValue object we previously created contains our image.

Now, we can finally evaluate the model against our input, by calling the EvaluateAsync() method exposed by the LearningModelSession object. As parameters, we must pass the LearningModelBinding object we just created and an identifier for the execution (we simply pass a fixed string).

When we call this method, WinML will start evaluating the model using the most appropriate technique, based on the hardware the application is running on. It might use the GPU, the CPU, or a combination of both. Once the operation is completed, we will find the results in the `Outputs` collection of the result. To decide which result is the correct one, we need to again use Netron. In the panel, we can see that the output is stored inside an item identified by the `squeezenet0_flatten0_reshape0` key.

We get back a `TensorFloat` object, which, by calling the `GetAsVectorView()` method, we can convert into an array of numbers. This result will contain exactly 1,000 items, one for each category supported by the model. For each category (identified by the index in the collection), we will get a number with the calculated probability, as you can see in the following screenshot:

Name	Value	Type
▲ 🔲 resultVector.ToList()	Count = 1000	System.Collections...
◈ [0]	45.87921	float
◈ [1]	127.02916	float
◈ [2]	76.03262	float
◈ [3]	36.4985237	float
◈ [4]	38.2322235	float
◈ [5]	48.30803	float
◈ [6]	44.6682854	float
◈ [7]	74.47798	float
◈ [8]	40.9283829	float
◈ [9]	108.549141	float
◈ [10]	109.729279	float
◈ [11]	125.945442	float
◈ [12]	108.440437	float
◈ [13]	74.41786	float
◈ [14]	50.14369	float
◈ [15]	82.57404	float
◈ [16]	167.0952	float
◈ [17]	62.8335381	float
◈ [18]	61.51476	float
◈ [19]	80.40032	float

Figure 10.2 – The list of results with the probability calculated by WinML

As the next step, as such, we have to sort the collection so that we can get the items with a higher probability first; they will tell us the category of the object that has been recognized in the image.

Sorting the results

To sort the results, we can rely on the powerful features offered by C# and .NET when it comes to collection manipulation. Here is how we can achieve our goal:

```
private List<(int index, float probability)>
  SortItems(IReadOnlyList<float> resultVector)
{
        // Find the top 3 probabilities
        List<(int index, float probability)> indexedResults =
    new List<(int, float)>();

        for (int i = 0; i < resultVector.Count; i++)
        {
                indexedResults.Add((index: i, probability:
      resultVector.ElementAt(i)));
        }

        // Sort the results in order of highest probability
        indexedResults.Sort((a, b) =>
        {
        if (a.probability < b.probability)
        {
            return 1;
        }
        else if (a.probability > b.probability)
        {
            return -1;
        }
        else
        {
            return 0;
        }
        });

        return indexedResults;
}
```

We have created a new `List` that will store a collection of tuples. Each item will have an index and a probability value. As the first step, we populate the list with all the items that are included in the vector that WinML has generated for us. Then, by using the `Sort ()` method, we reorder the collection using the probability as a condition to evaluate. The output of this method will be the same vector we received as input but ordered starting from the item with the highest probability to the one with the lowest probability. This is what the final collection looks like:

Name	Value	Type
▲ ⬙ indexedResults	Count = 1000	System.Collections...
▲ [0]	(817, 473.8865)	(int, float)
⬙ Item1	817	int
⬙ Item2	473.8865	float
▸ ⬙ [1]	(751, 448.814667)	(int, float)
▸ ⬙ [2]	(436, 444.7551)	(int, float)
▸ ⬙ [3]	(511, 438.104)	(int, float)
▸ ⬙ [4]	(479, 437.485016)	(int, float)
▸ ⬙ [5]	(581, 410.307129)	(int, float)
▸ ⬙ [6]	(468, 400.940277)	(int, float)
▸ ⬙ [7]	(717, 392.350677)	(int, float)
▸ ⬙ [8]	(654, 388.0898)	(int, float)
▸ ⬙ [9]	(584, 383.902649)	(int, float)
▸ ⬙ [10]	(477, 376.494)	(int, float)
▸ ⬙ [11]	(656, 374.6264)	(int, float)
▸ ⬙ [12]	(518, 367.143616)	(int, float)
▸ ⬙ [13]	(889, 361.33902)	(int, float)
▸ ⬙ [14]	(784, 359.392456)	(int, float)
▸ ⬙ [15]	(407, 356.72168)	(int, float)
▸ ⬙ [16]	(830, 356.073273)	(int, float)
▸ ⬙ [17]	(740, 354.302856)	(int, float)

QuickWatch dialog — Expression: `new System.Collections.Generic.ICollectionDebugView<(int, float)>(indexedResults).Items[0]`

Figure 10.3 – The list of results, this time ordered based on the probability

Each item in the collection has two values: `Item1` is the index of the element in the categories list and `Item2` is the probability that the image is described by that category.

There's a last step we must take care of: we know what the index is, but we don't know what the actual name of the category is. Remember, in fact, that the model returns just the index, not the label. As such, now we must reconcile the list that WinML has generated for us with the list of labels we populated at the beginning starting from the text file.

We can do this very easily. Since the collection of the labels contains the index of the category, we just need to extract the items from the `indexedResults` collection with the same index. This is the code we can use to achieve this goal:

```
for (int i = 0; i < 3; i++)
{
        Console.WriteLine(
          $"\"{labels[indexedResults[i].index]}\" with
            probability of {indexedResults[i].probability}");
}
```

With this snippet, we are displaying on the console output the top three items in the collection, which means the three categories that have the highest chances of matching the content of the image we have uploaded.

Putting everything together

Now that we have defined all the methods to work with WinML, we can put everything together in a single method that, for example, we can invoke when the user presses a `Button` control in the application. This is what it looks like:

```
private async void OnRecognize(object sender,
   RoutedEventArgs e)
{
        LearningModelSession session = await
        CreateLearningSession();

        Dictionary<long, string> labels = await LoadLabels();

        ImageFeatureValue image = await PickImageAsync();

        IReadOnlyList<float> resultVector= await
        EvaluateModelAsync(session, image);

        List<(int index, float probability)> indexedResults =
        SortItems(resultVector);

        // Display the results
        for (int i = 0; i < 3; i++)
```

```
        {
                    Debug.WriteLine($"\"{labels[indexedResults[i].
                    index]}\" with confidence of
                    {indexedResults[i].probability}");
        }    }
    }
```

This flow recaps everything we have done so far:

1. We have loaded the model starting from the ONNX file and we have created an evaluation session starting from it, using the `LearningModelSession` class offered by WinML.

2. We have loaded the list of labels of the categories supported by the model, starting from a text file provided with the ONNX model.

3. We have enabled users to pick an image from their computer. Then, we have gone through the process of converting the selected file into an `ImageFeatureValue` object, which is the type required as input by WinML.

4. We have run the evaluation, which returns, as a result, a list of the categories and, for each of them, a probability value, which tells us how many chances there are that the content of the image falls into that category.

5. We order the results starting from the one with the highest probability value.

6. Since the model doesn't return, as output, the category labels, we have to manually match the category index returned by the evaluation process with the list that we loaded at the beginning starting from the text file.

7. Finally, we display the first three items in the list, which are the three categories that have the greatest chances of matching the content of the image of the user.

The following screenshot shows what happens when you use WinML to evaluate the ONNX model we downloaded from GitHub, passing, as input, the photo of a car:

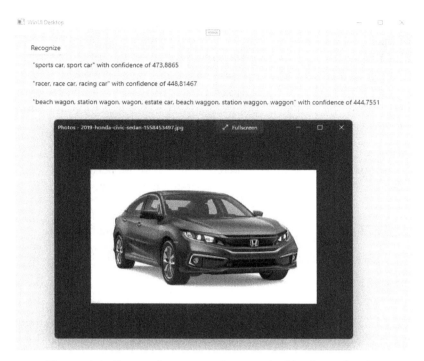

Figure 10.4 – Image of a car properly recognized by the application

You can see how the object recognition works well; the result with the highest probability matches the content of the image (sport car), followed by two other categories that are still a good match (race car and station wagon).

Exciting, isn't it? However, what if we can't find any model in ONNX format that can be a good fit for our scenario? Let's explore how we can train our own one and use it with WinML.

Training and using your own machine learning model

In this section, we're going to reuse the WinML APIs we have learned about so far, but on a different ONNX model. Instead of downloading it from the internet, we're going to create one. Thanks to the power of the cloud, there are many services that we can use to train our models without having to be data scientists. One of these services, provided by Microsoft, is called **Custom Vision**. It's based on Azure, and it requires you to have an Azure subscription. However, the service comes with a free tier, which is a great fit for testing scenarios.

Thanks to Custom Vision, we can train our own model without having to master all the inner details of machine learning algorithms. We will provide a series of pictures and tag them. Then, Custom Vision will train the model based on the information we have provided, enabling it to recognize the objects we have also tagged in every photo, even the ones that aren't part of the original training set. Custom Vision is a great companion for WinML because, among the available options, it supports exporting a trained model to an ONNX file, which we can import into our Windows application.

Let's start by going to `https://www.customvision.ai/` and signing in with a Microsoft account or work account connected to an Azure subscription. If you don't have one, you can create a trial starting from `https://azure.microsoft.com/en-us/offers/ms-azr-0044p/`.

Once you are logged in, click on **New project** and set the following fields:

- **Name**: Choose a name for your project.

- **Description**: This is optional; feel free to leave it empty or fill it with a description of your project.

- **Resource**: Choose the Azure resource that will host the service. If you don't have one, you can choose **Create new** to create a new service, by providing the following:

  - The name.

  - The Azure subscription where you want to host it.

  - The resource group that will contain the service.

  - For **Kind**, choose `CustomVision.Training`.

  - The Azure region that is going to host the service. Choose the one closest to your location.

  - **Pricing tier**: Choose **F0** if you want to use the free tier.

- **Project Types**: Choose **Classification** since we're going to train a model to recognize the content of an image.

- **Classification Types**: Choose **Multiclass** (single tag per image).

- **Domains**: Select **General (compact) [S1]**. It's important to choose one of the compact models since they are the ones that can be exported as ONNX. Once you choose this option, a new field will appear called **Export Capabilities**. Make sure to keep the default value, which is **Basic** platforms. Other domains can give you better results, but they can be consumed only through REST APIs, so they require an internet connection.

Once you have created a new project, you will have access to the main dashboard of the app, which is split into the following three sections:

- **Training Images**: We're going to use these to provide the training data for the model.

- **Performance**: This will tell us how reliable the generated model is.

- **Predictions**: We can use this to test our model.

Let's start with the training!

Training the model

To train the model, we must provide a set of images that describe the object (or the objects) that we want to recognize. In this example, we're going to build a model that can recognize planes in a photo. By clicking the **Add images** button on the dashboard, the tool will give you the option to look for training images on your computer. In the GitHub repository connected to this chapter, you will find a folder with various plane images; otherwise, you can look for them on the web.

Make sure to provide at least five images; it's the minimum requirement to start the training. Of course, the more images you provide, the better the accuracy will be. Once you have selected the images, the tool will ask you to tag them.

We're going to tag them with the `plane` keyword, as shown in the following screenshot:

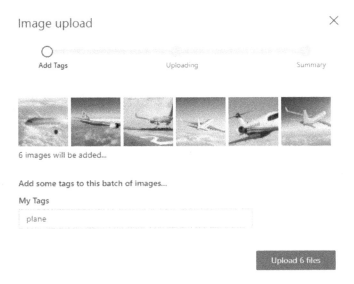

Figure 10.5 – The set of plane images we're providing as the training set for our model

Click on **Upload X files** (where X is the number of images you have uploaded) and wait for the process to finish.

Now, provide a few images that, instead, don't include a plane by again clicking on **Add new images** and, this time, looking for five random pictures you have on your computer, as long as they don't represent a plane. This time, like tags, we're going to use a special one called **Negative**, which means that the image doesn't include any visual subject of any other tag.

In the following screenshot, you can see an example with a set of photos of cars. Of course, I could have used the car tag to identify these photos; this way, the final model would be able to recognize both photos of planes and cars. However, for the sake of simplicity, we're going to create a model that is able to recognize only planes, so I'm marking the car photos as **Negative**:

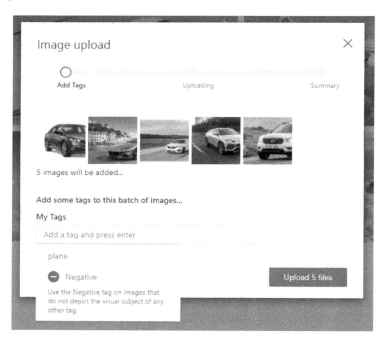

Figure 10.6 – The set of car images is tagged with the special Negative keyword

Now, we have everything we need to start basic training: five images with the content we want to recognize (planes) and five images with different content (cars). Click on the green **Train** button at the top and choose to perform a quick training.

Figure 10.7 – The button to start the training of the model

The training will take a few minutes. The duration of the training is proportional to the number of images you have uploaded. Once the training is completed, the portal will update to show you some statistics about the quality of the model that has been generated.

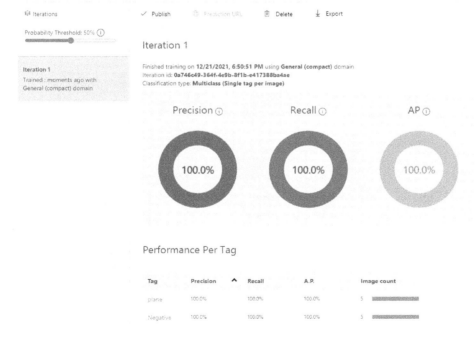

Figure 10.8 – The outcome of the training performed with the provided data

Now that we have a model, we can test it before exporting it thanks to a dedicated feature offered by the Custom Vision portal.

Testing the model

To test the model you have generated, you can use the **Quick Test** button, which is placed near the **Train** button one. You can use this tool to either load an image from the disk or by directly sharing the URL on the internet. Either way, Custom Vision will analyze the image against the model and calculate the probability for each of the tags. In the following screenshot, you can see how the tool has identified that the provided image doesn't contain a plane since **Probability** is **0%**:

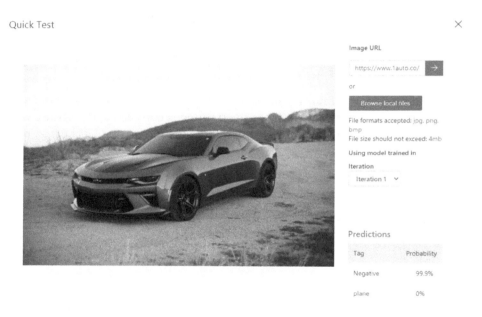

Figure 10.9 – Testing the model with the Quick Test tool

All the images you use for testing will be stored in the **Predictions** section of the portal, which you can use to further improve the model. For example, you can see in the following screenshot that the model doesn't work quite right with the GitHub logo, since it has assigned almost a 90% probability that the image includes a plane. As such, we can assign to the image the **Negative** tag, then press **Save and close**. The image will be added to the ones used as training data under the **Training Images** section:

Figure 10.10 – The model didn't properly classify the GitHub logo, so we can fix it by assigning the right tag

Now, if we want, we can again press the **Train** button to run a new iteration, which will take into consideration all the new images that we used for testing the model and, as such, will be more reliable.

Once we are satisfied with the model, we are ready to export it.

Exporting the model

To export the model, you must move to the **Performance** section of the dashboard, where you will find at the top the **Export** button. When you click it, the Custom Vision portal will offer you a series of formats that are supported. Choose **ONNX** and click **Export** to start the process. Once the model has been generated, click on **Download** to get the ONNX file on your local disk.

The model will be included in a ZIP file; so, before using it, extract the file with the `.onnx` extension in a folder. Then, import it into your project in Visual Studio, in the same way you imported the ONNX model you downloaded from GitHub at the beginning of this chapter.

Before testing it, however, we must make a few changes to our application:

- We must open the model with Netron to see whether the inputs and the outputs still match. We can observe that, in both cases, the expected type is the same. The input is an image (represented in BGRA8 format), while the output is an array of numbers. However, this time, the size of the array is only 2, since our model can recognize only two categories: `plane` and `negative`. Additionally, the name of the output is different; it's `model_output` now. As such, we have to change the `EvaluateModelAsync()` method, as follows:

```
private async Task<IReadOnlyList<float>>
  EvaluateModelAsync(LearningModelSession session,
    ImageFeatureValue image)
{
      LearningModelBinding bind = new
    LearningModelBinding(session);

      bind.Bind("data", image);

      var results = await session.EvaluateAsync(bind,
    "0");

      // Retrieve the results of evaluation
      var resultTensor = results.Outputs["model_
        output"]
    as TensorFloat;
      var resultVector = resultTensor.
        GetAsVectorView();

      return resultVector;
}
```

- The list of labels we are loading from the text file doesn't match the categories supported by our model anymore. Since, in our case, we support only two categories, we can change the `LoadLabels()` method by manually loading our list of tags in alphabetical order:

```
private Task<Dictionary<long, string>> LoadLabels()
  {
```

```
        var labels = new Dictionary<long, string>();
        labels.Add(0, "negative");
        labels.Add(1, "plane");

        return Task.FromResult(labels);
}
```

Now, we can run our application and pick any image of a plane. You should get the following result:

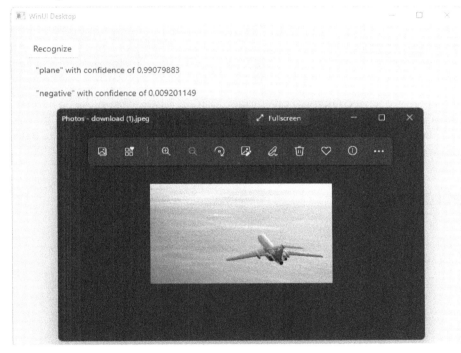

Figure 10.11 – The evaluation result returned by the model we have trained with Custom Vision

As you can see, our model is working. Through the evaluation performed by WinML, our application has recognized that the image is a photo of a plane.

Custom Vision is just an example of a way to train your model. In the official WinML documentation, available at `https://docs.microsoft.com/en-us/windows/ai/windows-ml/`, you will find many other examples, based on other frameworks that can also train models locally, such as ML.NET or PyTorch.

Summary

In this chapter, we have learned about another great feature provided by Windows, WinML, which enables our computers and tablets to become part of the Edge ecosystem. Thanks to WinML, we can evaluate machine learning models without the need for an internet connection and by using the power of the CPU and GPU of a computer. Object detection, object recognition, and sentiment analysis are just a few of the scenarios that we can enable in situations where we might not have a reliable internet connection, or where we need to process a lot of data in a very short amount of time.

WinML is another example of the new direction of the Windows development ecosystem. It started as a feature that could be used only in Universal Windows Platform applications and it gradually evolved to a platform that can be integrated into any kind of Windows application. In this chapter, we have seen a practical example with a WinUI application, but the same exact code would have worked in a WPF or Windows Forms application, enabling developers to add machine learning to many of the enterprise applications on the market.

With this chapter, we have concluded our modernization journey from the developer aspect of an application. In the remaining chapters, we'll explore another topic that doesn't involve code but is equally important for Windows developers, how to distribute a desktop application, starting in the next chapter.

Questions

1. What is ONNX?

2. WinML uses the power of the cloud to evaluate machine learning models and return the results back to a Windows application. True or false?

3. What is Custom Vision?

Section 4: Distributing Your Application

Section 4 will be dedicated to the multiple ways you can deploy a Windows application. The chapters will cover the new Microsoft Store, technologies such as App Installer to deploy your app via a website, and enterprise deployment and CI/CD for Windows applications.

This section contains the following chapters:

- *Chapter 11, Publishing Your Application*
- *Chapter 12, Enabling CI/CD for Your Windows Applications*

11
Publishing Your Application

When you build Windows desktop applications, there's another stage of the project that is equally important to the development phase: the deployment. If you build the best application in the world, but the deployment process is complicated and prone to errors, your customers will be equally dissatisfied, and they will just give up.

However, the deployment technology is only one side of the coin: once you have found the right one, you need a platform to distribute it.

In this chapter, we're going to address all these challenges by going through the following topics:

- Understanding the MSIX packaging technology
- Creating and signing an MSIX package
- Publishing your application to Microsoft Store
- Publishing your application to a website using AppInstaller
- Publishing your application to the Windows Package Manager repository

We have mentioned MSIX and the Windows App SDK package model multiple times in this book. It's time to dive a little bit deeper into the topic and gain a better understanding of what MSIX can do for us.

Understanding the MSIX packaging technology

MSIX is a new packaging technology introduced by Microsoft with the goal to solve many of the challenges that the previous installation technologies (such as MSI or ClickOnce) have created over the years. Some of these challenges include the following:

- The need for enterprises to repackage applications at every update when they have to make any customization to satisfy the company's requirements

- The requirement of using third-party tools or Visual Studio customizations to create an installer as part of the build process

- The need to use third-party services (or build your own custom one) if you want to provide automatic updates

- The inability to completely uninstall an application often leaves orphan registry keys or files, which, over time, slow down the operating system

The key difference between MSIX and other technologies is that it's completely integrated into Windows. This means that many of the challenges that we have outlined are solved without requiring any extra work from the developer. Let's briefly look at all of the main advantages of this technology.

Built-in optimizations around network bandwidth and disk space

Thanks to the Windows integration, MSIX can provide two features that are especially helpful in enterprise environments, where applications are deployed across thousands of devices scattered across different locations in the world.

The first one is differential updates. Whenever you release an update of your application, as a developer, you just need to create a new MSIX package with all of the content. The package can either be used for the first installation or to update an existing installation. In the second scenario, Windows won't download the entire content of the package, but only the files that have changed from the previous version. This feature introduces advantages for both users and developers:

- Users will save time and bandwidth since they won't have to redownload the entire application from scratch.

- Developers don't need to create a separate packaging and deployment process for the main application and updates. They can just create a single MSIX package that is optimized for both scenarios.

The second optimization is around disk space. If you install an MSIX package that contains a file, with the same hash of another file included in another MSIX package, Windows won't download it again and won't store a second copy. However, it will create a hard link so that the same file can be used by both applications. If one of the applications gets uninstalled, Windows will move the file so that it doesn't get lost.

Deployment flexibility

MSIX is the successor of AppX, the original packaging format introduced in Windows 8 for Store applications. The technology has evolved over the years, and while in the beginning it was only supported by Microsoft Store, today, MSIX can be deployed with a wide range of consumer and enterprise technologies, including the following:

- Sideloading, either locally or from a website or network share
- Microsoft Endpoint Manager
- Microsoft Intune
- Windows Package Manager

Also, MSIX has recently become the leading technology for deploying applications in cloud environments. Thanks to a feature called **MSIX App Attach**, today, you can use the same MSIX package to deploy an application either in a local environment or on Azure Virtual Desktop, that is, the Microsoft platform to create remote desktop environments, which makes it possible to access your workstation from anywhere. You can learn more about this feature at `https://docs.microsoft.com/en-us/azure/virtual-desktop/what-is-app-attach`.

Additionally, MSIX enables automatic updates not only in managed environments such as Microsoft Store but also in unmanaged ones like a website. We'll learn about this in more detail in the *Publishing your application to Microsoft Store* section.

Tamper protection and signature enforcement

MSIX packaged applications are installed inside a special folder, `C:\Program Files\WindowsApps`, which is a system-protected folder. Only Windows can modify or delete the content of this folder. Additionally, when an application is installed or updated, Windows Stores the hash of each file that is included inside the package. When the application starts, Windows double-checks that the hash of these files hasn't changed. If that happens, it means that a malicious actor has changed the content of the package; therefore, Windows will block the execution. The user must repair the application before using it again, which can be done in different ways based on the way it was deployed:

- If the application has been installed using Microsoft Store or a website, it will be automatically repaired by redownloading the package.

- If the application has been installed using an enterprise tool, it will be up to the IT Pro department to push the application to the affected devices again.

Additionally, certificates play a critical role to ensure the reliability of an MSIX package. A package must be signed with a certificate trusted by the machine; otherwise, Windows will block the installation. Unlike other technologies (such as MSI), there is no way to install a package that isn't signed or that is signed with an untrusted certificate. We'll learn more about certificates in the *Publishing your application to Microsoft Store* section.

Registry virtualization

The Windows registry is one of the areas that is heavily affected by bad installation experiences. It's very common for applications to create new hives and keys in the registry, which are left in the system even when the application is uninstalled. This leads the registry to become fragmented, which tends to slow down Windows over time. To overcome this limitation, MSIX introduces the concept of registry virtualization, which is implemented in the following way:

- When the application creates or updates a registry key, the operation is performed against a virtual copy of the registry that only belongs to your application. If you open the **Registry Editor** screen, you won't see these keys.

- When the application reads a registry key, Windows will merge the system registry with the local virtual copy for your application. This approach enables your application to read both keys that you have created and keys that belong to the system configuration.

This feature helps to deliver a clean uninstallation experience. When the application is uninstalled, Windows will simply delete the virtual copy of the registry, avoiding leaving any orphan registry keys on the system.

However, this feature also introduces one of the potential limitations that you must consider when you plan to adopt MSIX. Since the virtualized registry is only visible to your application, this means that you can't create keys to share information with another application, since they won't be able to see them.

Local application data virtualization

One of the best practices for Windows development is to use the special `AppData` folder to Store data that is tightly coupled with the application. Configuration files, databases, and log files are just a few examples of data that only makes sense for your application, and, as such, they should be removed when the application is uninstalled. The `AppData` folder is the right place to Store these files since it lives inside the user-space; therefore, no matter whether the user is an administrator or has regular permissions, you can be safe in the knowledge that your application will always have the permission it requires to work with these files.

However, having the `AppData` folder in a different location from the installation leads to the same problems that we have seen with the registry. It's very common for an application to create some files in the `AppData` folder, just to leave them even after it has been uninstalled.

When an application is packaged with MSIX instead, all the reading and writing operations against the `AppData` folder are automatically redirected to the local storage of the application, which we learned about in *Chapter 8, Integrating Your Application with the Windows Ecosystem*. The goal is the same as the registry virtualization: when the application is uninstalled, the local storage is automatically deleted, making sure that any unnecessary files aren't left behind, wasting space on the disk.

The Virtual File System (VFS)

Inside an MSIX package, you can create a special folder, called VFS (which stands for **Virtual File System**), that you can use to virtualize most of the system folders, such as C:\Windows, C:\Program Files\, and more. When your application runs and it looks for files in one of these folders (for example, it's very common when you have a dependency on a runtime or a framework), first, Windows will search for them inside the VFS folder. If it doesn't find them, it will fall back to the actual system folders. Thanks to the VFS, you can solve two familiar challenges around application deployment:

- **Conflicts with existing dependencies**: Often, a problem encountered here is **DLL Hell**. Let's consider the following scenario: you have an application A, which requires framework XYZ to properly work. This framework is installed system-wide, and it's shared by all the applications that require it. At some point in time, you install application B, which requires a newer version of framework XYZ that overwrites the existing one. Application B will run fine, but the new version of the framework contains multiple breaking changes that will prevent application A from running properly. Thanks to VFS, you can solve this problem by including the framework inside the package. Windows will automatically use the embedded version instead of the system-wide one, preventing any conflict.

- **Creating self-contained packages**: Thanks to VFS, you can include all of the dependencies that you need inside the package, removing the requirement for your users to manually install runtimes and frameworks to install your application.

On the following page, https://docs.microsoft.com/en-us/windows/msix/desktop/desktop-to-uwp-behind-the-scenes, you can find a table with all of the system folders supported by VFS and the name you must use to map them inside the VFS folder.

Now that we have learned all the main features of MSIX, let's see the things we have to bear in mind before choosing to adopt MSIX.

MSIX limitations

At the time of writing, MSIX doesn't provide all the capabilities that are supported by traditional MSI installers. Some of them will likely never be supported since they are incompatible with the goal of MSIX of providing reliable and safe installation technologies. Let's see the most important limitations to bear in mind and examine how we can overcome some of them:

- Since the registry and the application data folder are virtualized, you can't use them to share data across multiple applications. If you have this requirement, there are a few available options:

- Starting from Windows 11, you can enable flexible virtualization. You can specify a list of folders under `AppData` and registry keys under the HKCU (Current User) hive that won't be virtualized but will be created in the system. You can learn more about this feature at `https://docs.microsoft.com/en-us/windows/msix/desktop/flexible-virtualization`.

- You can disable virtualization completely by using a restricted capability called `unvirtualizedResources`. However, this capability will add some restrictions to the way you can deploy this package. You won't be able to publish it to Microsoft Store or via AppInstaller, but only using PowerShell.

- If you are an IT pro, Windows 11 has introduced a feature called **Shared Package Container**, which you can use to share the same content across multiple applications. This way, they will be able to access the same registry keys and the same files, even if virtualized. You can learn more about this feature at `https://docs.microsoft.com/en-us/windows/msix/manage/shared-package-container`.

- You can't install software or hardware drivers. Drivers must be certified and published through Windows Update.

- You can't make changes to the Windows configuration. Since the application runs inside a virtualized environment, operations such as creating environment variables or enabling a Windows feature aren't supported.

- You can't run scripts or custom tasks. The MSIX deployment process can only copy the package content on the disk. If you need to run scripts, you can evaluate the adoption of the Package Support Framework, which is a library that you can integrate into your package to overcome some of the MSIX limitations. One of the features supported by this library is to run scripts, with the following limitations:

  - Scripts can be run only when the application starts or quits. They can't be executed as part of the deployment.

  - Scripts will run inside the virtualized environment, so you still won't be able to perform tasks such as setting an environment variable.

  You can learn more about this option at `https://docs.microsoft.com/en-us/windows/msix/psf/run-scripts-with-package-support-framework`.

- You can't deploy files outside of the installation folder. The content of an MSIX package is automatically copied inside a dedicated folder in `C:\Program Files\ WindowsApps`. If your application needs to access files outside the installation folder, you can do the following:

  - Change the code of your application so that the files are copied at the first execution.

  - Use the Package Support Framework to run a script that copies the files you need outside of the installation folder.

  - Starting from Windows 11, you can use an extension called `MutablePackageDirectories`, which you can use to project a file into a system folder: `https://docs.microsoft.com/en-us/windows/msix/ manage/create-directory`.

    - If it's a read-only operation, you can include the file in one of the special folders supported by the VFS.

- You can't write any file in the installation folder since it is system-protected. The best solution to overcome this limitation is to change your application's code to avoid using the installation folder to store files. If you can't make this change, you can use the Package Support Framework to redirect all the writing operations against the installation folders to the local storage of the application. This solution is documented at `https://docs.microsoft.com/en-us/windows/msix/ psf/psf-filesystem-writepermission`.

Now you have all the tools that you need to understand whether MSIX is the right technology for your application. It's time to learn more about how an MSIX package is structured.

The anatomy of an MSIX package

Compared to other deployment technologies, MSIX is a quite simple format. In fact, an MSIX package is just a compressed file, which you can open with any popular compression utility such as 7-zip or WinRAR. Inside an MSIX package, you will find the following:

- The files that make up your application, such as binaries, executables, DLLs, and assets.

- The manifest file, called `AppxManifest.xml`. We have met this file multiple times across the book. It describes the application, the extensions, and the features it uses. It's essential for Windows to manage the life cycle of the application. For example, through the manifest, Windows can detect whether it's the first installation or whether it's an upgrade.

- The blockmap file, called `AppxBlockMap.xml`. This contains the list of all the files included in the package along with their hashes. This file is used to light up some of the features we have described so far, such as tamper protection, differential updates, and disk space optimization.

The metadata inside the manifest file is used to determine two key information instances to identify an MSIX package:

- **Package Family Name** (**PFN**): This identifies the lineage of the package. For example, the PFN of Microsoft Store, one of the built-in Windows apps, is `Microsoft.WindowsStore_8wekyb3d8bbwe`. It's composed of the publisher's name (`Microsoft`), the application identifier (`WindowsStore`), and a hash, which is calculated based on the publisher. This makes it impossible for two different companies to create an application that has the same PFN. The folder where Windows creates the local storage of the application uses the PFN as the name.

- **Full Package Name**: This identifies a specific version of the package installed on the machine. Other than the information already described by the PFN, it also includes the version number and the CPU architecture. For example, at the time of writing, the full package name of Microsoft Store is `Microsoft.WindowsStore_22112.1402.3.0_x64__8wekyb3d8bbwe`. Windows uses the full package name for the name of the folder where it Stores the content of the package (inside `C:\Program Files\WindowsApps`).

The identity of an application can be customized by using the Visual Studio manifest editor, which is automatically opened when you double-click on the `Package.appxmanifest` file. You will find all the options under the **Packaging** section, as you can see in the following screenshot:

Application	Visual Assets	Capabilities	Declarations	Content URIs	Packaging

Use this page to set the properties that identify and describe your package when it is deployed.

Package name:	WinUINotifications			
Package display name:	WinUI Notifications			
	Major:	Minor:	Build:	
Version:	1	0	0	More information
Publisher:	CN=Matteo Pagani, O=Matteo Pagani, L=Como, S=Como, C=IT			Choose Certificate...
Publisher display name:	Matteo Pagani			
Package family name:	WinUINotifications_e627vcndsd2rc			

Figure 11.1 – The Packaging section of the manifest editor in Visual Studio

The fields that are used to calculate the PFN are **Package name** and **Publisher**. When you change them, you will observe that the **Package Family Name** field at the bottom will also change. Additionally, notice how the **Publisher** field can't be freely edited, but there's a **Choose Certificate…** button on the right-hand side. As we're going to learn when we talk about signing, this is one of the MSIX requirements to satisfy: the **Publisher** field in the manifest must always match the subject of the certificate that you're going to use to sign the package.

Now that we have learned all the features of an MSIX package, it's time to see how to create one.

Creating and signing an MSIX package

There are three approaches that you can use to create an MSIX package:

- **Using Visual Studio**: This is the approach that we have seen in all the various chapters of the book so far. If you're either using the MSIX single-project or the Windows Application Packaging Project, you can quickly generate an MSIX package from your Windows application directly from Visual Studio. This is the approach we're going to explore in this chapter since it's the most suitable one for a developer.

- **Using the MSIX Packaging Tool**: This is a tool that has been released by Microsoft. It enables you to repackage any kind of application as MSIX. Using a special driver, the application captures any changes made to the system by a traditional installer and Stores them inside the MSIX package. It's mainly aimed at IT pros since you can use it to repackage applications for which you don't own the source code. You can learn more about this tool (including the link to download it) at `https://docs.microsoft.com/en-us/windows/msix/packaging-tool/tool-overview`.

- **Using a third-party tool**: In the last few years, most of the popular tools to create installers, such as Advanced Installer, InstallShield, or Wix, have added support for MSIX. By using the same project you use today to generate a traditional installer with MSI, you can also create an MSIX package. You can find the list of all the partners at `https://docs.microsoft.com/en-us/windows/msix/partners`.

If you have followed the book so far, your WinUI, WPF, or Windows Forms application should already be set up for MSIX packaging. We explored the various options in *Chapter 1, Getting Started with the Windows App SDK and WinUI*:

- In the case of a WinUI application, the MSIX generation is provided by the single-project MSIX, which is used when you create a new packaged application using the Visual Studio template.

- In the case of a WPF application or a Windows Forms application, you can package it with MSIX using the Windows Application Packaging Project.

Regardless of your choice, if you right-click on the project in Visual Studio, you will find an option called **Package and Publish** (if you're using the single-project MSIX template) or just **Publish** (if you're using the Windows Application Packaging Project). Inside this menu, you will find the **Create App Package** option, which will start a wizard that we can use to generate an MSIX package. Let's see the various steps in more detail.

Choosing a distribution method

The first choice in the wizard is how you want to distribute your application:

- **Microsoft Store**: This option generates a package optimized for Microsoft Store. This means the following:

    - The signing step will be skipped since Microsoft Store will automatically take care of the signing process as part of the package submission.

    - In the next step, we'll be asked to log in with the Microsoft account associated with our developer account in Microsoft Store. Then, Visual Studio will show a list of the names we have reserved, along with the option to create a new one so that the generated identity can be automatically injected inside the manifest.

    We'll learn more about Store publishing later in this chapter.

- **Sideloading**: This option is for all the other scenarios: distributions via a website, a network share, an enterprise tool, or even just mailing the MSIX package to a coworker.

Then, you can move on to choose the signing method.

Choosing a signing method

Thanks to this step, we can choose the certificate we want to use to sign our package. As mentioned earlier, this step will be skipped if we have chosen Microsoft Store as a distribution method. We'll be able to pick a certificate from multiple sources such as the following:

- Azure Key Vault: This is a cloud service where we can safely host protected information, such as secrets, passwords, and certificates. You can learn more at `https://azure.microsoft.com/en-us/services/key-vault/`.

- The local Windows certificate Store.

- A PFX file with the certificate.

- A new self-signed certificate.

    You can observe these options in the following screenshot:

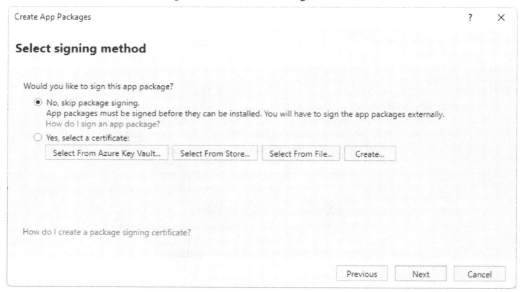

Figure 11.2 – The available signing options during the MSIX generation

We'll talk about the signing options, in more detail, later.

The next step will help us to define the content of the package.

Selecting and configuring packages

This section is very important as it defines the content of the package. This is what the section looks like:

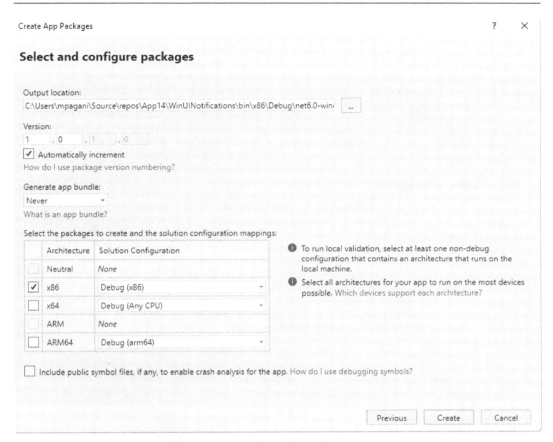

Figure 11.3 – The option to configure the packages you want to generate

From the preceding screenshot, we observe that we can first configure the following settings:

- **Version**: This is the version number associated with this release, which is stored in the manifest. The version must follow the X . Y . Z . 0 syntax: you can change the first three digits, but the last one must always be equal to 0. It's important to always increase the version number when you generate a new package since it enables Windows to use it to update an existing installation. For this reason, you also have an option (**Automatically increment**) that will take care of this for you.

- **Generate app bundle**: In this book, we learned that the Windows App SDK is a native runtime. As such, applications that use it must be compiled for a specific CPU architecture. This means that if you want to build an optimized version for each CPU, you must create and distribute multiple packages. By generating an app bundle, you can generate a single package that supports all CPU architectures. The package size will be bigger, but Windows will automatically download only the binaries targeting the CPU architecture of the current device.

The second section is needed to define what packages you want to create, based on the CPU architecture. The Windows App SDK supports x86, x64, and ARM64. For each of them, you can choose the configuration mode (either **Debug** or **Release**).

You have reached the end of the wizard: now you can hit **Create**, and Visual Studio will build your application and create the MSIX package. Once the operation has finished, a prompt will give you a link to directly access the folder where the package has been created:

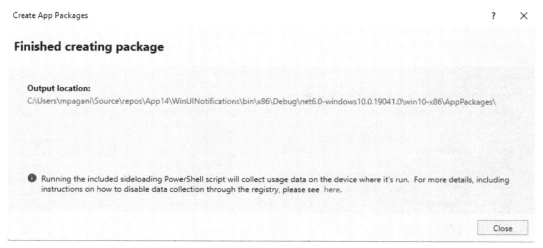

Figure 11.4 – The wizard is complete: the MSIX package has been generated

If you double-click on the MSIX package, Windows will open the user interface (provided by the AppInstaller application, which we're going to see in more detail later) that you can use to install the application on the machine:

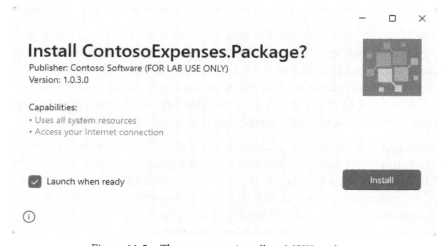

Figure 11.5 – The prompt to install an MSIX package

However, the application can be installed only if the package is signed with a trusted certificate. Let's learn more.

Signing an MSIX package

As already mentioned, signing is a key operation when you work with MSIX packages. If the package is not signed, users won't be able to install it.

This is the experience that users will get when they try to install a package that isn't signed or has been signed with a certificate that is not trusted by the current machine:

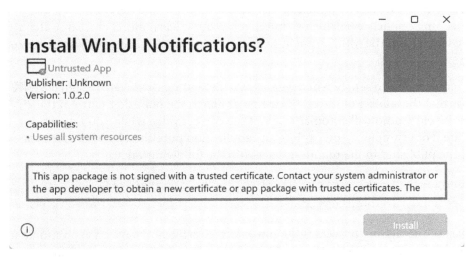

Figure 11.6 – The prompt to install an MSIX package that isn't signed or using a certificate that isn't trusted

There are three ways to obtain a code signing certificate:

- **A public certificate authority** (such as Digicert or Comodo): These certificates are the best fit for applications that must be publicly distributed since they are automatically considered to be trustworthy by Windows. When you buy one of these certificates, you must go through a strict process that verifies your identity. Thanks to this, the user won't have to take any special actions to install your MSIX package.

- **An enterprise certificate authority**: This is the most common scenario for enterprise distribution. These certificates are released to all the devices of the company, which makes them a good fit for signing applications that must be deployed internally.

- **With a self-signing certificate**: These certificates can be manually created on your local machine, without involving a certificate authority. However, they can only be used for testing scenarios since, by default, they aren't considered to be trustworthy by Windows. Unlike the ones released by a public certificate authority, you can create them with any random name, since the identity isn't validated. Users must manually add them to the **Trusted People** category in the certificate store, which requires administrative rights.

Microsoft is also working on a new service, called **Azure Code Signing Service**, which will make the signing experience easier and cheaper. One of the downsides of public certificates is that they can be quite expensive, and they must be renewed periodically. The service isn't publicly available yet, but if you want to learn more, you can watch the following video from the MSIX team at https://www.youtube.com/watch?v=Wi-4WdpKm5E.

The key requirement to satisfy when you use a code signing certificate to sign a MSIX package is that the subject of the certificate must match the publisher entry in the manifest. Visual Studio will automatically set the correct value when you import a certificate. For example, if you explore any application published by Microsoft, you will find that the publisher in the manifest is set with the following value:

```
CN=Microsoft Corporation, O=Microsoft Corporation,
   L=Redmond, S=Washington, C=US
```

A vital part of the signing process is the timestamp. Certificates have an expiration date, which can lead to a package to stop working after it's been passed. By specifying a timestamp server, you'll be able to continue installing and using the package even after the certificate has expired. You can find a list of available time stamp servers at https://gist.github.com/Manouchehri/fd754e402d98430243455713efada710.

If you're a developer working with Visual Studio, the easiest way to sign an MSIX package is to choose your certificate in the manifest editor or in the wizard we have previously seen to create an MSIX package, which supports all the options we have seen so far, including setting a timestamp server. The manifest editor will also give you the option to generate a self-signing certificate on the fly, if you don't have one, by clicking on the **Create...** button:

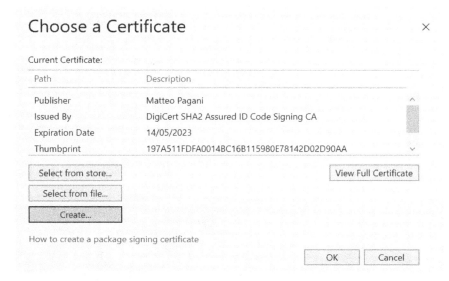

Figure 11.7 – Choosing a certificate

After you have chosen the certificate, Visual Studio will automatically sign the package as part of the build process when you create an MSIX package for distribution using the Create App Packages wizard.

Other options to sign an MSIX package include the following:

- **Using the signtool utility**: This is included in the Windows SDK. It's a command-line utility whose usage is described at https://docs.microsoft.com/en-us/windows/msix/package/sign-app-package-using-signtool.

- **Using a third-party tool such as MSIX Hero**: This is an application developed by Marcin Otorowski, which gives you multiple tools to work with MSIX, such as browsing the catalog of packages deployed on your machines, unpacking an MSIX package, and more. Among the available options, you have also the opportunity to sign an MSIX package given a valid certificate. You can download this from https://msixhero.net/.

Signing the package with Visual Studio is a good solution when you're in the development and testing phases. However, it isn't suitable for more professional scenarios in which you want to automate the building and deployment of your application. This is because it forces you to share the certificate as part of the project, which might lead to identity theft.

In the next chapter, we'll explore other ways to sign your package. Now, it's time to talk about the distribution options.

Publishing your application to Microsoft Store

Microsoft Store is the most efficient and straightforward way with which to distribute your application to a broader audience. It's a catalog of applications and games that is preinstalled on every Windows computer, which greatly simplifies the experience of downloading and installing applications on your machine. Users no longer have to open a browser, search for an application with a search engine, find the right website (avoiding, in the meantime, potentially malicious ones), download the installer, and execute it. With Microsoft Store, they can simply search for the app they need and, by clicking on a button, trigger the installation. The following screenshot shows Microsoft Store in the browser:

Figure 11.8 – The Microsoft Store main page

As a developer, Microsoft Store greatly simplifies the distribution experience thanks to the following features:

- You don't need to buy a certificate and manually handle the signing. The Store will automatically sign all the applications with a trusted certificate issued by Microsoft.

- By default, the Store offers automatic updates. You don't have to implement a custom solution, but it's enough to publish an update on the Store to automatically distribute it to your customers.

- You can leverage a built-in monetization infrastructure. You can sell your application or add-ons through Microsoft Store, rather than having to build your own licensing infrastructure. You just need to set your price during the submission process; users will pay the application using the payment systems that are connected to their Microsoft account, such as a credit card. Microsoft will keep a fee of 15% for every purchase. If you want to keep 100% of the income, or if you have your own licensing system, you aren't obligated to use the one provided by Microsoft.

- You have access to advanced deployment features, such as gradual rollout and package flights to deploy different versions of the same app to different users (such as alpha, beta, and more).

- You have built-in support for marketing activities, such as a dedicated window for your application with descriptions, screenshots, and videos; the ability to generate promotional codes to give away an application for free; and support for a sales campaign in which the application price is automatically reduced during a given period of time.

- You can engage with your customers by collecting feedback, reviews, and ratings.

The starting point to publish your Windows applications to Microsoft Store is by creating a developer account at `https://developer.microsoft.com/en-us/microsoft-store/register/`.

There are two types of accounts, as follows:

- **Personal**: These accounts are linked to your personal identity. To open it, you are asked to pay a one-time fee of $19. At the end of the process, the account will be ready to use immediately.

- **Business**: These accounts are linked to a business entity, such as a company. In this case, the one-time fee is $99. This type of account takes a bit longer to be opened since a vetting company will verify your identity and will make sure that the company exists and that you can operate on its behalf.

Regardless of the type you choose, a developer account must be linked to a personal Microsoft account. You can't open it using a work account, such as the one provided by Microsoft 365. However, once the developer account has been created, you can link it to an Azure Active Directory tenant (such as the one that belongs to your company). This is so that you can enable coworkers to access the developer portal. Additionally, you can assign them distinct roles: for example, a developer can only submit apps, whereas a finance contributor can see all the financial reports. You can learn more about this configuration at `https://docs.microsoft.com/en-us/windows/uwp/publish/manage-account-users`.

Once your account has been set up, you can log in to the **Partner Center** portal at `https://partner.microsoft.com/dashboard` and reserve a new name, which must be unique. The name will also be used to generate the PFN, which defines the application's identity. Once the name has been reserved, you can create a new submission, where you can provide all the information that will be made visible to users: pricing, distribution markets, description, age rating, screenshots, and more. You can find a detailed description of the submission flow at `https://docs.microsoft.com/en-us/windows/uwp/publish/`.

However, there's one step that is significantly different based on the deployment technology you choose. In fact, Microsoft Store supports submitting both packaged applications and unpackaged applications. This is a new feature that has been introduced by the new Microsoft Store version launched in Windows 11, which is now also available on Windows 10.

Let's see the specific details about how to manage both scenarios.

Submitting an MSIX packaged application

The MSIX packaged model is the most powerful one since it gives you access to all the features offered by Microsoft Store. In fact, an MSIX packaged application is hosted directly by the Microsoft infrastructure, which enables all the features highlighted at the beginning of this section regarding deployment, such as a built-in signing process, automatic updates, and gradual rollouts.

To generate an MSIX package for the Store, you can use the Visual Studio wizard that you can trigger by right-clicking on your project and choosing **Publish | Create app packages**. In the first step, choose Microsoft Store under a new app name. You will be asked to log in with your developer account so that Visual Studio can list all the names you have reserved. Choose the one you have created for your application and continue the wizard, following the steps we described earlier in this chapter.

The outcome will be an unsigned MSIX package (remember that the Store will take care of it for you) and the identity assigned by Partner Center. The package will have a special extension— `.msixupload`. Now you can navigate to the **Packages** section of the submission process and upload it. The Store will ingest it and make it available through its platform.

Now, let's see how you can handle unpackaged applications.

Submitting an unpackaged application

In the case of an unpackaged application, the Store will function mainly as a window for your software. Users will still be able to find your application in the catalog, discover information about it, and install it with a single click. However, the application won't be hosted by the Store. Instead, it must be hosted by your own infrastructure, such as a website or cloud storage. Since the Store doesn't handle the distribution, many of the features won't be available since it behaves like a sideloading scenario: you must take care of signing or delivering automatic updates. However, you will still be able to use the monetization services and the feedback system provided by the Store.

> **Note**
>
> At the time of writing, publishing unpackaged applications is in preview. Before getting access to this feature, you must submit your nomination at `https://aka.ms/storepreviewwaitlist`.

When you choose to publish an unpackaged application, the **Packages** section of the submission process will be different. Instead of uploading a package, you must provide the following information:

- The package URL: This is the URL that points to the MSI or EXE installer of your application. Ideally, you should use a CDN to make sure that you can manage the traffic that will be generated by the Store.

- The languages that your application supports: You will have the option to provide different metadata (such as description, screenshots, and more) for each language.

- The CPU architecture is supported by your installer.

- The Store must be able to install your application in a silent way, without user intervention. This is typically achieved by passing a parameter such as `/s`, `/q`, or `/quiet`. As part of the submission, you must specify this parameter.

Regardless of the deployment technology you choose, the application will go through a certification process.

The certification process

Once you submit the application, it won't be immediately available, but it must pass a certification process. The Store will run a series of automated tests that will make sure your application behaves properly: for instance, it doesn't crash at startup, it doesn't contain malicious code, it's reliable, and more. Additionally, your application might also be selected for manual testing, which will validate not only the technical side but the content policies too. In fact, some content isn't allowed in the Store, such as pornography, blasphemy, and the like.

You can review all the policies at `https://docs.microsoft.com/en-us/windows/uwp/publish/store-policies`.

Once your app has been certified, it's made available to all Windows users. At this point, you will also have the option to submit updates.

Submitting updates

To submit an update, you must create a new submission starting from an existing application. The submission process is the same as the one you have followed first the first release, with the difference that all the various steps (for instance, pricing, markets, metadata, and more) will be already filled with the information you have submitted the first time.

The way you submit an updated package changes based on the deployment scenario:

- For MSIX packaged apps, you have built-in support for automatic updates. You just need to create a new package from Visual Studio, set a higher version number, and upload it to the **Packages** section of the submission process.

- For unpackaged apps, the Store isn't able to provide automatic updates. You can only use the **Packages** section if you need to change the package URL. You must provide a built-in update feature in your application if you want to automatically deliver updates to your customers.

Microsoft Store is a powerful distribution system for consumer applications. However, for enterprise applications, it might not be the best option, since many of the features it provides are tied to the consumer ecosystem. For instance, the payment system is tied to the personal Microsoft account and enterprises are unable to build their own catalog of applications.

For all the other scenarios, you can opt for sideloading distribution from a website, cloud storage, or an internal share.

In the next section, we're going to learn how, thanks to MSIX and AppInstaller, we can retain some of the features that we have seen so far.

Deploying your applications from a website using AppInstaller

If you are building a Windows application, there are many reasons why you might want to distribute your application using a website. If you are an **Independent Software Vendor (ISV)**, you might only need to protect the download of the application to your registered users by using a reserved area on your website; if you are an enterprise, you might build an internal website or an internal network share where employees can download the applications they need.

Of course, if you prefer to use the unpackaged model, you can keep using traditional deployment technologies such as MSI or ClickOnce and add a link on your website to download it. However, thanks to MSIX, you can enable a more streamlined deployment experience using AppInstaller. This enables you to do the following:

- Deploy dependencies and optional packages automatically.
- Deliver automatic updates through your website or network share.
- Support automatic remediation if the application is corrupted.

AppInstaller is an application included in Windows 10 that enables all these scenarios other than providing the installation experience for MSIX packages. The user experience you see when you double-click on an MSIX package to start the installation is managed by AppInstaller.

Additionally, AppInstaller supports a special file, with the `.appinstaller` extension, which describes the installation experience of your MSIX package. You can specify the URL where the package can be downloaded, where to download eventual dependencies, such as the Windows App SDK runtime, and how you want to handle updates. This file can be published together with your MSIX package or embedded directly inside it.

Let's see what an AppInstaller file looks like:

```
<AppInstaller
 Version="1.0"
 Uri="https://www.contoso.com/ContosoExpenses
   .appinstaller "
     xmlns="http://schemas.microsoft.com/appx/appinstaller
       /2018">
   <MainPackage
     Name="ContosoExpenses"
     Publisher="CN=Matteo Pagani, O=Matteo Pagani, L=Como,
```

```
      S=Como, C=IT"
    Version="1.0.5.0"
    ProcessorArchitecture="x86"
    Uri="https:\\www.contoso.com\ContosoExpenses.msix" />
  <Dependencies>
    <Package
      Version="0.319.455.0"
      Uri="https://www.contoso.com/Microsoft.WindowsApp
        Runtime.1.0.msix"
      Publisher="CN=Microsoft Corporation, O=Microsoft
        Corporation, L=Redmond, S=Washington, C=US"
      ProcessorArchitecture="x86"
      Name="Microsoft.WindowsAppRuntime.1.0" />
  </Dependencies>
  <UpdateSettings>
    <OnLaunch
      HoursBetweenUpdateChecks="0"
      ShowPrompt="true" />
  </UpdateSettings>
</AppInstaller>
```

Let's break it down into the main four sections:

- The first key element is the `Uri` attribute of the main `AppInstaller` entry, which is the root of the XML file. This `Uri` attribute must point to the website or the network share where you are going to publish this AppInstaller file.

- The `MainPackage` entry describes the main MSIX package with its name, publisher, version number, and CPU architecture. Also, in this scenario, a key element is the `Uri` attribute, which must point to the location where you're going to publish your MSIX package. If you are distributing an app bundle instead of a single package, you can replace `MainPackage` with `MainBundle`.

- The `Dependencies` entry is used when an application has a dependency from a runtime or a library distributed with MSIX. The Windows App SDK runtime is a good example of such a dependency. Also, in this case, the key property is the `Uri` attribute, which points to the location of the dependency. Thanks to this section, we can install dependencies together with the application from a custom location, also supporting scenarios where Microsoft Store is blocked, so the dependency can't be downloaded automatically.

- The `UpdateSettings` section can be used to control the update logic. In this example, we are configuring our application to check for updates every time the application starts and, if an update is available, to show a notification prompt. Other available options are checking for updates in the background or making updates mandatory (which blocks the ability to run the application until the application has been updated).

Additionally, the AppInstaller file can specify other types of dependencies, such as optional packages (special packages that contain add-ons for the main application) or modification packages (special packages that can customize the existing installation of the application).

The following documentation gives you a detailed overview of the AppInstaller schema and all of the available options: `https://docs.microsoft.com/en-us/windows/msix/app-installer/how-to-create-appinstaller-file`.

However, there are better ways than having to generate the file manually. Let's explore the options we have next.

Generating an AppInstaller file with Visual Studio

Visual Studio can generate an AppInstaller file for you as part of the build process that generates an MSIX package. The starting point is the same wizard we looked at earlier, that is, the one that you can trigger by choosing **Publish | Create a package** when you right-click on your project.

> **Note**
>
> At the time of writing, this option is only available for the Windows Application Packaging Project. If you're building a WinUI application using the single-project MSIX approach, you will still see the option to generate an AppInstaller file, but it won't have any effect. At the end of the build process, the AppInstaller file will be missing.

When you choose **Sideloading** as a distribution method, you can also enable a flag called **Enable automatic updates**, as you can see in the following screenshot:

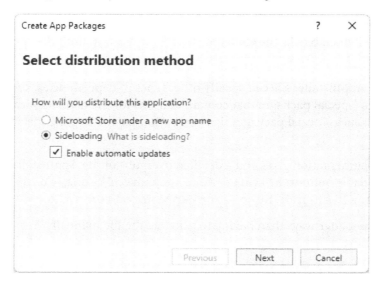

Figure 11.9 – The option to enable the creation of the AppInstaller file during the MSIX generation

When you check this option, you will see a new step of the wizard, titled **Configure update settings**, at the end of the process. It will appear before the creation of the package:

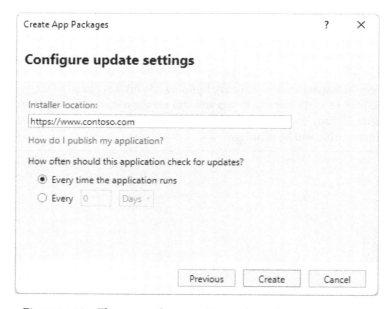

Figure 11.10 – The step in the wizard to configure the update settings

In this step, you must provide the URL of the network share path where you're planning to make your application available and how often Windows should check for updates. Then, you can hit **Create** to start the creation of the MSIX package.

At the end of the process, in the same folder where Visual Studio has generated the MSIX package, you will also find the following two additional files:

- One with the .appinstaller extension, already configured in the proper way by combining the information you have provided in the wizard (such as the installer location) and the ones generated by Visual Studio (such as the name of the MSIX package).

- Another is called index.html. This is a static web page that you can use to trigger the installation of the application. Other than displaying information such as the logo, name, and version number, it includes a button that uses the ms-appinstaller protocol (we'll learn more about this later) to start the installation of the application.

However, this wizard has a limitation: the AppInstaller file has evolved over time by supporting new features such as update prompts or mandatory updates. The wizard didn't follow the same evolution and, as you have seen in *Figure 11.9*, it doesn't support these new features.

To work around this limitation, you can add an AppInstaller template to your project. This is a special file that will be used as a template during the generation of the package, and it contains placeholders for the key information, such as the installer location or the application's version. During the build process, Visual Studio will replace them with the actual data generated during the build. To add this file, you must right-click on your project and choose **Add | New item**. Choose the AppInstaller template, leave the default name (Package.appinstaller), and press **Add**. This is what the file looks like:

```
<AppInstaller Uri="{AppInstallerUri}"
    Version="{Version}"
    xmlns="http://schemas.microsoft.com/appx/
    appinstaller/2017/2">

    <MainPackage Name="{Name}"
                Version="{Version}"
                Publisher="{Publisher}"
                Uri="{MainPackageUri}"/>

    <UpdateSettings>
```

```
    <OnLaunch HoursBetweenUpdateChecks="0"/>
  </UpdateSettings>

</AppInstaller>
```

This is very similar to the file we have seen before, except that most of the information is described with a placeholder and wrapped inside curly braces (such as {AppInstallerUri} or {Publisher}). Being a template, you can customize it as you prefer, for example, by adding new entries and attributes in the UpdateSettings section to enable new features.

> **Note**
>
> Visual Studio generates the AppInstaller template using an old xmlns schema, which doesn't support many of the new features. As such, it's important to replace the current one with the following:
>
> xmlns="http://schemas.microsoft.com/appx/appinstaller/2021"

When you generate a package using the wizard, Visual Studio will use this file as a base template. In fact, in the **Configure update setting** section of the wizard, you won't be able to customize the settings (other than the location URL) anymore since they will be taken directly from the template.

But what if you want more flexibility when creating an AppInstaller file? Let's explore another option next.

Generating an AppInstaller file with AppInstaller File Builder

AppInstaller File Builder is an application developed by Microsoft, which is part of the MSIX Toolkit—a series of utilities to support working with MSIX packages. You can download it from https://github.com/microsoft/MSIX-Toolkit/releases/.

You can use this tool to generate an AppInstaller file more easily, thanks to its visual interface:

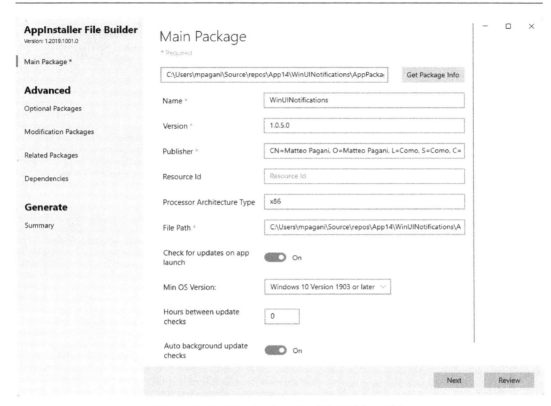

Figure 11.11 – The AppInstaller File Builder utility

What makes this a powerful tool is that instead of manually filling in all the required information, you can just click on the **Get Package Info** button to choose the MSIX package you want to publish from your hard disk. The tool will automatically extract all the information for you. Thanks to this tool, you can define the following:

- The main package
- The dependent packages (such as the optional packages, modification packages, related packages, and dependencies)
- The update conditions, which are based on the Windows version that you're targeting

Once you have filled in all of the information, you can move on to the **Summary** section. Here, you'll be asked to specify the URL where you're going to publish the AppInstaller file, the URL of the various packages, and the version number:

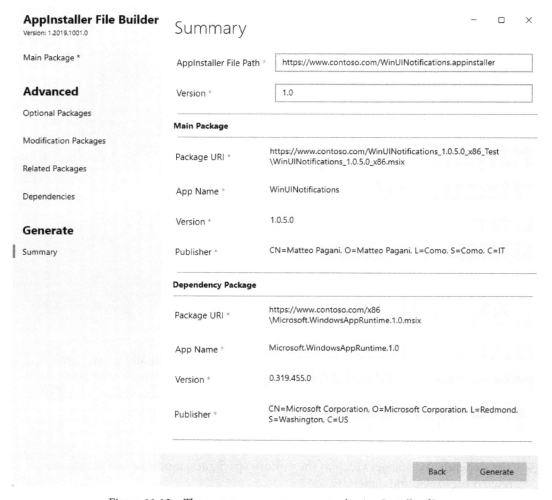

Figure 11.12 – The summary page to generate the AppInstaller file

By clicking on **Generate**, the tool will generate the AppInstaller file for you, all ready to be used. But how can we use it?

Deploying an AppInstaller file

Now that we have an AppInstaller file, all we need to do is deploy it together with our packages (the main one and the eventual dependencies) in the location we have chosen, which can either be an HTTP endpoint or a network share.

Once we have deployed them, we can use one of the following two ways to trigger the installation of the package:

- Add a link to our web page to download the AppInstaller file locally. The user must double-click on it to start the installation process. Windows will automatically download the MSIX package and its dependencies from the URLs that are specified inside the file.

- Using the `ms-appinstaller` protocol, which is the default approach used by Visual Studio. In this scenario, the link you add to your web page won't download the AppInstaller file locally, but it will point to it using the `ms-appinstaller` protocol (for example, `ms-appinstaller:?source=https://www.contoso.com/ContosoExpenses.appinstaller`). This approach provides the best user experience because the installation will be directly triggered from the website, saving the extra step of downloading the file locally and launching it.

> **Note**
>
> At the time of writing, Microsoft has discovered a vulnerability related to the `ms-appinstaller` protocol, which is described at `https://msrc.microsoft.com/update-guide/vulnerability/CVE-2021-43890`. For this reason, option 2 isn't currently available, but it might be restored in the future once the vulnerability has been addressed.

Starting from Windows 11, you can also choose to embed the AppInstaller file inside your MSIX package. This doesn't replace the need to publish your AppInstaller file to a website or network share: if you want to deliver automatic updates, you will still need to make it available through an HTTP endpoint. However, by embedding the AppInstaller file inside the package, you can enable automatic updates even if the application has been manually installed the first time; you can do this by directly using the MSIX package rather than the AppInstaller file.

If you want to use this possibility, first, you'll have to copy the same AppInstaller file you previously generated and deployed inside the Visual Studio project. Then, you must declare it in the application's manifest.

First, you'll need to add an extra namespace called `uap13`:

```
<Package
    xmlns:uap13="http://schemas.microsoft.com/appx
    /manifest
```

```
                 /uap/windows10/13" IgnorableNamespaces="uap13">
...
<Package>
```

Then, inside the `Properties` section of the manifest, you must add an `AppInstaller` entry with the path of the AppInstaller file inside the package:

```
<Properties>
    <uap13:AutoUpdate>
        <uap13:AppInstaller File="Update.appinstaller" />
    </uap13:AutoUpdate>
</Properties>
```

Now that we have deployed and installed our application via an AppInstaller file, let's see how we can apply updates.

Updating an application automatically

Thanks to the `UpdateSettings` section, we have added to our AppInstaller file. Now, we can enable automatic updates without having to write any other code in our application. We just need to generate an updated AppInstaller file and MSIX package, making sure that we set a higher version number for both. Then, we need to upload them in the same HTTP or network location where we deployed the original version of the application.

Now, based on the update logic that we have defined in the AppInstaller file, Windows will automatically pick the update and install it. Users won't need to come back to the website or the network share to download and install the package again.

For example, if you are using the `OnLaunch` configuration and you have enabled the `ShowPrompt` option, users will see the following message the next time they open your application:

Figure 11.13 – The prompt that is displayed to the user when an update is available

In other scenarios, you might want more granular control over the updates. Let's explore this scenario by using the Package APIs.

Updating an application from code

After you have deployed an application using AppInstaller, you can leverage which APIs belong to the `Windows.Management.Deployment` namespace to check for available updates and install them. This approach requires more work since you'll have to write some additional code, but it enables you to further customize the update logic since you are in full control of the update process.

Let's see a code example that we can use to trigger the update:

```
private async Task CheckUpdatesAsync()
{
    PackageManager pm = new PackageManager();
    Package package = pm.FindPackageForUser(string.Empty,
        Package.Current.Id.FullName);

    PackageUpdateAvailabilityResult result = await
        package.CheckUpdateAvailabilityAsync();
```

```
switch (result.Availability)
{
    case PackageUpdateAvailability.Available:
    case PackageUpdateAvailability.Required:
        var installTask =
            pm.AddPackageByAppInstallerFileAsync(
        new Uri("https://www.contoso.com/
            ContosoExpenses.appinstaller"),
                AddPackageByAppInstallerOptions
                    .ForceTargetAppShutdown, pm.GetDefault
                        PackageVolume());
        installTask.Progress += (installResult,
            progress) =>
            {
                Console.WriteLine(progress);
            };

        var installResult = await installTask;
        if (!installResult.IsRegistered)
        {
            Console.WriteLine(installResult.ErrorText);
        }

        break;
    case PackageUpdateAvailability.Unknown:
    case PackageUpdateAvailability.NoUpdates:
        MessageBox.Show("No updates available");
        break;
    }
}
```

Here, we use the `PackageManager` class to retrieve a reference to the current instance of the application by calling the `FindPackageForUser()` method. As parameters, we must pass the identifier of the user (we can pass an empty string if we want to retrieve the packages for all of the users) and the Package Family Name. Since our application is packaged, we can retrieve it using the `Package.Current.Id.FullName` property.

Once we have a reference to the package, we can invoke the
CheckUpdateAvailabilityAsync() method, which will use the information
stored inside the AppInstaller file that we used to install the application, to check whether
there's a new update available. If there's indeed a new version, we will get back the
Available value or the Required value of the PackageUpdateAvailability
enumerator (this depends on whether we marked the update as optional or required).
If that's the case, we can move on with the installation process by calling the
AddPackageByAppInstallerFileAsync() method of the PackageManager
class. This requires the following parameters:

- The URL of the AppInstaller file.

- The behavior to apply when the update is installed. In this example, by using the
  ForceTargetAppShutdown option, we force the application to be closed so that
  the update can be applied immediately.

- The volume where the application has been installed can be retrieved by calling the
  GetDefaultPackageVolume() method of the PackageManager class.

The result of the method is represented by a special Windows Runtime interface
called IAsyncOperationWithProgress, which can be used to track the
progress of asynchronous operations. In the previous snippet, you can see an
example of how you can use it: we don't immediately execute the method, but
we just Store a reference to the Task object that represents it (notice how we call the
AddPackageByAppInstallerFileAsync() method without prefixing it with
the await keyword). Then, we subscribe to the Progress event, which we can use to
monitor the progress of the operation.

Following this, we start the task by invoking it with the await prefix. We can use the
IsRegistered property of the result to determine whether the operation has been
completed successfully; otherwise, we can use the ErrorText and ErrorCode
properties to understand what went wrong.

So, what if you have opted to not use the AppInstaller file, but you still want to provide
a way to deliver updates? In this case, you can use the AddPackageAsync() method
offered by the PackageManager class, directly passing the URL of the updated MSIX
package as a parameter. This is shown in the following snippet:

```
PackageManager packagemanager = new PackageManager();
await packagemanager.AddPackageAsync(new
    Uri("https://www.contoso.com/ContosoExpenses.msix")
     ,null, AddPackageOptions.ForceApplicationShutdown);
```

Even if this code might look simpler than the AppInstaller scenario, it actually requires more work from your side. In fact, if the application hasn't been installed with an AppInstaller file, Windows doesn't have a way to understand whether there's a new package available. As such, you must implement this logic on your own, for example, by exposing this information through a REST API that you must call before installing the update.

Let's wrap up the AppInstaller overview by taking a quick look at two new features that have been added to Windows 11.

Supporting updates and repairs

Starting from Windows 11, AppInstaller supports two new features:

- The ability to specify multiple update URLs: This way, if one of them should go down, Windows will automatically fall back to one of the others.

- The ability to specify one or more repair URLs: If the application has been tampered with and the AppInstaller URI isn't available, Windows will use these URLs to retrieve the latest version of the package and repair the broken installation.

The first feature is enabled through the `UpdateUri` section, which looks like this:

```
<UpdateUris>
        <UpdateUri>https://www.contosobackup.com/
        ContosoExpenses.AppInstaller</UpdateUri>
        <UpdateUri>\\MyNetworkShare\\ContosoExpenses
        .AppInstaller</UpdateUri>
</UpdateUris>
```

The second one is enabled by the `RepairUris` section, which is defined as follows:

```
<RepairUris>
        <RepairUri>https://www.contosobackup.com/
        ContosoExpenses.AppInstaller</RepairUri>
        <RepairUri>\\MyNetworkShare\\ContosoExpenses.
        AppInstaller </RepairUri>
</RepairUris>
```

We have completed the exploration of another way to distribute our Windows applications. This is via a website or a network share, which is a scenario made easier by MSIX and AppInstaller. Now, let's see the last option we're going to discuss in this chapter: Windows Package Manager.

Publishing your application to the Windows Package Manager repository

Installing applications can be a long and tedious process, especially if you're building a new machine and you have to reinstall all the tools you need for your work and personal file. What if you could somehow automate this process? What if you can make it faster rather than having to pick the applications you need one by one? These are the reasons that led Microsoft to create a new utility called **Windows Package Manager** (the short name is `winget`), which is inspired by other popular tools such as **Advanced Package Tool** (**APT**) and Chocolatey.

Note

Windows Package Manager is deployed through the AppInstaller application, which is built inside Windows. As such, you will find this utility already available on every Windows installation starting from Windows 10 1709. Preliminary versions of the new releases are available through the official GitHub repository at `https://github.com/microsoft/winget-cli/releases`.

Windows Package Manager is a command-line tool that you can use to quickly install and upgrade applications on your machine from a variety of sources, such as websites and Microsoft Store. The tool is backed up by a repository, which is hosted on GitHub at `https://github.com/microsoft/winget-pkgs`. It contains a manifest for each available application. The manifest is a YAML file that describes the application, including the URL where it can be downloaded from. For example, this is an excerpt of the manifest of Microsoft Edge—the popular Microsoft browser:

```
# yaml-language-server: $schema=https://aka.ms/winget-
    manifest.singleton.1.0.0.schema.json
PackageIdentifier: Microsoft.Edge
Publisher: Microsoft
PackageName: Microsoft Edge
Author: Microsoft
ShortDescription: World-class performance with more
    privacy, more productivity, and more value while you
        browse.
Moniker: msedge
Tags:
...
```

```
MinimumOSVersion: 10.0.0.0
PackageVersion: 86.0.622.38
InstallerType: msi
Installers:
- Architecture: x64
  InstallerUrl: https://msedge.sf.dl.delivery.mp
    .microsoft.com/filestreamingservice/files/53a5f508-
      44da-4f9d-85d9-312fb8f92f4b/MicrosoftEdge
        EnterpriseX64.msi
  InstallerSha256: D49104F182675701423CB774
    CD2120B3A3FF63565661E4364D1169F64B9019EC
  Scope: machine
PackageLocale: en-US
ManifestType: singleton
ManifestVersion: 1.0.0
```

The key information elements in the manifest are as follows:

- `PackageIdentifier`: This is the unique ID that we can use to install/upgrade/ remove the application.

- `InstallerUrl`: This is the URL where the installer is available. In the preceding example, it's an MSI installer, but it could have also been an EXE package or an MSIX package.

If you want to install this application, you just need to open a Command Prompt or Windows Terminal and type in the following command:

```
winget install Microsoft.Edge
```

Windows Package Manager will display a progress bar, and it will let you know when the application has been installed. Thanks to `winget`, you can manage the full life cycle of the application. For example, you can use the following command to update Edge:

```
winget upgrade -q Microsoft.Edge
```

Also, you can use `winget` to remove an application by invoking the following command:

```
winget uninstall Microsoft.Edge
```

As a user, the whole `winget` backend is completely transparent. You don't have to know how and where manifests are stored. In fact, you can directly search whether an application is in the catalog with the `search` command, as shown in the following example:

```
winget search edge
```

The tool will give you a list of packages that matches the criteria and their identifiers so that you can install the one you're looking for, as shown in the following screenshot:

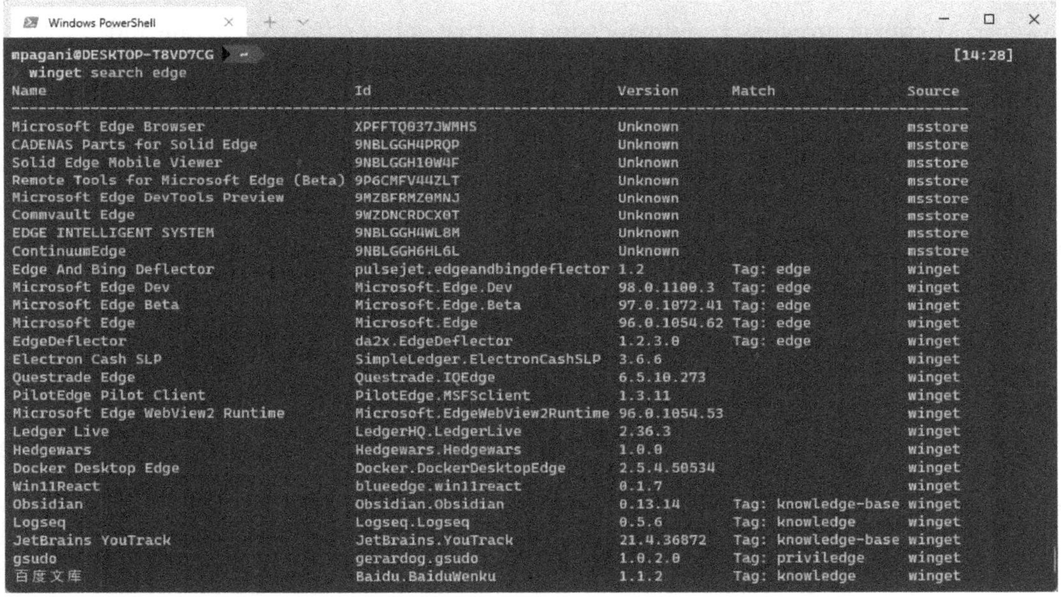

Figure 11.14 – The list of applications returned by a search in Windows Package Manager

As a developer, making your application available through Windows Package Manager is a worthwhile investment since the following can occur:

- You can make the life of users who want to automate the installation of applications much easier.

- Soon, Microsoft is going to enable the possibility of hosting your own private Windows Package Manager repository, which you'll be able to connect to enterprise tools such as Microsoft Endpoint Manager. This scenario will enable enterprises to manage their internal catalog of applications more efficiently since they won't be forced to host the various installers and packages internally anymore. Instead, they will just need to host the manifests. This solution is going to replace the Microsoft Store for Business as the official Microsoft solution to host an internal catalog of applications for enterprises. Therefore, by submitting your manifest, you will make the lives of companies interested in acquiring your application easier.

In the next section, let's see how we can create a manifest.

Creating a manifest for Windows Package Manager

As mentioned earlier, a manifest is just a YAML file that contains the metadata of the application. As such, you could create it manually with any text editor by following the schema described at `https://docs.microsoft.com/en-us/windows/package-manager/package/manifest`.

However, Microsoft provides a command-line tool that you can use to generate and update manifests more easily. The tool is called **WinGetCreate**, and guess what? It's available through Windows Package Manager. To install it, just open a Terminal and run the following command:

```
winget install wingetcreate
```

Once the installation is complete, you can start the creation of a new manifest by calling the following command:

```
wingetcreate new
```

A wizard will guide you through the flow so that you can specify the following information:

1. The first information you must provide is the URL of your installer or MSIX package. The tool will download it and parse it, to try to automatically extract as much metadata as possible.

2. The package identifier must be unique. The tool will propose you one based on the publisher and the name of the application.

3. The version number.

4. The default language of the application (for example, `en-US`).

5. The publisher's name.

6. The package name.

7. The package license.

8. A short package description.

The tool will automatically generate the manifest file in a path that follows the same rule adopted by the official repository, which is as follows:

```
manifests\<first letter of the publisher>\<name of the
   publisher>\<name of the application>\<version number>
```

For example, if you explore the official repository on GitHub, you will find that the Microsoft Edge manifest is stored in the following path:

```
Manifests\m\Microsoft\Edge\86.0.622.38\Microsoft.Edge.yaml
```

Once the manifest has been generated, the tool will give you the option to submit it to the Windows Package Manager repository. This is so that your application can be made available in the catalog. If you have created the manifest manually, you can validate it before submitting it, to make sure you haven't made any mistakes, by using the following command:

```
winget validate <path of the YAML file>
```

Regardless of the way you have created the manifest, to submit it, the `wingetcreate` tool will first ask you to log in to GitHub with your account.

The mechanism behind the manifest submission is the `Pull Request`. Once the manifest has been submitted, you will find a new pull request in the Windows Package Manager Community repository, which will add the manifest you have just created to the catalog. The pull request will trigger a validation process, which will verify that the metadata is correct, the installer URL is valid, the binaries are safe, and more. Once the pull request has been approved, the new manifest will be merged inside the main repository, and you'll be able to start using the `winget install` command to install your application.

The `wincreate` tool can also be used to create an update for an existing manifest and submit it to the repository, as shown in the following example:

```
wingetcreate update MatteoPagani.ContosoExpenses -v
   1.0.201.0 -u https://www.contoso.com/ContosoExpenses
      102010.msix
```

Unlike the create option, the updated one isn't interactive, which makes it a good fit to be integrated inside a CI/CD pipeline. We'll learn more about this scenario in *Chapter 12, Enabling CI/CD for Your Windows Applications*.

Summary

Finding the right technology and platform to distribute your application is essential for building a successful product. However, once the application has been installed, your work as a developer doesn't stop there: you must manage the licensing, the monetization, the feedback, and the automatic updates. All of these tasks are equally important to continuously deliver value to your users.

In this chapter, we have explored some of the technologies (such as MSIX) and the platforms (such as Microsoft Store, sideloading with AppInstaller and Windows Package Manager) that you can use to plan a successful deployment and, at the same time, simplify many of the challenges that you face as a developer. Examples of this include setting up automatic updates with AppInstaller, delivering a clean and reliable clean installation and uninstallation experience with MSIX, and more.

This chapter serves as a foundation for the next and concluding chapter: we're going to see how, thanks to the technologies we have learned, we'll be able to automate building and deploying our Windows desktop application, thanks to the adoption of DevOps best practices.

Questions

1. Windows Package Manager is an alternative distribution channel if you don't want to use Microsoft Store or AppInstaller. Is this true or false?

2. To enable automatic updates with AppInstaller, you don't need to change the code of your application. Is this true or false?

3. Microsoft Store is the best fit to deploy applications in an enterprise environment. Is this true or false?

12
Enabling CI/CD for Your Windows Applications

In the last few years, DevOps has become one of the most important trends in tech companies, and for a very good reason. There are multiple research reports and studies (please refer to the report here from **DevOps Research and Assessment (DORA)** team, *Accelerate: State of DevOps: Strategies for a New Economy* by *N. Forsgren, J. Humble, and G. Kim*) that demonstrate how companies that adopt a DevOps approach to software development are able to do the following:

- Reach the market faster
- Deploy changes and updates more often
- Recover faster in the case of bugs or issues
- Adapt more quickly to changes in requirements, either within the ecosystem or the market

DevOps is a complex topic that isn't just technical. In fact, DevOps isn't only about automating some operations or testing your code. Above all, it's about changing the mindset of how you envision, plan, and build your software projects, by introducing concepts such as the following:

- **Continuous planning**: By shifting to a different planning methodology (such as Agile), you can work on small-batch releases, increasing the opportunity of reducing any risks. Additionally, you can react better to changes in the ecosystem, customer requirements, or new technologies that are being released.

- **Continuous integration**: By continuously building and testing your code every time you make changes, you can find errors earlier in the development process, eliminate integration issues, and facilitate the collaboration and parallel development of new features.

- **Continuous delivery**: By automating the deployment of your software project, you can reduce the risk of errors. Additionally, you can quickly create environments to test your work before moving into production, and you can run experiments to validate theories.

Learning all the aspects of DevOps is beyond the scope of this book. In this chapter, we'll focus on two specific topics: **Continuous Integration** (**CI**) and **Continuous Delivery** (**CD**). We're going to learn how we can automate building, testing, and deploying Windows desktop applications by going through the following topics:

- Introducing CI/CD pipelines
- Building a Windows application in a CI/CD pipeline
- Supporting versioning
- Handling signing
- Automating the deployment

Let's get started!

Technical requirements

You need a GitHub account, which you can get for free at `https://www.github.com`.

The free account includes the following:

- Unlimited public and private repositories
- Unlimited automation minutes per month for public repositories and 2,000 minutes per month for private ones

- Unlimited storage for packages for public repositories and 500 MB for private repositories

The sample code can be found at the following URL:

```
https://github.com/PacktPublishing/Modernizing-Your-Windows-
Applications-with-the-Windows-Apps-SDK-and-WinUI/tree/main/
Chapter12
```

Introducing CI/CD pipelines

Let's start with the basic concepts. What is a pipeline? It's a series of actions that are executed in an automated way on a build machine, which can be hosted on the cloud or on-premises.

The goal of a pipeline is to automate all the steps that you typically do on your own development machine when you want to release your software, such as building the solution with Visual Studio, generating a distributable package, publishing it on a website, and more. Typically, in DevOps, you have two types of pipelines:

- The CI pipeline takes care of building and testing your application, to make sure that the code can be built successfully and that you haven't introduced any regression.

- The CD pipeline takes care of taking the output of the CI pipeline (such as an MSIX package) and deploying it on the distribution platform of your choice (such as a web server or Microsoft Store).

In both cases, we're calling them *continuous* because these operations aren't manually triggered. Instead, they are automatically performed every time you take a specific action, such as committing new code to your source control repository. In most cases, these two pipelines are combined (hence, the usage of the term CI/CD): if the CI pipeline is successful, the CD pipeline is automatically triggered to deploy the artifacts that the CI pipeline has just produced. In this way, you can focus on the developmental aspect of your project rather than wasting resources and time every time you deploy an updated version of your application to your testers and your customers. Additionally, by doing it in an automated way, you reduce the chances of introducing human error.

These pipelines are executed and managed by a CI/CD platform, which takes care of orchestrating all of the operations. There are many solutions on the market; in this chapter, we're going to focus on the most popular development platform in the world: **GitHub**. This is a complete platform to manage the whole life cycle of a software project: from hosting the code to planning and from tracking issues to hosting packages. Of course, it also includes a CI/CD solution, called **GitHub Actions**.

What makes GitHub Actions (and, overall, all cloud-based solutions) so powerful is that, as part of the platform, you get access to hosted machines to perform the build, called **GitHub-hosted runners**. You can think of them as virtual machines, which are created on the fly whenever you execute a CI or CD pipeline. Additionally, they come with all the software you might need to build your application: Visual Studio, Java, the Windows 10 SDK, Node.js, and .NET are just a few examples. GitHub-hosted agents also come with a variety of operating systems: Windows, Linux, and macOS. This solution helps you to mitigate one of the most time and resource-consuming tasks of setting up a CI/CD environment: building and maintaining a pool of machines to build your software.

If, for any reason, you need tools that aren't available, or you need to access resources that you can't expose on the cloud, GitHub also gives you the option to use self-hosted runners. Thanks to a service provided by GitHub for Linux, Windows, and macOS, you can turn any machine (such as a physical machine, a virtual machine, or a machine in the cloud) into an agent. The software will manage the communication between the machine and GitHub so that it can execute your workflows. You can learn more about this option at `https://docs.github.com/en/actions/hosting-your-own-runners/about-self-hosted-runners`.

Another powerful feature offered by cloud services is the availability of premade tasks (called **actions** in GitHub), which can ease the execution of specific actions within your project. For example, you can find an MSBuild action that you can use to compile software that has been built with Visual Studio; alternatively, you can use an Azure App Service action that you can use to deploy a web application to an instance of Azure App Service and more. GitHub comes with many built-in actions, but there's a rich ecosystem of third-party actions available at `https://github.com/marketplace?type=actions`.

All the modern CI/CD platforms, including GitHub, have adopted YAML as the language to define a pipeline (which is also called a workflow on GitHub). This is a data serialization language that you can use to configure all the aspects of a pipeline, such as the following:

- The runner to use
- The various stages that make up the whole build and deployment process
- The tasks to perform and their sequence at each stage

What makes YAML powerful is that, with this approach, the whole configuration of the pipeline is stored in a plain text file, which becomes part of your project. This makes it easy to redistribute, branch, or fork the pipeline.

This is an example of a simple workflow for GitHub:

```
name: CI
on:
```

```
push:
  branches: [ main ]
pull_request:
  branches: [ main ]
workflow_dispatch:

jobs:
  build:
    runs-on: windows-2022
    env:
      message_name: world

    steps:
      - name: Checkout
    uses: actions/checkout@v2

      - name: Run a one-line script
        run: echo Hello, ${{ env.message_name} !

      - name: Run a multi-line script
        run: |
          echo Add other actions to build,
          echo test, and deploy your project.
```

The workflow is defined with the following properties:

- name: This is the name of the workflow.

- on: This defines the condition that will execute the workflow. In this example, we specify that we want to run it every time there's a new push or a new pull request executed on the main branch. However, GitHub Actions can be used to automate everything, not just to implement CI/CD scenarios. You can have workflows that are triggered when a new issue is created or when a new package gets released. You can find a full list at https://docs.github.com/en/actions/learn-github-actions/events-that-trigger-workflows. By adding the workflow_dispatch condition, we are also setting up the option to manually run the workflow.

- jobs: This is a way to logically split your workflow into multiple stages. Each job is independent: it has its own set of actions, variables, and runners. By default, jobs run in parallel, but you can also define a dependency to run them in sequence (for example, a deployment job will need the build job to be completed before running). A workflow must have at least one job, such as in this example: here, we have a single job called build.

Each job can have multiple properties. In the preceding example, we can see the most important ones, as follows:

- runs-on: This defines the runner you want to use to execute the workflow. You can find the list of available runners at https://github.com/actions/virtual-environments. On the same website, for each runner, you can also find the list of tools that are installed on it, so you can determine which is the best platform for your project. For example, in our case, we can see that the runner called windows-2022 includes .NET 6, the Windows 10 SDK, and Visual Studio 2022, so it's the best fit in which to build a Windows application created with the technologies that we discussed in this book.

- env: You can use this to define the environment variables that can be referenced across the workflow. In this example, we have stored a string in a variable called message_name, which we can reference across the steps using the ${{ env. message_name }} syntax.

- steps: This is the list of actions that the job will perform one after the other. Each step can be defined by an action provided by GitHub, a third-party developer, or it can simply execute a script:

  - Actions can do all the heavy lifting for you. You just need to include them using the uses keyword, followed by their identifier. Optionally, if the action can be customized, you can supply parameters using the with keyword. For example, the step called checkout uses an action, which we reference using its full identifier (actions/checkout@v2). The identifier uses the <company or developer name>/<action name>@<version> syntax.

- Scripts can execute any arbitrary operation. They are defined with the `run` keyword, followed by the script you want to execute. In the preceding example, there are two steps that execute a single line and a multiline script.

The starting point to create an action is the **Actions** tab on GitHub. GitHub will suggest a few starter templates (which provide a built-in configuration, series of steps, and more) based on the content of your repository:

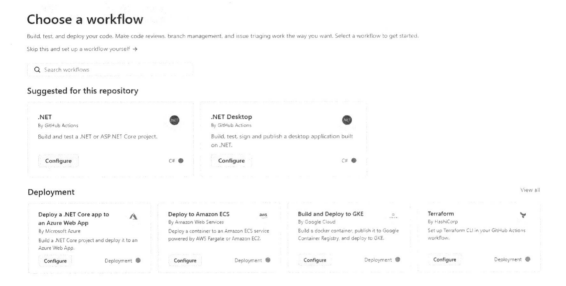

Figure 12.1 – The starting point to create a new workflow in GitHub

For our scenario, we can simply click on the **set up a workflow yourself** link, which will create a basic workflow, similar to the one we saw in the previous example. GitHub will open the editor, which is a simple text editor but with a twist: on the right-hand side, you will see a panel titled **Marketplace**. You can use it to browse the catalog of first-party and third-party actions. For each action, you will get quick access to the documentation and the YAML snippet, which you must copy inside your workflow to use:

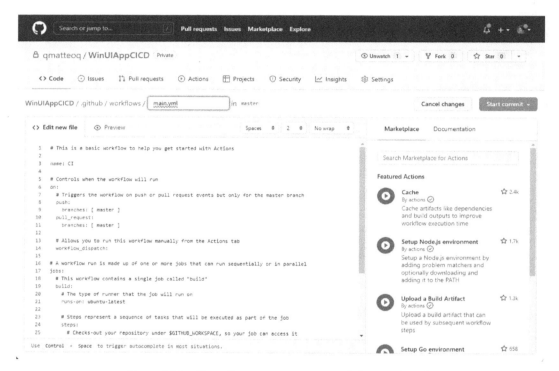

Figure 12.2 – The editor to create a new workflow in GitHub

Once you have completed the work, you can press **Start commit** to add your workflow file to your repository. If you are triggering the workflow automatically whenever there's a new commit, this action will also lead to the execution of the workflow, which you can monitor from the **Actions** tab of the repository.

For each execution, GitHub will provide a real-time logger, which displays the execution of the various steps and the output from each step. This is critical, especially when you have errors, to understand what went wrong:

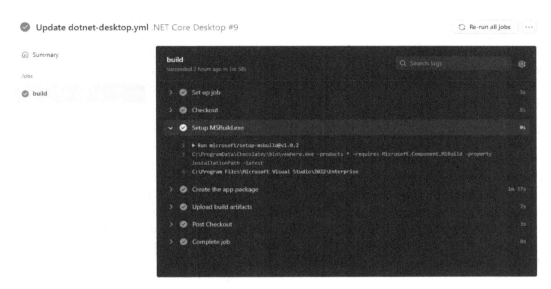

Figure 12.3 – The execution of a workflow on GitHub

Now that we have learned about the basic concepts of using GitHub Actions, let's see how we can create a workflow to build our Windows applications.

Building a Windows application in a CI/CD pipeline

Most of the documentation and stories related to CI/CD that you read on the internet are about server-side applications: cloud services, APIs, and web applications. One of the reasons (of course, other than because they're very popular and widely adopted) is that they're a perfect fit to build a solid CI/CD story, especially when it comes to deployment. These kinds of applications are hosted on a server, which is exposed through an HTTP endpoint. With the power of the cloud, you can quickly create new servers as needed; for instance, PaaS services such as Azure App Services give you advanced scaling and staging capabilities. Thanks to all of these features, it's quite straightforward to define a pipeline that can automatically deploy a new version of the software in one or more environments, such as a staging server, a production server, and more.

When it comes to Windows desktop applications, it's more challenging. This is because they don't sit on a server but are distributed on each computer of our userbase. If our application is installed across 10,000 users, that means we must update 10,000 instances whenever we have a new version.

To solve these challenges, we're going to reuse the knowledge we acquired in *Chapter 11, Publishing Your Application*. We learned MSIX provides many features (such as more reliability, automatic updates, differential updates, and more) that can help us to define a solid CI/CD story for our desktop applications. In this chapter, we're going to use a WinUI application as a starting point, which we're going to package as MSIX and deploy on a website using App Installer.

Note that the same steps we're going to see in this chapter can also be applied to WPF and Windows Forms applications. As we learned in *Chapter 1, Getting Started with the Windows App SDK and WinUI*, we can just add a Windows Application Packaging Project to our solution to enable Visual Studio to create an MSIX package as part of the building project. This is another feature that makes MSIX a great fit for CI/CD scenarios: unlike other installer technologies (such as MSI), which require third-party tools and special Visual Studio extensions to be generated, MSIX can be created with a standard Visual Studio build. As we learned in the previous section, the GitHub-hosted runner template for Windows already comes with everything we need.

Let's take a look at the general configuration of our workflow:

```
name: WinUI application
on:
  push:
    branches: [ master ]
  pull_request:
    branches: [ master ]
jobs:
  build:
    runs-on: windows-2022
    env:
      Solution_Name: WinUIAppCICD.sln
  steps:
    . . .
```

This is the same setup as the template we saw earlier. There are only two important things to highlight:

- We must run this workflow on the GitHub-hosted runner called `windows-2022`, which already comes with the .NET 6.0 SDK and Visual Studio 2022 and all the workloads available, including the one for building .NET and Universal Windows Platform apps. Don't use the `windows-latest` runner; this is because, at the time of writing, it still refers to the `windows-2019` image, which wouldn't work since it's based on the .NET 5.0 SDK and Visual Studio 2019.

- We set up an environment variable with the name of the solution so that we can easily change it later if we need to.

Now, let's see, step by step, the actions we need to perform in our job. We're going to add them under the steps section of the YAML file.

Pulling the source code

The first action to add is the following:

```
- name: Checkout
  uses: actions/checkout@v2
  with:
    fetch-depth: 0
```

This is required for each workflow related to CI since it will pull the latest version of the code from the repository and will save it on the runner in a path that is exposed by the `${{ github.workspace }}` environment variable.

Adding MSBuild to the system's path

The second action to add is the following:

```
- name: Setup MSBuild.exe
  uses: microsoft/setup-msbuild@v1.0.2
```

Here, we're using one of the actions available in the GitHub marketplace, provided by Microsoft, called `setup-msbuild`. This action will locate the path of the MSBuild executable (that is, the build engine provided by Visual Studio), and it will add it to the global path of the machine. In this way, the next actions will be able to use MSBuild just by invoking the `msbuild.exe` executable, without having to specify the full path.

Building the Windows application

The third action is the most important one since it's the one that will actually build our solution and generate the MSIX package:

```
- name: Create the app package
  run: msbuild ${{ env.Solution_Name }} /restore
/p:Platform=${{ env.Appx_Bundle_Platforms }}
/p:AppxBundlePlatforms=${{ env.Appx_Bundle_Platforms }}
/p:Configuration=${{ env.Configuration }}
```

```
/p:UapAppxPackageBuildMode=${{ env.Appx_Package_Build_Mode
}} /p:AppxBundle=${{ env.Appx_Bundle }}
/p:AppxPackageDir=${{github.workspace}}\AppPackages\
/p:GenerateAppxPackageOnBuild=true
  env:
    Appx_Bundle: Never
    Appx_Bundle_Platforms: x64
    Appx_Package_Build_Mode: SideloadOnly
    Configuration: Release
```

In this case, we aren't using an action, but we're just running a script. We're executing MSBuild by passing a series of parameters that will lead to the generation of an MSIX package with our application. The parameters we set are as follows:

- `Platform`: This is the CPU architecture that we want to target.

- `Configuration`: This is the configuration we want to use to build the project (**Debug** or **Release**). We set it to **Release** since we're building a package for distribution.

- `AppxBundlePlatforms`: In this scenario, it's also the CPU architecture that we want to target. This is a specific parameter required for generating the MSIX.

- `UapAppxPackageBuildMode`: This is the type of package that we want to generate. By setting it to **SideloadOnly**, we're generating a package for sideloading; if we're planning to publish the application in Microsoft Store, we can set this parameter to **StoreUpload**, which will also lead us to generate the `.msixupload` file.

- `AppxBundle`: This defines whether we want to generate a single package for each architecture or a bundle. In this scenario, we're setting it to **Never**, which will generate a single package.

- `AppxPackageDir`: This is the path where the MSIX package will be generated. To define it, we can use the environment variable we learned about earlier, `${{ github.workspace }}`, which is automatically translated with the path where the repository has been cloned.

In this step, notice how variables can be defined not only at the workflow level but also at the step level. In this example, we're setting the MSBuild parameters with variables rather than a hardcoded string. We don't need them at the workflow level (since this is the only step that uses them), but having them as a variable will make it easier to change them at a later stage if we need them.

Publishing the artifact

Once the build has been completed, we can generate an artifact. An artifact is the output of the build process that we want to make available to the developer (via the GitHub dashboard) or other jobs of the same workflow. In our case, it's the folder that contains the MSIX package and all of its dependencies, so we set up our step as follows:

```
- name: Upload build artifacts
  uses: actions/upload-artifact@v2
  with:
    name: MSIX Package
    path: ${{github.workspace}}\AppPackages
```

Here, we use a predefined action, this time provided directly by GitHub (the ones with `actions` as the company name are built into the platform). Compared to the other actions we have seen, in this scenario, we need to configure it. So, we use the `with` keyword to set up its properties:

- `name`: This is used to set the artifact name.

- `path`: This is used to set the path of the folder that contains the files we want to expose as a package. In our scenario, we set the same path that we have passed to the `AppxPackageDir` parameter of the MSBuild execution.

Testing the workflow

Once you commit to your updated workflow, it will be triggered immediately. If the configuration is correct, the workflow will complete successfully, and you will be able to download, as an artifact, the MSIX package that has been generated, as shown in the following screenshot:

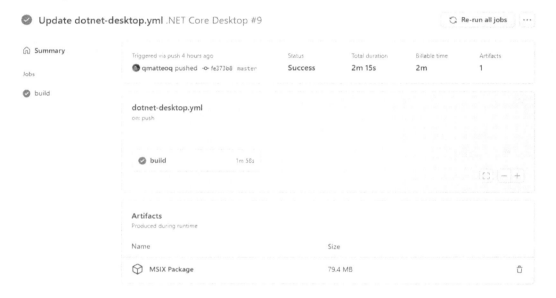

Figure 12.4 – The workflow has been completed successfully, and the MSIX package is available as an artifact

However, if you try to download and install the MSIX package, you will fail, since it isn't signed. This is a task that we'll take care of during the deployment process. Now, let's move on to handle another important requirement for when you are working with MSIX packages: versioning.

Supporting versioning

Versioning is a critical part of a software release. When it comes to MSIX, it becomes even more important because it ensures that Windows can effectively manage updates. As such, supporting versioning as part of our CI/CD pipeline is critical: if every execution of the workflow could generate an MSIX package with the same version, we wouldn't be able to deploy it if the application was already installed, we would break the auto-update feature if using App Installer, and Microsoft Store wouldn't be able to accept it.

One of the most interesting tools that you can use to support versioning is called **Nerdbank.GitVersioning**. This is an open source project that is part of the .NET Foundation, which is hosted on GitHub at `https://github.com/dotnet/Nerdbank.GitVersioning`.

This tool has the following three features that make it a great fit for enabling versioning:

- Version numbers are generated using Git information, such as the Git height or the Git commit ID, which ensures that every generated version will be higher than the previous one. You can find out more information at `https://github.com/dotnet/Nerdbank.GitVersioning#what-is-git-height`.

- Version numbers aren't only calculated automatically; they take into account the information provided in the file that is included in the solution called `version.json`. Thanks to this file, you can decide which numbers should be updated manually and which ones should be automatically generated. This is especially helpful if you adopt semantic versioning as every number has a precise meaning (for example, you should increase the major number only when you introduce breaking changes or new features, so you only want your workflow to automatically increase the minor number or the revision number).

- It offers built-in support for cloud CI/CD environments. When you run it on a platform such as GitHub or Azure DevOps, it will automatically create a series of environment variables that store the generated version number so that you can reuse it across your workflow.

The best way to start using `Nerdbank.GitVersioning` is to install it as a .NET CLI tool. First, you must install it on your own development machine: here, we're going to use some of the CLI commands to set it up in our solution.

Open Windows Terminal (or Command Prompt on your machine) and run the following command:

```
dotnet tool install -g nbgv
```

Then, in the same Terminal, move to the folder that contains your solution and run the following command:

```
nbgv install
```

The tool will perform two actions:

- It will add a `Directory.build.props` file into the root of your solution, which will ensure that the `Nerdbank.GitVersioning` NuGet package is installed in all your projects.

- It will add the `version.json` file to the root of your solution.

Let's take a look at the content of the file:

```
{
    "$schema": "https://raw.githubusercontent.com/AArnott
        /Nerdbank.GitVersioning/master/src/NerdBank
            .GitVersioning/version.schema.json",
    "version": "1.0-beta",
    "publicReleaseRefSpec": [
        "^refs/heads/master$",
        "^refs/heads/v\\d+(?:\\.\\d+)?$"
    ],
    "cloudBuild": {
        "buildNumber": {
            "enabled": true
        }
    }
}
```

The key property is called `version`, and it's the one that enables the flexible scenario we described earlier. The version we set in this property will never change automatically: the tool will only generate numbers for the subsequential numbers. For example, if we set it to 1.0, the tool will automatically generate version numbers such as 1.0.1, 1.0.2, 1.0.3, and more.

The default value is `1.0-beta`, but we must change it to `1.0` to support our scenario: MSIX requires the version number to be `x.y.z.0`, and as such, we can't include alphabetical characters.

Instead, the `publicReleaseRefSpec` collection is used to specify which branches of our repository can be used to generate packages that are publicly distributed. Version numbers generated outside these branches will automatically include, in the end, the Git commit ID, such as in the following example:

```
1.0.24-g9a7eb6c819
```

Make sure that you match the branches you have on your repository; otherwise, you might generate a version number that isn't valid for the MSIX manifest.

The last section, called `cloudBuild`, enables one of the features I previously highlighted: the option to automatically create environment variables in a CI/CD environment with the generated version number.

This is what the `version.json` file we made:

```
{
    "$schema": "https://raw.githubusercontent.com/AArnott
        /Nerdbank.GitVersioning/master/src/NerdBank
            .GitVersioning/version.schema.json",
    "version": "1.0",
    "publicReleaseRefSpec": [
        "^refs/heads/master$",
        "^refs/heads/main$",
        "^refs/heads/v\\d+(?:\\.\\d+)?$"
    ],
    "cloudBuild": {
        "buildNumber": {
            "enabled": true
        }
    }
}
```

Now, it's time to go back to the workflow file to change it. You can go back to GitHub and use the visual interface. However, if you prefer, you can also use your favorite code editor. Remember that, thanks to YAML, the workflow is now a file that is stored inside the `.github\workflows` folder of your project. If you use Visual Studio Code, you can find some interesting extensions that will make your authoring experience easier, for instance, by providing IntelliSense or the option to monitor a workflow execution directly from the editor. Here is an example: `https://marketplace.visualstudio.com/items?itemName=cschleiden.vscode-github-actions`.

Let's see the steps we need to add.

Installing the tool and generating the version number

The first step of our workflow is to install `Nerdbank.Gitversioning` on the runner, similarly to how we installed it on our development machine. However, we don't have to do this manually. This is because the author of the library has created a dedicated GitHub action that takes care of installing the tool on the runner and setting the environment variables. As such, you'll just need to add the following step at any point in the `steps` section, as long as it's placed before the execution of the MSBuild task:

```
- name: Use Nerdbank.GitVersioning to set version variables
  uses: dotnet/nbgv@v0.4.0
  with:
    setAllVars: true
```

Setting the version number

The next step is to set the version number, which must be changed in the manifest. The easiest way to do this is with a PowerShell script as this scripting language provides a powerful set of APIs to manipulate XML files. This is what our step looks like:

```
- name: Update manifest version
  run: |
    [xml]$manifest = get-content ${{env.manifestPath}}
    $manifest.Package.Identity.Version =
      "${{env.NBGV_SimpleVersion}}.0"
    $manifest.save("${{env.manifestPath}}")
  env:
    manifestPath: .\WinUIAppCICD\Package.appxmanifest
```

First, we define a local environment variable with the full path of the manifest file, which is called `Package.appxmanifest` and is stored inside the folder of the main project (if we're using the single-project MSIX approach) or inside the Windows Application Packaging Project.

Using the PowerShell APIs, we read the whole XML, and then we change the value of the `Package.Identity.Version` node to the newer version number, which we can find in the `NBGV_SimpleVersion` variable that was generated in the previous task. However, we must add `.0` at the end: remember that the MSIX versioning must always follow the `x.y.z.0` syntax. Then, we use the `save()` method to save our changes, passing again the path of the manifest file so that it can be overwritten.

That's all we need to do. Just remember to add this step after the installation of the `Nerdbank.GitVersioning` tool and before the execution of the MSBuild task. With these changes, the MSBuild task will now generate an MSIX package using the version number we have just injected with our PowerShell script.

To check whether you have made the changes in the right way, you just need to commit them and trigger a new execution of the workflow. In the end, download the generated artifact and look inside the ZIP file: the filename of the MSIX package will include a higher version number instead of the default `1.0.0.0` version number:

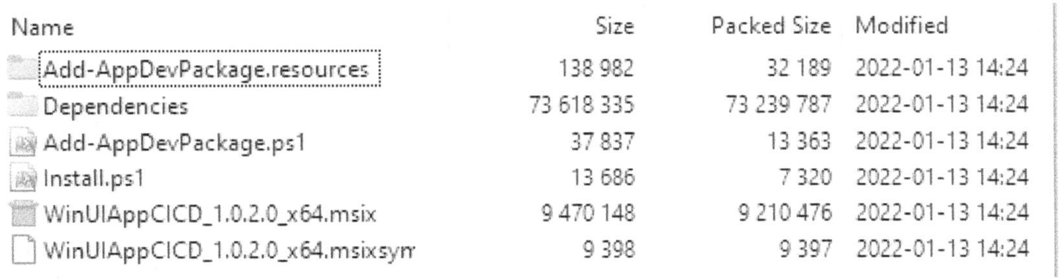

Name	Size	Packed Size	Modified
Add-AppDevPackage.resources	138 982	32 189	2022-01-13 14:24
Dependencies	73 618 335	73 239 787	2022-01-13 14:24
Add-AppDevPackage.ps1	37 837	13 363	2022-01-13 14:24
Install.ps1	13 686	7 320	2022-01-13 14:24
WinUIAppCICD_1.0.2.0_x64.msix	9 470 148	9 210 476	2022-01-13 14:24
WinUIAppCICD_1.0.2.0_x64.msixsym	9 398	9 397	2022-01-13 14:24

Figure 12.5 – The generated MSIX package has a higher version number

This step concludes our CI pipeline. Now we have an MSIX package that is ready to be deployed, so let's explore the CD pipeline.

Handling signing

So far, we have succeeded in creating an MSIX package, but we still can't use it. In *Chapter 11*, *Publishing Your Application*, we learned that an MSIX package should be as follows:

- It must be signed with a certificate.
- The certificate must be trusted by the machine.
- The **Publisher** defined in the manifest must match the subject of the certificate.

As such, we can't really use the package created by our CD pipeline since it isn't signed. If we try to install it, Windows will show the following error:

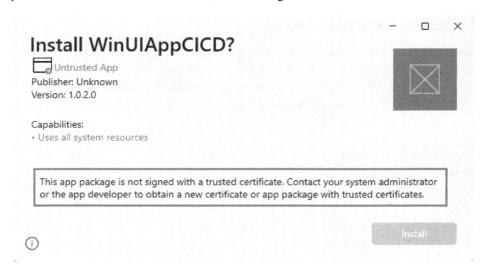

Figure 12.6 – The error displayed by Windows when we try to install an unsigned MSIX package

To solve this problem, we must sign the package as part of our workflow. Signing is a very delicate task as we must find a way to sign the package without exposing our certificate. If we don't protect it, a malicious developer could steal it and use it to sign other applications using our identity.

Visual Studio supports the possibility of signing the package as part of the build process, but this is a very fragile approach since it would require storing the certificate on the repository and making it accessible to every developer on the team (or to the broader developer community if we're talking about an open source project). Additionally, you might want to use different certificates based on the version of the application you're building: for instance, testing, production, and more.

Consequently, it's good practice to perform the signing task in the CD pipeline, right before the deployment. Let's see how we can do that by using the features provided by GitHub.

Storing the certificate on GitHub

This approach requires a few steps to be implemented since, at the time of writing, GitHub doesn't support safely storing files. Instead, it supports storing secrets, which are strings that are protected behind a repository: you can use their value inside a workflow, but you can't see them once you have set them.

As such, the first task is to convert our certificate into a base64 string that we can store as a secret. The starting point is a PFX file (which is protected by a password), which contains our certificate. If you don't have one, you can find many tutorials on the internet regarding how to export a certificate from the Windows certificate store into a PFX file.

These are the steps to follow:

1. Open a PowerShell terminal and launch the following command:

   ```powershell
   $pfx_cert = Get-Content '.\MyCertificate.pfx' -
     Encoding Byte
   [System.Convert]::ToBase64String($pfx_cert) | Out-File
     'MyCertificate_Encoded.txt'
   ```

   Assuming that your PFX file is called MyCertificate.pfx, this command will generate a text file called MyCertificate_Encoded.txt with the base64 version of the certificate.

2. Go back to your repository on GitHub, and navigate to the **Settings** tab.

3. Move to the **Secrets** section, and click on **New repository secret**.

4. Set the name as CERTIFICATE and, for the value, copy and paste the content of the MyCertificate_Encoded.txt file.

5. Click on **New repository secret** again to create a new secret.

6. Set PASSWORD as the name and, for its value, use the password you just encoded to open the PFX certificate.

Now it's time to add the first task. Since we're moving on to the CD phase, we must first add a new job to our workflow, as shown in the following example:

```yaml
name: WinUI application
on:
  push:
    branches: [ master ]
  pull_request:
    branches: [ master ]
  workflow_dispatch:

jobs:
  build:
    runs-on: windows-2022
    env:
```

```
        Solution_Name: WinUIAppCICD.sln
    steps:
        . . .
  deploy:
    runs-on: windows-2022
    needs: [ build ]
    steps:
        . . .
```

We have defined a new job called deploy, which will also be executed on a windows-2022 runner. By using the needs keyword, we specify that this job must only run when the build job has been completed successfully. If we don't set up this relationship, GitHub will execute the two jobs in parallel.

Now, let's add the steps. The first one downloads the artifact that was generated during the build job:

```
- uses: actions/download-artifact@v2
  with:
    name: MSIX Package
    path: MSIX-package
```

We are downloading the artifact called MSIX Package (this is the name we used in the Upload build artifact step in the build job) into a folder called MSIX-package.

The next step is to convert the base64 string that we stored as a secret back into a PFX file:

```
- name: Decode the pfx
  run: |
    $pfx_cert_byte = [System.Convert]::FromBase64String
      ("${{ secrets.CERTIFICATE }}")
    [IO.File]::WriteAllBytes("${{ github.workspace}}\MSIX-
      package\MyCertificate.pfx", $pfx_cert_byte)
```

The secret can be retrieved using the ${{ secrets.CERTIFICATE }} syntax. By using the .NET APIs available via PowerShell, we convert the base64 string back into a file called MyCertificate.pfx.

Now that we have the PFX file stored on the runner, we can use it to sign the MSIX package thanks to the signtool utility:

```
- name: Sign package
```

```
run:  |
      Get-ChildItem -recurse -Include *.msix | ForEach-
         Object {
      $msixPath = $_.FullName
      & "C:/Program Files (x86)/Windows Kits/10
         /bin/10.0.22000.0/x64/signtool.exe" sign /f "${{
         github.workspace}}\MSIX-package\MyCertificate.pfx
         " /p ${{ secrets.PASSWORD }} /fd SHA256 $msixPath
      }
```

We're using the `Get-ChildItem` and `ForEach-Object` commands in PowerShell to iterate through all the files inside the `MSIX-package` folder. This is so that we only extract the ones with `.msix` as an extension. When we find one, we use the `signtool` utility that is included in the Windows 10 SDK by invoking the following command:

```
C:/Program Files (x86)/Windows Kits/10/bin/10.0.22000.0/x64
/signtool.exe" sign /f <path of the PFX file> /p <password>
/fd SHA256 <path of the MSIX file>
```

Again, we use the secrets feature offered by GitHub to retrieve the password that we saved previously by using the `${{ secrets.PASSWORD }}` syntax.

The last step, for security reasons, is to delete the PFX file. In theory, this isn't necessary because runners are automatically discarded at the end of the workflow, but being very sensitive information, it's better to be safe:

```
- name: Remove the pfx
  run: Remove-Item -path "${{ github.workspace}}\MSIX-
  package\MyCertificate.pfx"
```

That's it! If you want to test that your CD pipeline works as expected, you can add another action to upload the signed MSIX package as an artifact:

```
- name: Upload build artifacts
  uses: actions/upload-artifact@v2
  with:
    name: Signed MSIX Package
    path: MSIX-package
```

This time, when the workflow completes, you will be able to find two different artifacts: **MSIX Package** (which is the original unsigned one) and **Signed MSIX Package** (which is the signed one). When you download and open the signed one, this time the **Install** button should be enabled:

Figure 12.7 – This time the package is signed, so Windows allows us to install it

If you want to adopt a safer approach to store your certificate, you can use a service provided by Azure called Key Vault, which you can use to protect secrets, passwords, and certificates. By supporting advanced security features, such as encryption and FIPS 140-2 Level 2 and Level 3-validated HSMs, this is the best way to store sensitive information such as a certificate. Additionally, the protected data never leaves the service, but the communication happens through a series of APIs. In our scenario, this approach adds an extra layer of security because the certificate is never stored on the repository or the runner.

The perfect companion for Azure Key Vault is **Azure SignTool**, which is an open source tool that is available at https://github.com/vcsjones/AzureSignTool and developed by Kevin Jones, who is part of the Security and Cryptography team at GitHub. This tool works like the original SignTool utility we learned about earlier; however, instead of requiring the local path of the PFX file or the password, it needs the URL, the client ID, and the client secret to connect to your Azure Key Vault instance. It's a command-line tool that is also available as a .NET global tool, so it can be easily installed and used in your CD pipeline to sign the MSIX packages you have generated.

In this chapter, we won't explain this approach in detail, but you can find a detailed step-by-step guide in the official documentation at https://docs.microsoft.com/en-us/windows/msix/desktop/cicd-keyvault.

Now that our package is signed, we can plan the deployment.

Automating the deployment

One of the key features of MSIX that can make it easier to deploy a Windows desktop application is the App Installer technology. In *Chapter 11*, *Publishing Your Application*, we learned how, thanks to an App Installer file, we can support features such as automatic updates, dependencies installation, and more.

Let's see how we can create it as part of our workflow.

Generating the App Installer file

The easiest way to generate the App Installer file is by leveraging Visual Studio, which can generate the file as part of the build, as we learned in the previous chapter. However, here, we are building a CI/CD pipeline, so we can't use the **Publish** wizard provided by Visual Studio to configure App Installer. As such, we need to add some extra parameters to the MSBuild step in the build job, as shown in the following example:

```
- name: Create the app package
  run: msbuild ${{ env.Solution_Name }} /restore
/p:Platform=${{ env.Appx_Bundle_Platforms }}
/p:AppxBundlePlatforms=${{ env.Appx_Bundle_Platforms }}
/p:Configuration=${{ env.Configuration }}
/p:UapAppxPackageBuildMode=${{ env.Appx_Package_Build_Mode
}} /p:AppxBundle=${{ env.Appx_Bundle }}
/p:AppxPackageDir=${{github.workspace}}\AppPackages\
/p:GenerateAppxPackageOnBuild=true
/p:GenerateAppInstallerFile=true /p:AppInstallerUri=${{
env.AppInstallerUrl}} /p:HoursBetweenUpdateChecks=${{
env.UpdateFrequency }}
  env:
    Appx_Bundle: Never
    Appx_Bundle_Platforms: x64
    Appx_Package_Build_Mode: SideloadOnly
    Configuration: Release
    AppInstallerUrl: https://www.mywebsite.com
    UpdateFrequency: 0
```

Here, we have added three parameters compared to the original implementation:

- `GenerateAppInstallerFile`: We set this to `true`.

- `AppInstallerUri`: We set this with the URL of the website or network share from which we're going to deploy our application. We set this information with an environment variable.

- `UpdateFrequency`: This indicates how often we want to check updates. By setting it to `0`, Windows will check for new versions every time the application starts. This information is also set with an environment variable.

If you commit these changes and run your workflow again, this time you will notice that the generated artifact will include two additional files:

- One with `extension.appinstaller`, which has been configured using the information we passed as MSBuild parameters. If you're building a WinUI application, the App Installer file will already include all of the required dependencies, such as the Windows App SDK runtime.

- The other is called `index.html`, which is a web page that you can publish on your website to trigger the installation of the package.

If you need to further customize the App Installer file (for example, by enabling additional features such as the option to show a prompt when there's a new update available), you can add an App Installer template to your project, as we explained in *Chapter 11, Publishing Your Application*.

Before adopting this solution, there are a few important things to bear in mind:

- As mentioned in the previous chapter, at the time of writing, if you're building a WinUI application using the single-project MSIX approach, the App Installer file won't be generated even if you specify the dedicated parameters. Until this issue is fixed, the workaround is to create a WinUI application using a template called **Blank App, Packaged with Windows Application Packaging Project** (WinUI 3 in Desktop). If you're building a Windows Forms application or a WPF application, you should already have a Windows Application Packaging Project in your solution to enable MSIX packaging, so you don't have to take any extra steps.

- The `index.html` file that is generated as part of the build uses the `ms-appinstaller` protocol to enable the installation of the application. As explained in *Chapter 11, Publishing Your Application*, this protocol has been disabled in the recent releases of App Installer due to a security issue. Until the issue is resolved, you can't use the `index.html` file as it is, but you can change the installation URL to directly download the App Installer file on the user's machine.

For example, let's assume that the link on the page is defined as the following:

```
<a href='ms-appinstaller:?source=https://
www.mywebsite.com/WinUIAppCICD_ x64.appinstaller'>
<button>Install for x64</button></a>
```

You must change it into the following:

```
<a href='https://www.mywebsite.com/WinUIAppCICD
_x64.appinstaller'><button>Install for
x64</button></a>
```

Now that we have an App Installer file together with our MSIX package, we can deploy it on a website or network share.

Deploying our Windows application

Now that we have an MSIX package and an App Installer file, we can deploy it to the distribution platform of our choice. All you need is a static website that is accessible via HTTPS or a network share, so there are multiple options available. In this chapter, we won't go into the specific implementation details, since each option has its own different approach. However, thanks to GitHub Actions, you can easily hook your distribution platform into your CD pipeline, thanks to a wide range of available deployment actions.

For example, if you're using Azure as a cloud platform, there are multiple tutorials that cover the integration with GitHub Actions, including the following:

- Azure Storage with a static website: `https://docs.microsoft.com/en-us/azure/storage/blobs/storage-blobs-static-site-github-actions`

- Azure Static Web Apps: `https://azure.microsoft.com/en-us/services/app-service/static/`

- Azure App Service: `https://docs.microsoft.com/en-us/azure/app-service/deploy-github-actions`

If you aren't using Azure, you will find many actions to deploy your package to different cloud providers, such as AWS or Google Cloud. And, in the end, if you aren't using any cloud provider, you also have actions to deploy files via FTP, enabling the option to deploy your package virtually on any web server.

Thanks to MSIX and App Installer, we can enable a true CD story. Every time we push new changes to the repository, GitHub will create a new MSIX package with an updated App Installer file that we can push to our deployment website. Thanks to App Installer, whenever we deploy an updated package, all the users who have installed the application will automatically receive the update, without taking any manual action.

Improving the deployment story

Thanks to GitHub, you can improve the deployment story by adding extra features such as the following:

- A workflow can have as many jobs as you want. As such, you can define multiple stages to support multiple deployment environments. Your workflow could deploy the MSIX package on a testing website, which is only accessible in your company internally. Then, later, you could deploy it on the production website once all the validation tests have passed.

- Thanks to the concept of environments, you can enable an approval workflow in your deployment stages. In this way, you can pause the execution of a deployment workflow until a specific list of people you have defined have approved it. This way you can enable, for example, a workflow where the deployment to a staging environment is performed automatically once the build has been completed, but the deployment to the production environment is paused until the PM of the project approves it. You can learn more about this feature at https://docs. github.com/en/actions/managing-workflow-runs/reviewing-deployments.

This way, you can support more complex workflows to better adapt to the DevOps implementation within your company.

Summary

The adoption of the DevOps mindset and practices is becoming increasingly important to deliver successful software projects. CI and CD are critical parts of a successful DevOps implementation since they help you to ensure the quality of the applications you build and enable you to release updates more often and with confidence.

However, when it comes to adopting CI and CD for Windows desktop applications, there are a few challenges we must consider: versioning, signing, the need to distribute updates to a wide range of machines running the application, and more.

In this chapter, we explored how we can create a CI/CD pipeline for a Windows desktop application using one of the most popular platforms in the world for developers: GitHub. Thanks to GitHub Actions and the provided runners, we can automate all the tasks that are needed to produce a deployable Windows application: for instance, generating an MSIX package, versioning it, signing it with a certificate, and then deploying it to a website or a network share.

All the techniques we have learned in this chapter are mostly based on the scripts and CLI tools, so you can easily adapt the workflow we have created for other CI/CD platforms. Additionally, if you are using Azure DevOps as a DevOps platform, the MSIX team has created a dedicated set of extensions for Azure Pipelines, which you can integrate into your workflow to perform MSIX-related tasks such as generating an MSIX package, creating an App Installer file, or signing the MSIX package. You can learn about them in more detail at `https://docs.microsoft.com/en-us/windows/msix/desktop/msix-packaging-extension?tabs=yaml`.

What if, for any reason, you can't use MSIX, but you must use an unpackaged deployment? In this case, the story is more complicated since Microsoft doesn't provide built-in support in Visual Studio to generate custom setups or MSI installers. Microsoft offers a technology called **ClickOnce** to deploy unpackaged .NET apps (you can learn more at `https://docs.microsoft.com/en-us/visualstudio/deployment/clickonce-security-and-deployment?view=vs-2022`). However, at the time of writing, it's only supported from the user interface. You can't generate a ClickOnce installer as part of a CI/CD pipeline. To achieve this goal, you can evaluate third-party solutions, such as Squirrel (`https://github.com/clowd/Clowd.Squirrel`) and **Wix** (`https://wixtoolset.org/`), which can generate an installer as part of a Visual Studio build, or Advanced Installer (`https://www.advancedinstaller.com/`), which offers tasks for CI/CD platforms to generate an installer starting from their projects without user intervention.

This chapter concludes our journey into developing modern Windows applications. Now we have all the knowledge we need to start building new applications with WinUI, modernizing existing applications with the Windows App SDK, and deploying them using MSIX and CI/CD pipelines.

Questions

1. If a GitHub-hosted runner doesn't satisfy the requirements to build your software, you can't use GitHub Actions, but you have to find an on-premises solution. Is this true or false?

2. You can reuse a GitHub workflow on every other CI/CD platform (such as Azure DevOps) since they are all based on YAML. Is this true or false?

3. You can use technologies other than MSIX to enable CI/CD for Windows desktop applications. Is this true or false?

Further reading

- *Accelerate: The Science of Lean Software and DevOps: Building and Scaling High Performing Technology Organizations* by Nicole Forsgren, Jez Humble, and Gene Kim.

- *The Phoenix Project*, by Gene Kim, Kevin Behr, and George Spafford.

- *The Unicorn Project*, by Gene Kim.

- *The DevOps Handbook: How to Create World-Class Agility, Reliability, and Security in Technology Organizations*, by Gene Kim, Jez Humble, Patrick Debois, and John Willis

Assessments

Chapter 1

1. No. The Windows App SDK is supported only by the new .NET runtime, so you must migrate your Windows Forms application to .NET 5.0 or .NET 6.0 first.

2. MSIX is the best deployment technology, not only for applications that integrate the Windows App SDK, but also for every Windows application. It delivers a clean install and uninstall experience, built-in features such as automatic updates, and network bandwidth optimization. However, if, for any reason, you can't adopt the MSIX technology, the Windows App SDK also supports the opportunity to work unpackaged, which means that you can deploy your apps with any existing technology.

3. No. The Windows App SDK includes other features, such as resource management, push notifications, and activations.

Chapter 2

1. The best approach is to implement the `INotifyPropertyChanged` interface, which gives you access to an event that you must raise every time a property value changes. Another supported approach is to use dependency properties, but they are harder to declare and maintain, so they should be used only when you're working with XAML controls (such as user control).

2. False. The goal of using styles is to maximize the re-use of resources, so they should be declared at the highest possible level, such as the `Page` container or the entire application. Even better, you can use resource dictionaries to centralize styles in a single place and organize them in a more structured way.

3. False. The default binding mode is `OneWay`, which means that only the target will receive notifications when the source changes and not the other way around. If you need to support two-way communication, you must set the binding mode to `TwoWay`.

Chapter 3

1. False. Thanks to the function support offered by `x:Bind`, you can also use system functions (such as `String.Format()`) directly in a markup expression.

2. False. With the help of the Windows Community Toolkit, you can introduce an extension method called `EnqueueAsync()` that supports the `async/await` pattern, enabling you to wait until the dispatched action is completed before performing additional tasks.

3. False. The Windows App SDK will take care of loading the proper resource file, based on the user configuration.

Chapter 4

1. False. It was true for UWP apps but, since Windows App SDK apps are based on the Win32 model, they have more flexibility. They will be automatically terminated only if the application raises an exception that isn't handled.

2. False. If packaged apps can support all the activation contracts offered by the UWP, at the time of writing, unpackaged apps can support only four types of activation: launch, file, protocol, and start up task.

3. True. This is the opposite behavior of UWP apps which, instead, are single-instance by default.

Chapter 5

1. False. `Canvas` uses absolute positioning to place the controls on a page, making it hard to react to changes in the size of the window. It's better to use controls such as `Grid` or `RelativePanel`, which better support layouts that can dynamically change.

2. False. The `NavigationView` control has a built-in `VisualStateManager` which takes care of changing its display mode (Full, Compact, or Minimized) automatically based on the available space.

3. It's called Visual Layer and it sits between the XAML layer and the DirectX one.

Chapter 6

1. False. Libraries such as MVVM Toolkit or Prism can help you implement the pattern in a faster and more reliable way, but it isn't an absolute requirement. You can choose to build the required building blocks on your own since the pattern is based on standard C# and XAML features.

2. One of the goals of the MVVM pattern is to decouple as much as possible all the components, so that you evolve, maintain, and test them independently. Thanks to messages, you can exchange data between two classes without coupling them.

3. Ideally, the answer should be false. The goal of the Model domain is to be platform-agnostic. By introducing the INotifyPropertyChanged implementation in classes that belong to the Model layer, you create a dependency with XAML-based technologies, making it hard to reuse the same classes in a different type of project, such as a web application or a REST API.

Chapter 7

1. False. Even though they are both based on XAML, and you can reuse most of the knowledge, they support different features, a different set of controls, and different namespaces. As such, some work is needed to migrate your UI layer.

2. False. The x:Bind markup expression uses the code-behind to resolve the binding channels. As such, you must expose your ViewModel through a property in the code-behind class.

3. True. Windows App SDK applications, being based on the Win32 ecosystem, have a simpler life cycle, so they support only a single activation point: the OnLaunched event of the App class. Multiple activation points are supported, but you must manage them directly in this event.

Chapter 8

1. False. The WebView2 control requires the WebView2 runtime, which is a special version of the Edge engine optimized for hybrid scenarios.

2. False. You just need to set the TargetFramework property of your project to use one of the available workloads for Windows 10 and the library will be installed automatically.

3. True. Only the use of local storage is supported just by packaged apps, but everything else also works with unpackaged apps, developed with any .NET UI platform.

Chapter 9

1. False. They are also supported by WPF and Windows Forms applications. However, based on the platform of your choice, you must place the code to handle the activation in a different location, since every platform has its own way of handling the startup phase.

2. False. It's required when you want to replace existing notifications, rather than adding a new one; or when you need to retrieve it at a later point in time, for example, to delete it. However, notifications are also displayed just fine without a tag.

3. False. A packaged application must register two extensions in the manifest so that Windows can properly manage the activation phase.

Chapter 10

1. ONNX is an open source standard to store an ML model in a file, which can be exchanged and consumed by different types of applications. Being portable, it's a good fit for Edge scenarios, where the model must be evaluated locally.

2. False. Thanks to WinML, an ML model can be evaluated on the local machine, without needing any internet connection and any communication with a cloud service.

3. Custom Vision is a service provided by Azure that removes the complexity of learning ML algorithms to train a model to recognize objects, sentiments, and so on. You just need to provide a series of images as input, classify them, and train the model. Custom Vision is a great companion for WinML since you can export the trained model to ONNX.

Chapter 11

1. False. Windows Package Manager is a tool that makes it more convenient to install applications from existing sources, such as the Microsoft Store or a website. The repository only contains manifests that describe the application and the URL from which to grab the installer, but it doesn't host the actual application.

2. True. If you want granular control over the update process, you can use a set of APIs to manage updates within your application. However, you aren't required to use them. You can just rely on the update settings you can define in the App Installer file.

3. False. The Microsoft Store is the perfect distribution platform for consumer apps since it's already installed in Windows and greatly simplifies the installation experiences for users. However, most of its features target a consumer audience, such as the requirement to use a personal Microsoft account to purchase and download applications. If you want to create a personal catalog of applications for your company (which can also include applications from the Microsoft Store), soon Windows Package Manager will enable you to host your own private repository.

Chapter 12

1. False. The easiest option is to install the missing requirements in the first steps in a workflow (such as a specific version of .NET or Java). If this isn't an option for you (for example, you need a tool that can't be installed in an unattended way), you can create your own self-runner and connect it to GitHub. This way, you can continue to use the powerful features offered by GitHub Actions, but the workflow will be executed on one of your machines.

2. False. Even if all the modern CI/CD platforms are using YAML, each of them has its own different syntax and actions available.

3. True, even if MSIX is the simplest one since it doesn't require any additional tool. You just need Visual Studio and the Windows 10 SDK, which is pre-installed on all the Windows build machines provided by the most important vendors.

Index

B

Packt.com

Subscribe to our online digital library for full access to over 7,000 books and videos, as well as industry leading tools to help you plan your personal development and advance your career. For more information, please visit our website.

Why subscribe?

- Spend less time learning and more time coding with practical eBooks and Videos from over 4,000 industry professionals

- Improve your learning with Skill Plans built especially for you

- Get a free eBook or video every month

- Fully searchable for easy access to vital information

- Copy and paste, print, and bookmark content

Did you know that Packt offers eBook versions of every book published, with PDF and ePub files available? You can upgrade to the eBook version at packt.com and as a print book customer, you are entitled to a discount on the eBook copy. Get in touch with us at customercare@packtpub.com for more details.

At www.packt.com, you can also read a collection of free technical articles, sign up for a range of free newsletters, and receive exclusive discounts and offers on Packt books and eBooks.

Other Books You May Enjoy

If you enjoyed this book, you may be interested in these other books by Packt:

Learn WinUI 3.0

Alvin Ashcraft

ISBN: 9781800208667

- Get up and running with WinUI and discover how it fits into the landscape of Project Reunion and Windows UI development
- Build new Windows apps quickly with robust templates
- Develop testable and maintainable apps using the MVVM pattern
- Modernize WPF and WinForms applications with WinUI and XAML Islands
- Discover how to build apps that can target Windows and leverage the power of the web
- Install the XAML Controls Gallery sample app and explore available WinUI controls

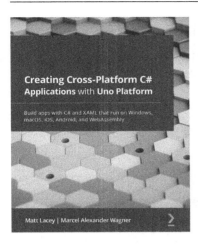

Creating Cross-Platform C# Applications with Uno Platform

Matt Lacey, Marcel Alexander Wagner

ISBN: 9781801078498

- Understand how and why Uno could be the right fit for your needs
- Set up your development environment for cross-platform app development with the Uno Platform and create your first Uno Platform app
- Find out how to create apps for different business scenarios
- Discover how to combine technologies and controls to accelerate development
- Go beyond the basics and create 'world-ready' applications
- Gain the confidence and experience to use Uno in your own projects

Packt is searching for authors like you

If you're interested in becoming an author for Packt, please visit `authors.packtpub.com` and apply today. We have worked with thousands of developers and tech professionals, just like you, to help them share their insight with the global tech community. You can make a general application, apply for a specific hot topic that we are recruiting an author for, or submit your own idea.

Share Your Thoughts

Now you've finished *Modernizing Your Windows Applications with the Windows App SDK and WinUI 3*, we'd love to hear your thoughts! Scan the QR code below to go straight to the Amazon review page for this book and share your feedback or leave a review on the site that you purchased it from.

https://packt.link/r/1803235667

Your review is important to us and the tech community and will help us make sure we're delivering excellent quality content.

www.ingramcontent.com/pod-product-compliance
Lightning Source LLC
Chambersburg PA
CBHW081453050326
40690CB00015B/2781